T0313738

PIEZOELECTRIC ENERGY HARVESTING

PIEZOELECTRIC ENERGY HARVESTING

Alper Erturk
Georgia Tech, USA

Daniel J. Inman
Virginia Tech, USA

WILEY

A John Wiley and Sons, Ltd., Publication

Registered office
John Wiley & Sons Ltd, The Atrium, Southern Gate, Chichester, West Sussex, PO19 8SQ, United Kingdom

For details of our global editorial offices, for customer services and for information about how to apply for permission to reuse the copyright material in this book please see our website at www.wiley.com.

Library of Congress Cataloging-in-Publication Data

Erturk, Alper.
  Piezoelectric energy harvesting / Alper Erturk, Daniel J. Inman.
    p. cm.
  Includes bibliographical references and index.
  ISBN 978-0-470-68254-8 (hardback)
  1. Piezoelectric transducers.   2. Electric generators.   3. Piezoelectricity.   I. Inman, D. J.   II. Title.
  TK7872.P54E78 2011
  621.31′3–dc22

                                                                              2010046394

A catalogue record for this book is available from the British Library.

Print ISBN: 978-0-470-68254-8
E-PDF ISBN: 978-1-119-99116-8
O-Book ISBN: 978-1-119-99115-1
E-Pub ISBN: 978-1-119-99135-9

Set in 10/12pt Times by Aptara Inc., New Delhi, India

These oscillations arise freely, and I have determined various conditions, and have performed a great many beautiful experiments on the position of the knot points and the pitch of the tone, which agree beautifully with the theory.

—Daniel Bernoulli (from a letter to Leonhard Euler)[1]

We have found a new method for the development of polar electricity in these same crystals, consisting in subjecting them to variations in pressure along their hemihedral axes.

—Pierre and Paul-Jacques Curie
(from the paper announcing their discovery)[2]

---

[1] In Timoshenko, S.P., 1953, *History of Strength of Materials (with a brief account of the history of theory of elasticity and theory of structures)*, McGraw-Hill, New York.

[2] In Cady, W.G., 1946, *Piezoelectricity: An Introduction to the Theory and Applications of Electromechanical Phenomena in Crystals*, McGraw-Hill, New York.

Contents

About the Authors

Dr. Alper Erturk is an Assistant Professor in the George W. Woodruff School of Mechanical Engineering at the Georgia Institute of Technology. Since 2008, he has published more than 50 papers in journals and conference proceedings on modeling and applications of piezoelectric materials with a focus on vibration-based energy harvesting. He is a member of the American Society of Mechanical Engineers, the American Institute of Aeronautics and Astronautics, the Society for Experimental Mechanics, and the International Society for Optical Engineering. Dr. Erturk received his PhD in Engineering Mechanics at Virginia Tech (2009).

Professor Daniel J. Inman is the Director of the Center for Intelligent Material Systems and Structures and the G.R. Goodson Professor in the Department of Mechanical Engineering at Virginia Tech and the Brunel Chair in Intelligent Materials and Structures Institute for Smart Technologies in the Department of Mechanical Engineering at the University of Bristol. Since 1980, he has published 8 books and more than 700 papers in journals and conference proceedings on subjects ranging from vibrations to structural health monitoring, graduated 50 PhD students, and supervised more than 75 MS degrees. He is a Fellow of the American Academy of Mechanics, the American Society of Mechanical Engineers, the International Institute of Acoustics and Vibration, the American Institute of Aeronautics and Astronautics, and the National Institute of Aerospace. Professor Inman received his PhD in Mechanical Engineering at Michigan State University (1980).

Preface

Energy harvesting from ambient waste energy for the purpose of running low-powered electronics has emerged during the last decade as an enabling technology for wireless applications. The goal of this technology is to provide remote sources of electric power and/or to recharge storage devices, such as batteries and capacitors. The concept has ecological ramifications in reducing the chemical waste produced by replacing batteries and potential monetary gains by reducing maintenance costs. The potential for enabling wireless monitoring applications, such as structural health monitoring, also brings an element of increasing public safety. With the previously mentioned potential as motivation, the area of energy harvesting has captivated both academics and industrialists. This has resulted in an explosion of academic research and new products. The evolution of low-power-consuming electronics and the need to provide wireless solutions to sensing problems have led to an emergence of research in energy harvesting. One of the most studied areas is the use of the piezoelectric effect to convert ambient vibration into useful electrical energy. Most products currently available for harvesting vibrational energy are based on this use of the piezoelectric effect. The focus of this book is placed on detailed electromechanical modeling of piezoelectric energy harvesters for various applications.

The area of vibration-based energy harvesting encompasses mechanics, materials science, and electrical circuitry. Researchers from all three of these disciplines contribute heavily to the energy harvesting literature. Due to the topic being spread over numerous different fields of study, many oversimplifications resulted from early attempts to understand and to develop predictive models. Our hope with the current volume is to provide reliable techniques for precise electromechanical modeling of piezoelectric energy harvesters and to understand the relevant phenomena. The term *energy harvester* is defined in this book as the generator device undergoing vibrations due to a specific form of excitation. The main focus is therefore placed on modeling the electromechanical response of the device for the respective form of excitation rather than investigating the storage components and the power electronics aspects. A brief review of the literature of piezoelectric energy harvesting circuits is also provided.

As far as the prerequisite of the material covered in this book is concerned, we have assumed that the reader is knowledgeable at the level of a BS degree in an engineering curriculum that includes a basic vibrations or structural dynamics course. Fundamental background in ordinary differential equations and partial differential equations is essential. Some of the topics in this book are related to subjects not necessarily covered in most undergraduate engineering curricula, such as random vibrations, nonlinear oscillations, and aeroelastic vibrations. However, references to excellent books and papers are provided as required. These aforementioned topics of vibrations and structural dynamics are coupled with the electrical domain throughout this text to formulate and/or to investigate various problems for vibration-based energy harvesting using piezoelectric transduction.

Several configurations ranging from conventional cantilevers to more sophisticated devices exhibiting nonlinear phenomena are modeled and tested for vibration-based energy harvesting. Both analytical and approximate analytical distributed-parameter electromechanical models are presented along with several case studies for experimental validation. Guidelines are provided for experimental modal analysis of a piezoelectric energy harvester in a laboratory environment.

The electromechanical response of a piezoelectric energy harvester and the amount of power it generates are completely dependent on the nature of the ambient energy. We consider harmonic as well as non-harmonic forms of ambient excitation. Moving-load excitations, transient vibrations, periodic inputs, and airflow-induced vibrations are also discussed. Each chapter ends with a notes section, which provides additional discussions, and references for further reading.

We are indebted to financial support for our research and experiments performed in energy harvesting. Specifically, some of the results presented here are referred from publications funded by energy harvesting grants from the Air Force Office of Scientific Research (under the programs and encouragement of Dr. "Les" Lee) and the National Institute of Standards and Technology (under the direction of Dr. Jean-Louis Staudenmann). We have also enjoyed collaboration with several colleagues on this subject, such as Dr. Carlos De Marqui Jr. (University of Sao Paulo, Sao Carlos, Brazil) on energy harvesting from aeroelastic vibrations and Dr. Didem Ozevin (University of Illinois, Chicago) along with the MISTRAS Group, Inc. Products and Systems Division on the energy harvesting potential of bridges as well as the bridge strain data. In addition, we have had the pleasure and support of communications with Professor Earl Dowell, Dr. Brian Mann, Mr. Sam Stanton (Duke University), Professor Ephrahim Garcia (Cornell University), Professor Yi-Chung Shu (National Taiwan University), Professor Niell Elvin (The City College of New York), Professor Dane Quinn (University of Akron), Dr. Mohammed Daqaq (Clemson University), Dr. Steve Burrow, Dr. David Barton (University of Bristol, UK), Professor Sondipon Adhikari, Professor Michael Friswell (Swansea University, UK), Dr. Andres Arrieta (TU Darmstadt, Germany), and Dr. Ho-Yong Lee (Ceracomp Co. Ltd., Korea). Professor Shashank Priya, Mr. Steve Anton, Ms. Na Kong, Mr. Justin Farmer, and other colleagues and graduate students in the Center for Intelligent Material Systems and Structures at Virginia Tech helped to form an atmosphere of discovery and collegiality, without which this effort would not be possible. In particular, we are indebted to Ms. Beth Howell for all of her help as our program manager. We are also indebted to Mr. Neville Hankins for copyediting and Ms. Shalini Sharma for typesetting the book. Lastly, we would like to thank the energy harvesting community at large for its contributions through the literature and many discussions at conferences and workshops for forming an intellectually stimulating environment.

A. Erturk
Atlanta, Georgia

D.J. Inman
Blacksburg, Virginia

1

Introduction to Piezoelectric Energy Harvesting

This chapter provides an introduction to vibration-based energy harvesting using piezoelec-tric transduction. Following a summary of the basic transduction mechanisms that can be used for vibration-to-electricity conversion, the advantages of piezoelectric transduction over the other alternatives (particularly electromagnetic and electrostatic transductions) are dis-cussed. Since the existing review articles mentioned in this chapter present an extensive review of the literature of piezoelectric energy harvesting, only the self-charging structure concept that uses flexible piezoceramics and thin-film batteries is summarized as a motivating example of multifunctional aspects. The focus is then placed on summarizing the literature of mathe-matical modeling of these devices for various problems of interest, ranging from exploiting mechanical nonlinearities to aeroelastic energy harvesting. Along with historical notes, the mathematical theory of linear piezoelectricity is briefly reviewed in order to derive the con-stitutive equations for piezoelectric continua based on the first law of thermodynamics, which are later simplified to reduced forms for use throughout this text. An outline of the remaining chapters is also presented.

1.1 Vibration-Based Energy Harvesting Using Piezoelectric Transduction

Vibration-based energy harvesting has received growing attention over the last decade. The research motivation in this field is due to the reduced power requirement of small electronic components, such as the wireless sensor networks used in passive and active monitoring applications. The ultimate goal in this research field is to power such small electronic devices by using the vibrational energy available in their environment. If this can be achieved, the requirement of an external power source as well as the maintenance costs for periodic battery replacement and the chemical waste of conventional batteries can be reduced.

As stated by Williams and Yates [1] in their early work on harvesting vibrational energy for microsystems, the three basic vibration-to-electric energy conversion mechanisms are the

Piezoelectric Energy Harvesting, First Edition. Alper Erturk and Daniel J. Inman.
© 2011 John Wiley & Sons, Ltd. Published 2011 by John Wiley & Sons, Ltd.

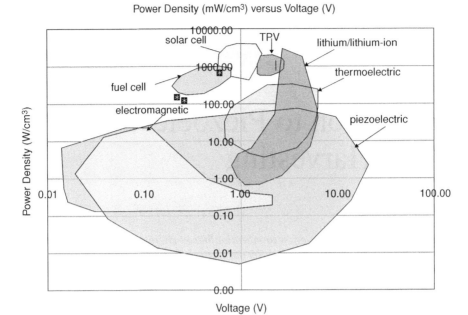

Figure 1.1 Power density versus voltage comparison of common regenerative and lithium/lithium-ion power supply strategies (from Cook-Chennault *et al.* [13], reproduced by permission of IOP © 2008)

electromagnetic [1–3], electrostatic [4,5] and piezoelectric [6,7] transductions.[1] Over the last decade, several articles have appeared on the use of these transduction mechanisms for low power generation from ambient vibrations. Two of the review articles covering mostly the experimental research on all transduction mechanisms are given by Beeby *et al.* [12] and Cook-Chennault *et al.* [13]. Comparing the number of publications that have appeared using each of these three transduction alternatives, it can be seen that piezoelectric transduction has received the greatest attention, especially in the last five years. Several review articles [13–16] have appeared in four years (2004–2008) with an emphasis on piezoelectric transduction to generate electricity from vibrations.

The main advantages of piezoelectric materials in energy harvesting (compared to using the other two basic transduction mechanisms) are their large power densities and ease of application. The power density[2] versus voltage comparison given in Figure 1.1 (due to Cook-Chennault *et al.* [13]) shows that piezoelectric energy harvesting covers the largest area in the graph with power density values comparable to those of thin-film and thick-film lithium-ion

---

[1] Other techniques of vibration-based energy harvesting include magnetostriction [8,9] and the use of electroactive polymers [10,11].

[2] The *power density* of an energy harvester is the power output divided by the device volume for a given input. In vibration-based energy harvesting, the input (excitation) is often characterized by the acceleration level and the frequency (usually a single harmonic is considered). Therefore, unless the input energy is provided, the power density is an insufficient parameter to compare different energy harvesters, although this has been frequently done in the literature.

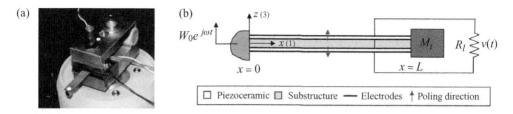

Figure 1.2 (a) A cantilevered piezoelectric energy harvester tested under base excitation and (b) its schematic representation

batteries and thermoelectric generators. As can be seen in Figure 1.1, voltage outputs in electromagnetic energy harvesting are typically very low and often multistage post-processing is required in order to reach a voltage level that can charge a storage component. In piezoelectric energy harvesting, however, usable voltage outputs can be obtained directly from the piezoelectric material itself. When it comes to electrostatic energy harvesting, an input voltage or charge needs to be applied so that the relative vibratory motion of the capacitor elements creates an alternating electrical output [4,5]. On the other hand, the voltage output in piezoelectric energy harvesting emerges from the constitutive behavior of the material, which eliminates the requirement of an external voltage input. As another advantage, unlike electromagnetic devices, piezoelectric devices can be fabricated both in macro-scale and micro-scale due to the well-established thick-film and thin-film fabrication techniques [7,17]. Poor properties of planar magnets and the limited number of turns that can be achieved using planar coils are some of the main practical limitations in enabling micro-scale electromagnetic energy harvesters [13].

Most piezoelectric energy harvesters are in the form of cantilevered beams with one or two piezoceramic layers (i.e., a unimorph or a bimorph).[3] The harvester beam is located on a vibrating host structure and the dynamic strain induced in the piezoceramic layer(s) results in an alternating voltage output across their electrodes. An example of a cantilever tested under base excitation is shown in Figure 1.2 along with its schematic. An alternating voltage output is obtained due to the harmonic base motion applied to the structure. In the mechanics research on piezoelectric energy harvesting as well as in the experimental research conducted to estimate the device performance for AC power generation, it is common practice to consider a resistive load in the electrical domain [6,7,20–27] to represent and electronic load as depicted in Figure 1.2b.

From an electrical engineering point of view, the alternating voltage output should be converted to a stable rectified voltage through a rectifier bridge and a smoothing capacitor (which constitute an AC–DC converter) for charging a small battery or a capacitor by using the harvested energy. Often a second stage (DC–DC converter) is employed to regulate the voltage output of the rectifier so that the power transfer to the storage device can be maximized (Figure 1.3). These electrical circuit and power electronics aspects [28–34] are addressed in

---

[3] Typical exceptions are the stack [18] and the cymbal [19] transducers used under direct force excitation in limited cases or patches excited by surface strain fluctuations (Chapter 7). Polyvinylidene fluoride (PVDF) membranes can also be used for piezoelectric power generation; however, they exhibit very low electromechanical coupling compared to piezoceramics and they are not discussed here.

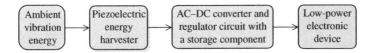

Figure 1.3 Schematic representation of the concept of a piezoelectric energy harvesting system

the last chapter of this book with a brief review of the relevant literature. The majority of this text takes a perspective from the mechanical engineering standpoint with a focus on detailed electromechanical modeling of piezoelectric energy harvesters (with simplified circuits) for various forms of excitation and engineering applications of practical importance.

1.2 An Example of a Piezoelectric Energy Harvesting System

Although examples from the literature of piezoelectric energy harvesting will not be reviewed here (the reader is referred to the existing review articles [13–16]), it is worth discussing a specific application of multidisciplinary aspects. The concept of *self-charging structures* refers to structures composed of elastic substructures (usually metallic or carbon fiber), flexible piezoceramics embedded in kapton layers, and flexible thin-film battery layers [35–38]. Proof-of-concept prototypes of self-charging structures with aluminum and carbon-fiber substructures are shown in Figure 1.4a. The goal of this concept is to use these structures in load-bearing applications to improve *multifunctionality* [39, 40] for low-power applications (such as powering a wireless sensor in the vicinity of the load-bearing structure using the harvested and stored energy). That is, if the existing load-bearing structure can be modified using flexible piezoceramics and thin-film batteries, electrical energy can be generated from the dynamic loads and stored inside the structure itself. As long as the load-bearing capacity of the modified structure is within an acceptable margin, this concept provides a multifunctional structure that can find applications for remote structural systems with battery-powered wireless electronic components. Dynamic excitation of a self-charging structure to charge its battery layers is shown in Figure 1.4b and its schematic view is depicted in Figure 1.4c. Energy-efficient charging of the battery layers (Figure 1.4d) by using the electrical output of the piezoceramic layers requires sophisticated considerations in the electrical domain. A regulator circuit that operates in the discontinuous conduction mode [34] for resistive impedance matching is shown in Figure 1.3e. In order to claim multifunctionality, it is required to test and characterize the load-bearing capacity of the multilayer structure. Three-point bending tests (Figure 1.4f) verify that the most critical layers are the brittle piezoceramics, although the kapton layers prevent disintegration of the structure after the fracture [41]. A primary motivation for self-charging structures is to use them for powering small electronic components in unmanned aerial vehicle (UAV) applications (Figure 1.4g).

More recently, flexible solar panels have also been combined with piezoceramics, thin-film batteries, and metallic layers as shown in Figure 1.5 [42]. The purpose is to utilize both sunlight and structural vibrations for charging the battery layers. Multifunctionality is again due to the load-bearing capability of the entire assembly with flexible layers. The electrical energy that can be harvested through the flexible solar cells depends mainly on the amount of the solar irradiance level, while the piezoelectric power output strongly depends on the electromechanical frequency response of the structure due to the *resonance* phenomenon.

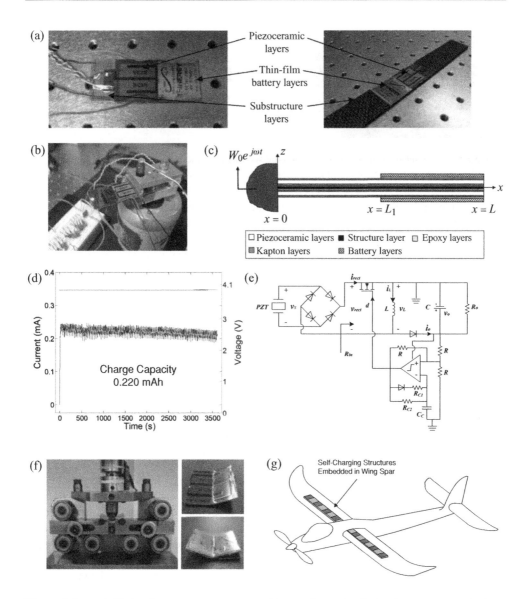

Figure 1.4 (a) Self-charging structure prototypes; (b) a cantilevered self-charging structure under base excitation; (c) schematic representation of the self-charging structure; (d) time history of charging a thin-film battery; (e) resistive impedance matching circuit; (f) experimental characterization of the mechanical strength; (g) an application for unmanned aerial vehicles

For a given vibration input (prescribed acceleration or dynamic strain at a certain frequency and amplitude or with certain statistical characteristics), the piezoelectric power output of the self-charging structure shown in Figure 1.5 is a complicated function of the geometric and material properties of its layers. Although the electrical response due to structural vibrations

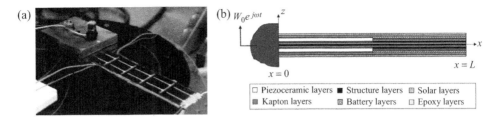

Figure 1.5 (a) A self-charging structure prototype with flexible metallic, piezoceramic, thin-film battery, solar, kapton, and epoxy layers, and (b) its schematic representation

is obtained from the piezoceramics layers, it is the dynamics of the entire composite structure that determine the resonance frequencies, that is, the effective frequencies of maximum power generation. Therefore, it is important to model the structure with substantial accuracy from the mechanical standpoint so that the resulting electromechanical model can provide reliable estimates of the acceleration level required to reach a desired voltage output as well as the matched resistance (for the circuit design) and the maximum acceleration that can be sustained by the piezoceramic layers (for the structural design), among other design parameters at the systems level. Moreover, often the vibratory motion of the energy harvester can be more sophisticated than simple harmonic oscillations at a single frequency; for example, it might involve nonlinear dynamical aspects or coupling with the surround airflow. The aim of this book is also to provide an introduction to such advanced topics in vibration-based energy harvesting.

1.3 Mathematical Modeling of Piezoelectric Energy Harvesters

As in the particular case of the self-charging structure concept summarized in Figures 1.4 and 1.5, research in the area of piezoelectric energy harvesting is strongly connected to various disciplines of engineering. Consequently, this promising way of powering small electronic components and remote sensors has attracted researchers from different disciplines of engineering, including mechanical, aerospace, electrical, and civil, as well as researchers from the field of materials science [13–16], and various modeling approaches have appeared as summarized in the following. A comprehensive mathematical model should be as simple as possible yet sophisticated enough to capture the important phenomena needed to represent and predict the dynamics of the physical system as required by the application of interest.

In the early mathematical modeling treatments, researchers [20, 21] employed lumped-parameter solutions with a single mechanical degree of freedom to predict the coupled system dynamics of piezoelectric energy harvesters (such as the one shown in Figure 1.2). Lumped-parameter modeling is a convenient modeling approach since the electrical domain already consists of lumped parameters: a capacitor due to the internal (or inherent) capacitance of the piezoceramic and a resistor due to an external load resistance. Hence, the only requirement is to obtain the lumped parameters representing the mechanical domain so that the mechanical equilibrium and the electrical loop equations can be coupled through the piezoelectric constitutive equations [43] and a transformer relation can be established. This was the main procedure followed by Roundy and Wright [20] and duToit *et al.* [21] in their lumped-parameter model derivations. Although lumped-parameter modeling gives initial insight into the problem by allowing simple expressions, it is an approximation limited to a single vibration mode and it lacks some important aspects of the coupled physical system, such as the dynamic mode shapes and accurate strain distribution as well as their effects on the electrical response.

Since cantilevered energy harvesters are basically excited under base motion, the well-known lumped-parameter harmonic base excitation relation taken from elementary vibration texts [44, 45] has been used widely in the energy harvesting literature for both modeling [21] and studying the maximum power generation and parameter optimization [46, 47]. In both lumped-parameter models [20, 21] (derived for the transverse vibrations and longitudinal vibrations, respectively), the contribution of the distributed mass (spring mass in the lumped-parameter sense) to the forcing amplitude in the base excitation problem is neglected. The inertial contribution of the distributed mass to the excitation amplitude can be important, especially if the harvester does not have a large proof mass [48].

As an improved modeling approach, the Rayleigh–Ritz type [44] of discrete formulation originally derived by Hagood *et al.* [49] for piezoelectric actuation (using the generalized Hamilton's principle for electromechanical systems given by Crandall *et al.* [50]) was employed by Sodano *et al.* [22] and duToit *et al.* [21,26] for modeling cantilevered piezoelectric energy harvesters (based on the Euler–Bernoulli beam theory). The Rayleigh–Ritz model gives a spatially discretized model of the distributed-parameter system and is a more accurate approximation compared to lumped-parameter modeling with a single mechanical degree of freedom. The Rayleigh–Ritz model gives an approximate representation of the distributed-parameter system (Figure 1.2) as a discretized system by reducing its mechanical degrees of freedom from infinity to a finite dimension and usually it is computationally more expensive than the analytical solution (if available).

In order to obtain analytical expressions, several others [23–25,27] used the vibration mode shapes obtained from the Euler–Bernoulli beam theory along with the piezoelectric constitutive equation that gives the electric displacement to relate the electrical outputs to the vibration mode shape. The important deficiencies in these mathematical modeling attempts and others were summarized in the literature [51] and they include the lack of consideration of the resonance phenomenon, ignorance of modal expansion, misrepresentation of the forcing due to base excitation, oversimplified modeling of piezoelectric coupling in the beam equation as viscous damping, and use of the static sensing/actuation equations in a fundamentally dynamic problem.

The analytical solutions based on distributed-parameter electromechanical modeling were given by Erturk and Inman [52, 53] along with experimental validations. Convergence of the Rayleigh–Ritz solution to the analytical solution was shown by Elvin and Elvin [54], who combined the lumped parameters obtained from the Rayleigh–Ritz formulation with circuit simulation software to investigate more sophisticated circuits in the time domain. Finite-element simulations given by Rupp *et al.* [55], De Marqui *et al.* [56], Elvin and Elvin [57], and Yang and Tang [58] were also shown to agree with the analytical solutions [52,53] and they were introduced for various purposes ranging from topology optimization [55] and added mass optimization [56] to analysis of nonlinear circuits [57,58].

Later, researchers focused on exploiting mechanical nonlinearities in vibration-based energy harvesting and the linear electromechanical models were modified accordingly. Erturk *et al.* [59] used the large-amplitude periodic attractor of a bistable *piezomagnetoelastic* device[4] and observed a substantially improved broadband energy harvesting performance that

---

[4] This device is based on the magnetoelastic structure introduced by Moon and Holmes [60,61] to study chaotic vibrations in structural mechanics.

had a power output an order of magnitude larger than the conventional cantilever configuration. The mechanical component of the nonlinear electromechanical equations [59] obeys the form of the *bistable* Duffing oscillator [62–65].[5] Contemporaneously, Stanton *et al.* [66] also combined the piezoelastic cantilever configuration with magnets to create softening and hardening stiffness effects in the form of the *monostable* Duffing oscillator.[6] In addition, they [70] studied an alternative bistable structural configuration focusing on chaotic vibrations as well as other dynamic responses emerging from the bifurcations of the system with respect to excitation frequency and base acceleration amplitude. Parametric excitation [63] of a nonlinear piezoelectric energy harvester was theoretically and experimentally investigated by Daqaq *et al.* [71].

The effect of inherent piezoelectric nonlinearity (formerly pointed out by Crawley and Anderson [72] and Crawley and Lazarus [73] for large electric fields in structural actuation) was implemented by Triplett and Quinn [74] for energy harvesting. Stanton and Mann [75] presented a Galerkin solution by taking into account the geometric and piezoelastic nonlinearities for weak electric fields [76] (since the electric field levels in energy harvesting are not as high as in actuation). For relatively stiff cantilevers, inherent piezoelastic nonlinearities can be pronounced even when the oscillations are geometrically linear.[7] Therefore the geometrically and piezoelectrically nonlinear modeling framework [75] was later simplified and used for quantitatively identifying the piezoelastic nonlinearities in the absence of geometric nonlinearity for a bimorph cantilever [77].

Another modeling problem of interest is the *stochastic* excitation of vibration-based energy harvesters since waste vibrational energy often appears in nondeterministic [78, 79] forms. Scruggs [80] investigated the optimal control of a linear energy harvester network for increased power flow to a storage system under stochastic excitation. An analysis of random vibrations of a linear piezoelectric energy harvester was carried out by Adhikari *et al.* [81] based on lumped-parameter modeling with a single mechanical degree of freedom. Daqaq [82] considered hardening stiffness in the lumped-parameter formulation (for electromagnetic energy harvesting) and showed that the monostable Duffing oscillator does not provide any enhancement over the typical linear oscillators under white Gaussian and colored excitations.[8] Cottone *et al.* [84] and Gammaitoni *et al.* [85] reported that nonlinear oscillators can outperform the linear ones under noise excitation both for bistable [84] and monostable [85] configurations. Utilization of nonlinear *stochastic resonance* [86–88] for vibration-based

---

[5] The terms *bistable* and *monostable* are used in this text for the *static* equilibrium of the respective configuration (i.e., statically bistable and statically monostable). In this context, for instance, the dynamic response of a monostable Duffing oscillator can have multiple (coexisting) stable solutions but it is statically monostable.

[6] Implementations of the monostable Duffing oscillator with hardening stiffness were formerly presented by Burrow *et al.* [67] and Mann and Sims [68] for electromagnetic energy harvesting. Use of the hardening stiffness in the monostable form of the Duffing oscillator was also discussed by Ramlan *et al.* [69] along with snap-through in a bistable mass–spring–damper mechanism.

[7] Recently, it has been observed that the nonlinearities in piezoelastic behavior and dissipation can become effective if the base excitation level exceeds a few hundreds of milli-*g* acceleration [75,77]. In the absence of other nonlinearities, the manifestation of piezoelastic nonlinearities is a softening stiffness effect; that is, the peak of the amplitude moves toward the left on the frequency axis with increasing base acceleration.

[8] His theoretical observation agrees with the experimental results contemporaneously reported by Barton *et al.* [83] for electromagnetic energy harvesting with stiffness nonlinearity of Duffing-type (where the majority of the high-energy branch obtained under periodic forcing is not reached under random forcing).

energy harvesting was formerly discussed by McInnes *et al.* [89] as well, without focusing on a specific transduction mechanism. More recently, Litak *et al.* [90] employed the lumped-parameter nonlinear equations given by Erturk *et al.* [59] and presented numerical simulations of stochastic resonance in the piezomagnetoelastic energy harvester under noise excitation.

Some of the interesting applications of piezoelectric energy harvesting involve fluid–structure interactions. For the piezoaeroelastic problem of energy harvesting from the airflow excitation of a cantilevered plate with embedded piezoceramics, De Marqui *et al.* [91, 92] presented finite-element models based on the vortex-lattice method [91] and the doublet-lattice method [92] of aeroelasticity [93–96]. Time-domain simulations [91] were given for a cantilevered plate with embedded piezoceramics for various airflow speeds below the linear flutter speed and at the flutter boundary. Frequency-domain simulations [92] considering resistive and resistive–inductive circuits were also presented focusing on the linear response at the flutter boundary. Bryant and Garcia [97, 98] studied the aeroelastic energy harvesting problem for a typical section by using the finite state theory of Peters *et al.* [99]. Erturk *et al.* [100] presented an experimentally validated lumped-parameter model for a wing section (airfoil) with piezoceramics attached onto plunge stiffness members using the unsteady aerodynamic model proposed by Theodorsen [101]. Piezoelectric power generation at the flutter boundary and its effect on the linear flutter speed have also been discussed [100]. In addition to their possible integration with aerospace structures, aeroelastic energy harvesters are considered as a scalable alternative to the conventional windmill design [102]. As an alternative to airfoil-based and wing-based configurations, St. Clair *et al.* [103] presented a design that uses a piezoelectric beam embedded within a cavity under airflow and described the resulting limit-cycle oscillations through the governing equations of the Van der Pol oscillator [63–65,104]. Vortex-induced oscillations of piezoelectric cantilevers located behind bluff bodies were investigated by Robbins *et al.* [105], Pobering *et al.* [106] and Akaydin *et al.* [107, 108] through experiments and numerical simulations. Other than airflow excitation, Elvin and Elvin [109] theoretically investigated the flutter response of a cantilevered pipe with piezoceramics for power generation from liquid flow and its effect on the flutter instability.

The foregoing summary includes some of the major papers on the mathematical modeling of piezoelectric energy harvesters for various applications where the presence of waste vibrational energy might have some value. Before reviewing the outline of the material covered in this text, it is useful to summarize the theory of linear piezoelectricity with brief historical notes.

1.4 Summary of the Theory of Linear Piezoelectricity

Piezoelectricity is a form of coupling between the mechanical and the electrical behaviors of ceramics and crystals belonging to certain classes. These materials exhibit the *piezoelectric effect*, which is historically divided into two phenomena as the *direct* and the *converse* piezoelectric effects. In simplest terms, when a piezoelectric material is mechanically strained, electric polarization that is proportional to the applied strain is produced. This is called the direct piezoelectric effect and it was discovered by the Curie brothers in 1880. When the same material is subjected to an electric polarization, it becomes strained and the amount of strain is proportional to the polarizing field. This is called the converse piezoelectric effect (sometimes called the "inverse" piezoelectric effect) and it was deduced mathematically from the fundamental principles of thermodynamics by Lippmann[9] in 1881 [110] and then confirmed experimentally by the Curie brothers in the same year. It is important to note that these two effects usually coexist in a piezoelectric material. Therefore in an application where the direct piezoelectric effect is of particular interest (which is the case in vibration-based energy harvesting), ignoring the presence of the converse piezoelectric effect would be thermodynamically inconsistent.[10]

Several natural crystals were observed to exhibit the piezoelectric effect in the first half of the last century, for example, Rochelle salt, quartz, and so on [111]. However, in order to use them in engineering applications, the *electromechanical coupling* between the mechanical and the electrical behaviors of the material has to be sufficiently strong. Piezoelectric ceramics were developed in the second half of the last century and they exhibit much larger coupling compared to natural crystals. The most popular of engineering piezoceramics, PZT (lead zirconate titanate), was developed at the Tokyo Institute of Technology in the 1950s and various versions of it (particularly PZT-5A and PZT-5H) are today the most commonly used engineering piezoceramics. As far as energy harvesting research is concerned, PZT-5A and PZT-5H are the most widely implemented piezoceramics according to the literature [13].

Decades after the fundamental contributions to the field of piezoelectricity (by Voigt [111], Cady [112], Heising [113], Mason [114], Mindlin and Tiersten [115–117], among others cited in their work), today's commonly used IEEE Standard on Piezoelectricity [43] was published.[11] It is useful to review the derivation of the linear constitutive equations of a piezoelectric material [43], which will later be simplified to reduced forms (Appendix A) for use in various chapters of this text.

The first law of thermodynamics (the principle of conservation of energy) for a linear piezoelectric continuum leads to [117]

$$\dot{U} = T_{ij}\dot{S}_{ij} + E_i\dot{D}_i \qquad (1.1)$$

where U is the stored energy density of the piezoelectric continuum, T_{ij} is the stress tensor, S_{ij} is the strain tensor, E_i is the electric field tensor, D_i is the electric displacement tensor and the overdot represents differentiation with respect to time.[12]

The electric enthalpy density H is given by

$$H = U - E_i D_i \qquad (1.2)$$

Substituting Equation (1.1) into the time derivative of Equation (1.2) gives

$$\dot{H} = T_{ij}\dot{S}_{ij} - D_i\dot{E}_i \qquad (1.3)$$

---

[9] Lippmann's relevant article [110] on the application of thermodynamic principles to reversible processes involving electrical quantities includes electrostriction and pyroelectricity as well (which also have their converse effects).

[10] Ignoring the effect of piezoelectric power generation on the dynamics of the energy harvester is a typical oversimplification repeated in the existing literature, as summarized by Erturk and Inman [51].

[11] This [43] refers to ANSI/IEEE Std 176-1987. An earlier version of it (ANSI/IEEE Std 176-1978) was published in 1978 (which followed various standards dating back to 1950s).

[12] The electric field and the electric displacement tensors are Cartesian tensors [118] of order 1 (i.e., they are vectors). Although, in principle, the stress and the strain tensors are of order 2, these are symmetric tensors and their subscripts can be relabeled based on Voigt's notation [111] so that they become vectors of six components (Appendix A).

which implies $H = H(S_{ij}, E_i)$ so that the components of the stress and the electric displacement tensors can be derived from the electric enthalpy density as [117]

$$T_{ij} = \frac{\partial H}{\partial S_{ij}}, \quad D_i = -\frac{\partial H}{\partial E_i} \tag{1.4}$$

The form of the electric enthalpy density in the linearized theory of piezoelectricity is

$$H = \frac{1}{2}c_{ijkl}^E S_{ij} S_{kl} - e_{kij} E_k S_{ij} - \frac{1}{2}\varepsilon_{ij}^S E_i E_j \tag{1.5}$$

where c_{ijkl}^E, e_{kij}, and ε_{ij}^S are the elastic, piezoelectric, and permittivity constants, respectively, while the superscripts E and S denote that the respective constants are evaluated at constant electric field and constant strain, respectively.

Using Equations (1.4) and (1.5) along with the relation $\partial S_{ij}/\partial S_{ji} = \delta_{ij}$ (where δ_{ij} is the Kronecker delta defined as being equal to unity for $i = j$ and equal to zero for $i \neq j$), one obtains [43,117]

$$T_{ij} = c_{ijkl}^E S_{kl} - e_{kij} E_k \tag{1.6}$$

$$D_i = e_{ikl} S_{kl} + \varepsilon_{ik}^S E_k \tag{1.7}$$

which is the form of the linear constitutive equations for the unbounded piezoelectric continuum. The following three pairs constitute alternative forms of piezoelectric constitutive equations and they are used for approximations under certain limiting circumstances [43] (usually for bounded piezoelectric media):

$$S_{ij} = s_{ijkl}^E T_{kl} + d_{kij} E_k \tag{1.8}$$

$$D_i = d_{ikl} T_{kl} + \varepsilon_{ik}^T E_k \tag{1.9}$$

and

$$S_{ij} = s_{ijkl}^D T_{kl} + g_{kij} D_k \tag{1.10}$$

$$E_i = -g_{ikl} T_{kl} + \beta_{ik}^T D_k \tag{1.11}$$

and

$$T_{ij} = c_{ijkl}^D S_{kl} - h_{kij} D_k \tag{1.12}$$

$$E_i = -h_{ikl} S_{kl} + \beta_{ik}^S D_k \tag{1.13}$$

where d_{kij}, g_{kij}, and h_{kij} are alternative forms of the piezoelectric constants, s_{ijkl}^E and s_{ijkl}^D are the elastic compliance constants, and β_{ik}^T and β_{ik}^S are the impermittivity constants. Here, the superscripts D and T denote that the respective constants are evaluated at constant electric displacement and constant stress, respectively. The transformations between the elastic, piezoelectric, and dielectric constants under different electrical and mechanical conditions can be found in the IEEE standard [43].

The electromechanical models derived in this book use the form of the constitutive equations given by Equations (1.8) and (1.9) for deriving reduced elastic, piezoelectric, and dielectric constants depending on the structural theory used and the locations of the electrodes (Appendix A). The surviving stress components and the electric displacement components are then used in model derivations as the dependent field variables in the form of Equations (1.6) and (1.7) while the strain and the electric field components are expressed in terms of the displacement field and the voltage output, respectively.

1.5   Outline of the Book

Vibrational energy is available in various environments as an alternative form of waste energy. The modeling treatment required to predict the coupled dynamics of a piezoelectric energy harvester changes dramatically depending on the application and the form of the excitation. For instance, although it is common practice to consider harmonic excitation to characterize the resonance behavior of a piezoelectric energy harvester, scavenging energy from vibrations of an aircraft wing requires a more involved analysis due to the coupling of the piezoelastic structure with the surrounding airstream. The motivation of this book is therefore to cover the basic mathematical modeling treatments for different problems and applications of piezoelectric energy harvesting. The problems discussed here range from nonlinear energy harvesting to energy harvesting from aeroelastic vibrations and the applications of interest include civil, mechanical, and aerospace engineering structures.

Chapter 2 reviews the formal treatment of the base excitation problem since piezoelectric energy harvesters are often designed as cantilevers operated under harmonic base excitation. After showing the inaccuracy of the frequently quoted lumped-parameter base excitation equation for a single mechanical degree of freedom, correction factors are derived to improve the lumped-parameter equations for both the transverse and the longitudinal vibration cases. Experimental validations are given and an amplitude correction factor is introduced to the lumped-parameter piezoelectric energy harvester equations. The distributed-parameter derivations given in this chapter constitute the mechanical background for the modeling of cantilevers excited under base motion in vibration-based energy harvesting.

Chapter 3 introduces electromechanically coupled analytical solutions of symmetric bimorph piezoelectric energy harvester configurations under base excitation for the series and parallel connections of the piezoceramics layers. The formulation is given based on the thin-beam theory since piezoelectric energy harvesters are typically thin structures. The multi-mode solutions (valid for arbitrary excitation frequencies) and the single-mode solutions (approximately valid at or near the resonance frequencies) are obtained and an extensive theoretical case study is provided. The analytical solutions obtained in this chapter are used for several purposes throughout the book.

Chapter 4 provides detailed experimental validations for the analytical expressions derived in Chapter 3. Three case studies are given with a focus on a brass-reinforced PZT-5H cantilever in the absence and presence of a tip mass and a brass-reinforced PZT-5A cantilever. The effect of the rotary inertia of a tip mass is discussed and the single-mode and multi-mode electromechanical frequency response functions (FRFs) are compared to the experimental measurements. Validations of the electromechanical FRFs are given along with the electrical performance diagrams. In addition to validation of the analytical solutions, this chapter provides a comprehensive discussion of the experimental testing and characterization of a piezoelectric energy harvester.

Chapter 5 presents dimensionless forms of the analytical expressions and detailed mathematical analyses of these equations along with experimental validations. The asymptotic forms of the voltage and displacement FRFs are obtained for the short-circuit and open-circuit conditions, which are then employed to derive closed-form expressions for identifying various system parameters. Analysis of the voltage asymptotes leads to a simple experimental technique for identifying the optimum electrical load using an open-circuit voltage measurement and a single resistor.

Chapter 6 presents approximate analytical solutions using an electromechanical version of the assumed-modes method for relatively complicated structural configurations which do not allow analytical solutions. The Euler–Bernoulli, Rayleigh, and Timoshenko types of solutions are presented. The derivations given in this chapter can be used for energy harvesters with varying cross-section and material properties, asymmetric laminates as well as moderately thick energy harvesters. The experimental cases given in Chapter 4 are revisited for validation of the approximate solutions using the electromechanical assumed-modes method. Convergence of the assumed-modes solution to the analytical solution with increasing number of modes is also shown. Another experimental case study is given for the modeling of a two-segment bimorph energy harvester.

Chapter 7 introduces distributed-parameter electromechanical models for piezoelectric energy harvesting under various forms of dynamic loading. The problems discussed in this chapter are periodic excitation, random excitation in the form of ideal white noise, moving load excitation of slender bridges, excitation due to strain fluctuations on large structures, and general transient base excitation. Detailed derivations are given for representing the input force to be used in the electromechanical problem and for predicting the electrical output under these various excitation forms. Two case studies are presented for the periodic excitation of a bimorph cantilever located on a four-bar mechanism and for strain fluctuations on a civil engineering structure.

Chapter 8 focuses on modeling and exploiting mechanical nonlinearities in piezoelectric energy harvesting. Enhancement of the bandwidth using the monostable form of the Duffing equation with softening and harvesting stiffness effects is discussed and a perturbation-based electromechanical formulation is given. The perturbation solution is verified against the time-domain numerical simulations. After the monostable case, the outstanding broadband power generation performance of a bistable piezomagnetoelastic energy harvester is investigated theoretically and experimentally. Chaotic vibration of the bistable piezoelectric energy harvester configuration is also discussed. Experimental performance results of a bistable carbon-fiber-epoxy plate with piezoceramics are summarized for nonlinear broadband piezoelectric energy harvesting.

Chapter 9 investigates the problem of piezoelectric energy harvesting from aeroelastic vibrations of structures with piezoceramic layers under airflow excitation. The linear piezoaeroelastic models discussed here include a two-degree-of-freedom formulation (for the typical section problem with plunge and pitch degrees of freedom) and assumed-modes formulation (for the distributed-parameter problem of a cantilever with bending and torsion modes). Finite-element piezoaeroelastic models using the vortex-lattice and the doublet-lattice methods are also reviewed in this chapter. Experimental validations are presented for the piezoaeroelastic typical section model and theoretical simulations are given for the finite element solutions. The effect of piezoelectric power generation on the linear flutter speed is also discussed.

Chapter 10 considers the effects of material constants and mechanical damping on piezo-electric energy harvesting. Various soft piezoelectric ceramics and single crystals are compared for power generation by using the experimentally validated model introduced in Chapter 3. Performance comparisons for resonance excitation are presented and the effect of damping is investigated along with other important material parameters in the beam-type plane-stress formulation. Soft and hard ceramics as well as soft and hard crystals are also compared for resonant and off-resonant energy harvesting.

Chapter 11 provides a brief review of the literature of electrical circuit papers used in piezoelectric energy harvesting. Lumped-parameter modeling of a piezoelectric energy har-vester with a standard AC–DC converter is summarized and simulation results are reviewed. Two-stage energy harvesting circuits combining AC–DC and DC–DC converters are also dis-cussed. Finally, the synchronous switch harvesting on inductor technique and its modeling are reviewed for performance enhancement in weakly coupled piezoelectric energy harvesters.

References

1. Williams, C.B. and Yates, R.B. (1996) Analysis of a micro-electric generator for microsystems. *Sensors and Actuators A*, **52**, 8–11.
2. Glynne-Jones, P., Tudor, M.J., Beeby, S.P., and White, N.M. (2004) An electromagnetic, vibration-powered generator for intelligent sensor systems. *Sensors and Actuators A*, **110**, 344–349.
3. Arnold, D. (2007) Review of microscale magnetic power generation. *IEEE Transactions on Magnetics*, **43**, 3940–3951.
4. Roundy, S., Wright, P., and Rabaey, J. (2003) *Energy Scavenging for Wireless Sensor Networks*, Kluwer Academic, Boston, MA.
5. Mitcheson, P., Miao, P., Start, B., Yeatman, E., Holmes, A., and Green, T. (2004) MEMS electrostatic micro-power generator for low frequency operation. *Sensors and Actuators A*, **115**, 523–529.
6. Roundy, S., Wright, P.K., and Rabaey, J.M. (2003) A study of low level vibrations as a power source for wireless sensor nodes. *Computer Communications*, **26**, 1131–1144.
7. Jeon, Y.B., Sood, R., Jeong, J.H., and Kim, S. (2005) MEMS power generator with transverse mode thin film PZT. *Sensors & Actuators A*, **122**, 16–22.
8. Wang, L. and Yuan, F.G. (2008) Vibration energy harvesting by magnetostrictive material. *Smart Materials and Structures*, **17**, 045009.
9. Adly, A., Davino, D., Giustiniani, A., and Visone, C. (2010) Experimental tests of a magnetostrictive energy harvesting device towards its modeling. *Journal of Applied Physics*, **107**, 09A935.
10. Koh, S.J.A., Zhao, X., and Suo, Z. (2009) Maximal energy that can be converted by a dielectric elastomer generator. *Applied Physics Letters*, **94**, 262902.
11. Aureli, M., Prince, C., Porfiri, M., and Peterson, S.D. (2010) Energy harvesting from base excitation of ionic polymer metal composites in fluid environments. *Smart Materials and Structures*, **19**, 015003.
12. Beeby, S.P., Tudor, M.J., and White, N.M. (2006) Energy harvesting vibration sources for microsystems applications. *Measurement Science and Technology*, **17**, R175–R195.
13. Cook-Chennault, K.A., Thambi, N., and Sastry, A.M. (2008) Powering MEMS portable devices – a review of non-regenerative and regenerative power supply systems with emphasis on piezoelectric energy harvesting systems. *Smart Materials and Structures*, **17**, 043001.
14. Sodano, H., Park, G., and Inman, D.J. (2004) A review of power harvesting from vibration using piezoelectric materials. *Shock and Vibration Digest*, **36**, 197–205.
15. Anton, S.R. and Sodano, H.A. (2007) A review of power harvesting using piezoelectric materials (2003–2006). *Smart Materials and Structures*, **16**, R1–R21.
16. Priya, S. (2007) Advances in energy harvesting using low profile piezoelectric transducers. *Journal of Electro-ceramics*, **19**, 167–184.

17. Choi, W.J., Jeon, Y., Jeong, J.H., Sood, R., and Kim, S.G. (2006) Energy harvesting MEMS device based on thin film piezoelectric cantilevers. *Journal of Electroceramics*, **17**, 543–548.

18. Feenstra, J., Granstrom, J., and Sodano, H. (2009) Energy harvesting through a backpack employing a mechanically amplified piezoelectric stack. *Mechanical Systems and Signal Processing*, **22**, 721–734.

19. Kim, H.W., Batra, A., Priya, S., Uchino, K., Markley, D., Newnham, R.E., and Hofmann, H.F. (2004) Energy harvesting using a piezoelectric "cymbal" transducer in dynamic environment. *Japanese Journal of Applied Physics*, **43**, 6178–6183.

20. Roundy, S. and Wright, P.K. (2004) A piezoelectric vibration based generator for wireless electronics. *Smart Materials and Structures*, **13**, 1131–1144.

21. duToit, N.E., Wardle, B.L., and Kim, S. (2005) Design considerations for MEMS-scale piezoelectric mechanical vibration energy harvesters. *Journal of Integrated Ferroelectrics*, **71**, 121–160.

22. Sodano, H.A., Park, G., and Inman, D.J. (2004) Estimation of electric charge output for piezoelectric energy harvesting. *Strain*, **40**, 49–58.

23. Lu, F., Lee, H.P., and Lim, S.P. (2004) Modeling and analysis of micro piezoelectric power generators for micro-electromechanical-systems applications. *Smart Materials and Structures*, **13**, 57–63.

24. Chen, S.-N., Wang, G.-J., and Chien, M.-C. (2006) Analytical modeling of piezoelectric vibration-induced micro power generator. *Mechatronics*, **16**, 397–387.

25. Lin, J.H., Wu, X.M., Ren, T.L., and Liu, L.T. (2007) Modeling and simulation of piezoelectric MEMS energy harvesting device. *Integrated Ferroelectrics*, **95**, 128–141.

26. duToit, N.E. and Wardle, B.L. (2007) Experimental verification of models for microfabricated piezoelectric vibration energy harvesters. *AIAA Journal*, **45**, 1126–1137.

27. Ajitsaria, J., Choe, S.Y., Shen, D., and Kim, D.J. (2007) Modeling and analysis of a bimorph piezoelectric cantilever beam for voltage generation. *Smart Materials and Structures*, **16**, 447–454.

28. Ottman, G.K., Hofmann, H.F., Bhatt, A.C., and Lesieutre, G.A. (2002) Adaptive piezoelectric energy harvesting circuit for wireless remote power supply. *IEEE Transactions on Power Electronics*, **17**, 669–676.

29. Ottman, G.K., Hofmann, H.F., and Lesieutre, G.A. (2003) Optimized piezoelectric energy harvesting circuit using step-down converter in discontinuous conduction mode. *IEEE Transactions on Power Electronics*, **18**, 696–703.

30. Guyomar, D., Badel, A., Lefeuvre, E., and Richard, C. (2005) Toward energy harvesting using active materials and conversion improvement by nonlinear processing. *IEEE Transactions of Ultrasonics, Ferroelectrics, and Frequency Control*, **52**, 584–595.

31. Guan, M.J. and Liao, W.H. (2007) On the efficiencies of piezoelectric energy harvesting circuits towards storage device voltages. *Smart Materials and Structures*, **16**, 498–505.

32. Lefeuvre, E., Audigier, D., Richard, C., and Guyomar, D. (2007) Buck-boost converter for sensorless power optimization of piezoelectric energy harvester. *IEEE Transactions on Power Electronics*, **22**, 2018–2025.

33. Shu, Y.C., Lien, I.C., and Wu, W.J. (2007) An improved analysis of the SSHI interface in piezoelectric energy harvesting. *Smart Materials and Structures*, **16**, 2253–2264.

34. Kong, N., Ha, D.S., Erturk, A., and Inman, D.J. (2010) Resistive impedance matching circuit for piezoelectric energy harvesting. *Journal of Intelligent Material Systems and Structures*, **21**, pp. 1293–1302.

35. Erturk, A., Anton, S.R., and Inman, D.J. (2009) Piezoelectric energy harvesting from multifunctional wing spars for UAVs – Part 1: coupled modeling and preliminary analysis. *Proceedings of SPIE*, **7288**, 72880C.

36. Anton, S.R., Erturk, A., and Inman, D.J. (2009) Piezoelectric energy harvesting from multifunctional wing spars for UAVs – Part 2: experiments and storage applications. *Proceedings of SPIE*, **7288**, 72880D.

37. Anton, S.R., Erturk, A., Kong, N., Ha, D.S., and Inman, D.J. (2009) Self-charging structures using piezoceramics and thin-film batteries. Proceedings of the ASME Conference on Smart Materials, Adaptive Structures and Intelligent Systems, Oxnard, CA, September 20–24, 2009.

38. Anton, S.R., Erturk, A., and Inman, D.J. (2010) Multifunctional self-charging structures using piezoceramics and thin-film batteries. *Smart Materials and Structures*, **19**, 115021.

39. Thomas, J.P. and Qidwai, M.A. (2005) The design and application of multifunctional structure-battery materials systems. *Journal of Minerals, Metals and Materials Society*, **57**, 18–24.

40. Snyder, J.F., Wong, E.L., and Hubbard, C.W. (2009) Evaluation of commercially available carbon fibers, fabrics, and papers for potential use in multifunctional energy storage applications. *Journal of the Electrochemical Society*, **156**, A215–A224.

41. Anton, S.R., Erturk, A., and Inman, D.J. (2010) Strength analysis of piezoceramic materials for structural considerations in energy harvesting for UAVs. *Proceedings of SPIE*, **7643**, 76430E.

42. Gambier, P., Anton, S.R., Kong, N., Erturk, A., and Inman, D.J. (2010) Combined piezoelectric, solar and thermal energy harvesting for multifunctional structures with thin-film batteries. Proceedings of the 21st International Conference on Adaptive Structures and Technologies, State College, PA, October 4–6, 2010.
43. Standards Committee of the IEEE Ultrasonics, Ferroelectrics, and Frequency Control Society (1987) *IEEE Standard on Piezoelectricity*, IEEE, New York.
44. Meirovitch, L. (2001) *Fundamentals of Vibrations*, McGraw-Hill, New York.
45. Inman, D.J. (2007) *Engineering Vibration*, Prentice Hall, Englewood Cliffs, NJ.
46. Stephen, N.G. (2006) On energy harvesting from ambient vibration. *Journal of Sound and Vibration*, **293**, 409–425.
47. Daqaq, F.M., Renno, J.M., Farmer, J.R., and Inman, D.J. (2007) Effects of system parameters and damping on an optimal vibration-based energy harvester. Proceedings of the 48th AIAA/ASME/ASCE/AHS/ASC Structures, Structural Dynamics, and Materials Conference, Honolulu, Hawaii, April 23–26, 2007.
48. Erturk, A. and Inman, D.J. (2008) On mechanical modeling of cantilevered piezoelectric vibration energy harvesters. *Journal of Intelligent Material Systems and Structures*, **19**, 1311–1325.
49. Hagood, N.W., Chung, W.H., and Von Flotow, A. (1990) Modelling of piezoelectric actuator dynamics for active structural control. *Journal of Intelligent Material Systems and Structures*, **1**, 327–354.
50. Crandall, S.H., Karnopp, D.C., Kurtz, E.F. Jr., and Pridmore-Brown, D.C. (1968) *Dynamics of Mechanical and Electromechanical Systems*, McGraw-Hill, New York.
51. Erturk, A. and Inman, D.J. (2008) Issues in mathematical modeling of piezoelectric energy harvesters. *Smart Materials and Structures*, **17**, 065016.
52. Erturk, A. and Inman, D.J. (2008) A distributed parameter electromechanical model for cantilevered piezoelectric energy harvesters. *ASME Journal of Vibration and Acoustics*, **130**, 041002.
53. Erturk, A. and Inman, D.J. (2009) An experimentally validated bimorph cantilever model for piezoelectric energy harvesting from base excitations. *Smart Materials and Structures*, **18**, 025009.
54. Elvin, N.G. and Elvin, A.A. (2009) A general equivalent circuit model for piezoelectric generators. *Journal of Intelligent Material Systems and Structures*, **20**, 3–9.
55. Rupp, C.J., Evgrafov, A., Maute, K., and Dunn, M.L. (2009) Design of piezoelectric energy harvesting systems: a topology optimization approach based on multilayer plates and shells. *Journal of Intelligent Material Systems and Structures*, **20**, 1923–1939.
56. De Marqui, C. Jr., Erturk, A., and Inman, D.J. (2009) An electromechanical finite element model for piezoelectric energy harvester plates. *Journal of Sound and Vibration*, **327**, 9–25.
57. Elvin, N.G. and Elvin, A.A. (2009) A coupled finite element – circuit simulation model for analyzing piezoelectric energy generators. *Journal of Intelligent Material Systems and Structures*, **20**, 587–595.
58. Yang, Y. and Tang, L. (2009) Equivalent circuit modeling of piezoelectric energy harvesters. *Journal of Intelligent Material Systems and Structures*, **20**, 2223–2235.
59. Erturk, A., Hoffmann, J., and Inman, D.J. (2009) A piezomagnetoelastic structure for broadband vibration energy harvesting. *Applied Physics Letters*, **94**, 254102.
60. Moon, F.C. and Holmes, P.J. (1979) A magnetoelastic strange attractor. *Journal of Sound and Vibration*, **65**, 275–296.
61. Holmes, P. (1979) A nonlinear oscillator with a strange attractor. *Philosophical Transactions of the Royal Society of London, Series A*, **292**, 419–449.
62. Duffing, G. (1918) *Erzwungene Schwingungen bei Veränderlicher Eigenfrequenz und ihre technische Bedeutung*, Vieweg, Braunschweig.
63. Nayfeh, A.H. and Mook, D.T. (1979) *Nonlinear Oscillations*, John Wiley & Sons, Inc., New York.
64. Guckenheimer, J. and Holmes, P. (1983) *Nonlinear Oscillations, Dynamical Systems, and Bifurcations of Vector Fields*, Springer-Verlag, New York.
65. Moon, F.C. (1987) *Chaotic Vibrations*, John Wiley & Sons, Inc., New York.
66. Stanton, S.C., McGehee, C.C., and Mann, B.P. (2009) Reversible hysteresis for broadband magnetopiezoelastic energy harvesting. *Applied Physics Letters*, **96**, 174103.
67. Burrow, S.G., Clare, L.R., Carrella, A., and Barton, D. (2008) Vibration energy harvesters with non-linear compliance. *Proceedings of SPIE*, **6928**, 692807.
68. Mann, B.P. and Sims, N.D. (2009) Energy harvesting from the nonlinear oscillations of magnetic levitation. *Journal of Sound and Vibration*, **319**, 515–530.
69. Ramlan, R., Brennan, M.J., Mace, B.R., and Kovacic, I. (2009) Potential benefits of a non-linear stiffness in an energy harvesting device. *Nonlinear Dynamics*, **59**, 545–558.

70. Stanton, S.C., McGehee, C.C., and Mann, B.P. (2010) Nonlinear dynamics for broadband energy harvesting: investigation of a bistable inertial generator. *Physica D*, **239**, 640–653.
71. Daqaq, M.F., Stabler, C., Qaroush, Y., and Seuaciuc-Osorio, T. (2009) Investigation of power harvesting via parametric excitations. *Journal of Intelligent Material Systems and Structures*, **20**, 545–557.
72. Crawley, E. and Anderson, E. (1990) Detailed models for the piezoelectric actuation of beams. *Journal of Intelligent Material Systems and Structures*, **1**, 4–25.
73. Crawley, E. and Lazarus, K. (1991) Induced strain actuation of isotropic and anisotropic plates. *AIAA Journal*, **29**, 944–951.
74. Triplett, A. and Quinn, D.D. (2009) The effect of non-linear piezoelectric coupling on vibration-based energy harvesting. *Journal of Intelligent Material Systems and Structures*, **20**, 1959–1967.
75. Stanton, S.C. and Mann, B.P. (2010) Nonlinear electromechanical dynamics of piezoelectric inertial generators: modeling, analysis, and experiment. *Nonlinear Dynamics* (in review).
76. von Wagner, U. and Hagedorn, P. (2002) Piezo-beam systems subjected to weak electric field: experiments and modeling of nonlinearities. *Journal of Sound and Vibration*, **256**, 861–872.
77. Stanton, S.C., Erturk, A., Mann, B.P., and Inman, D.J. (2010) Nonlinear piezoelectricity in electroelastic energy harvesters: modeling and experimental identification. *Journal of Applied Physics*, **108**, 074903.
78. Bendat, J.S. and Piersol, A.G. (1986) *Random Data Analysis and Measurement Procedures*, John Wiley & Sons, Inc., New York.
79. Newland, D.E. (1993) *Random Vibrations, Spectral and Wavelet Analysis*, John Wiley & Sons, Inc., New York.
80. Scruggs, J.T. (2009) An optimal stochastic control theory for distributed energy harvesting networks. *Journal of Sound and Vibration*, **320**, 707–725.
81. Adhikari, S., Friswell, M.I., and Inman, D.J. (2009) Piezoelectric energy harvesting from broadband random vibrations. *Smart Materials and Structures*, **18**, 115005.
82. Daqaq, M.F. (2010) Response of uni-modal Duffing-type harvesters to random forced excitations. *Journal of Sound and Vibration*, **329**, 3621–3631.
83. Barton, D.A.W., Burrow, S.G., and Clare, L.R. (2010) Energy harvesting from vibrations with a nonlinear oscillator. *ASME Journal of Vibration and Acoustics*, **132**, 021009.
84. Cottone, F., Vocca, H., and Gammaitoni, L. (2009) Nonlinear energy harvesting. *Physical Review Letters*, **102**, 080601.
85. Gammaitoni, L., Neri, I., and Vocca, H. (2009) Nonlinear oscillators for vibration energy harvesting. *Applied Physics Letters*, **94**, 164102.
86. Benzi, R., Sutera, A., and Vulpiani, A. (1981) The mechanism of stochastic resonance. *Journal of Physics A: Mathematical and General*, **14**, L453–L457.
87. Benzi, R., Parisi, G., Sutera, A., and Vulpiani, A. (1982) Stochastic resonance in climatic change. *Tellus*, **34**, 10–16.
88. Gammaitoni, L., Hanggi, P., Jung, P., and Marchesoni, F. (1998) Stochastic resonance. *Reviews of Modern Physics*, **70**, 223–287.
89. McInnes, C.R., Gorman, D.G., and Cartmell, M.P. (2008) Enhanced vibrational energy harvesting using non-linear stochastic resonance. *Journal of Sound and Vibration*, **318**, 655–662.
90. Litak, G., Friswell, M.I., and Adhikari, S. (2010) Magnetopiezoelastic energy harvesting driven by random excitations. *Applied Physics Letters*, **96**, 214103.
91. De Marqui, C. Jr., Erturk, A., and Inman, D.J. (2010) Piezoaeroelastic modeling and analysis of a generator wing with continuous and segmented electrodes. *Journal of Intelligent Material Systems and Structures*, **21**, 983–993.
92. De Marqui, C. Jr., Vieira, W.G.R., Erturk, A., and Inman, D.J. (2011) Modeling and analysis of piezoelectric energy harvesting from aeroelastic vibrations using the doublet-lattice method. *ASME Journal of Vibration and Acoustics*, **133**, 011003.
93. Bisplinghoff, R.L. and Ashley, H. (1962) *Principles of Aeroelasticity*, John Wiley & Sons, Inc., New York.
94. Fung, Y.C. (1969) *Introduction to the Theory of Aeroelasticity*, Dover, New York.
95. Dowell, E.H., Curtiss, H.C. Jr., Scalan, R.H., and Sisto, F. (1978) *A Modern Course in Aeroelasticity*, Sijthoff & Nordhoff, The Hague.
96. Hodges, D.H. and Pierce, G.A. (2002) *Introduction to Structural Dynamics and Aeroelasticity*, Cambridge University Press, New York.
97. Bryant, M. and Garcia, E. (2009) Development of an aeroelastic vibration power harvester. *Proceedings of SPIE*, **7288**, 728812.

98. Bryant, M. and Garcia, E. (2009) Energy harvesting: a key to wireless sensor nodes. *Proceedings of SPIE*, **7493**, 74931W.
99. Peters, D.A., Karunamoorthy, S., and Cao, W.M. (1995) Finite state induced flow models; Part I; two dimensional thin airfoil. *Journal of Aircraft*, **32**, 313–322.
100. Erturk, A., Vieira, W.G.R., Marqui, C. Jr., and Inman, D.J. (2010) On the energy harvesting potential of piezoaeroelastic systems. *Applied Physics Letters*, **96**, 184103.
101. Theodorsen, T. (1935) General theory of aerodynamic instability and mechanism of flutter. Langley Memorial Aeronautical Laboratory, NACA-TR-496.
102. Priya, S., Chen, C.T., Fye, D., and Zahnd, J. (2005) Piezoelectric windmill: a novel solution to remote sensing. *Japanese Journal of Applied Physics*, **44**, L104–L107.
103. St. Clair, D., Bibo, A., Sennakesavababu, V.R., Daqaq, M.F., and Li, G. (2010) A scalable concept for micropower generation using flow-induced self-excited oscillations. *Applied Physics Letters*, **96**, 144103.
104. Van Der Pol, B. (1920) A theory of the amplitude of free and forced triode vibrations. *Radio Review*, **1**, 701–710, 754–762.
105. Robbins, W.P., Morris, D., Marusic, I., and Novak, T.O. (2006) Wind-generated electrical energy using flexible piezoelectric materials. Proceedings of ASME IMECE 2006, Chicago, IL, November 5–10.
106. Pobering, S., Ebermeyer, S., and Schwesinger, N. (2009) Generation of electrical energy using short piezoelectric cantilevers in flowing media. *Proceedings of SPIE*, **7288**, 728807.
107. Akaydin, H.D., Elvin, N., and Andreopoulos, Y. (2010) Wake of a cylinder: a paradigm for energy harvesting with piezoelectric materials. *Experiments in Fluids*, **49**, 291–304.
108. Akaydin, H.D., Elvin, N., and Andreopoulos, Y. (2010) Energy harvesting from highly unsteady fluid flows using piezoelectric materials. *Journal of Intelligent Material Systems and Structures*, **21**, pp. 1263–1278.
109. Elvin, N.G. and Elvin, A.A. (2009) The flutter response of a piezoelectrically damped cantilever pipe. *Journal of Intelligent Material Systems and Structures*, **20**, 2017–2026.
110. Lippmann, G. (1881) Principe de la conservation de l'electricité. *Annales de chimie et de physique*, **24**, 145–178.
111. Voigt, W. (1910) *Lehrbuch der Kristallphysik*, Teubner, Leipzig.
112. Cady, W.G. (1946) *Piezoelectricity: An Introduction to the Theory and Applications of Electromechanical Phenomena in Crystals*, McGraw-Hill, New York.
113. Heising, R.A. (1946) *Quartz Crystals for Electrical Circuits*, Van Nostrand, New York.
114. Mason, W.P. (1950) *Piezoelectric Crystals and Their Applications to Ultrasonics*, Van Nostrand, New York.
115. Mindlin, R.D. (1961) On the equations of motion of piezoelectric crystals, In: *Problems of Continuum Mechanics*, Society for Industrial and Applied Mathematics, Philadelphia, pp. 282–290.
116. Tiersten, H.F. and Mindlin, R.D. (1962) Forced vibrations of piezoelectric crystal plates. *Quarterly Applied Mathematics*, **20**, 107–119.
117. Tiersten, H.F. (1969) *Linear Piezoelectric Plate Vibrations*, Plenum Press, New York.
118. Jeffreys, H. (1931) *Cartesian Tensors*, Cambridge University Press, New York.

2

Base Excitation Problem for Cantilevered Structures and Correction of the Lumped-Parameter Electromechanical Model

This chapter reviews analytical distributed-parameter modeling of the transverse and longitudinal vibrations for cantilevered beams and bars under base excitation. First the general solution of the base excitation problem is given for a cantilevered thin beam in transverse vibrations. The formal treatment of proportional mechanical damping as a combination of internal (strain-rate) and external (air) damping mechanisms is discussed and a possible effect of the external damping on the excitation force is addressed. Then, it is shown that the lumped-parameter base excitation model might yield highly inaccurate results in predicting the vibration response and an amplitude correction factor is derived for improving the lumped-parameter model predictions. Variation of the correction factor with the tip mass - to - beam mass ratio is also investigated and it is observed that the uncorrected lumped-parameter model can be accurate only when this ratio is sufficiently large. Case studies are presented for experimental validation of the dimensionless correction factor. The base excitation problem is summarized for the case of longitudinal vibrations and a correction factor is introduced for the lumped-parameter base excitation model of the longitudinal vibration problem as well. Finally, the amplitude correction factor is introduced to the electromechanically coupled lumped-parameter piezoelectric energy harvester equations and a theoretical case study is given. The distributed-parameter derivations given here constitute the mechanical background for the modeling of cantilevers excited under base motion in vibration-based energy harvesting.

Piezoelectric Energy Harvesting, First Edition. Alper Erturk and Daniel J. Inman.
© 2011 John Wiley & Sons, Ltd. Published 2011 by John Wiley & Sons, Ltd.

Figure 2.1 Cantilevered beam transversely excited by the translation and small rotation of its base

2.1 Base Excitation Problem for the Transverse Vibrations of a Cantilevered Thin Beam

2.1.1 Response to General Base Excitation

The most commonly used energy harvester design is based on the cantilevered (clamped–free) beam configuration as illustrated in Figure 2.1. Furthermore, the length-to-thickness aspect ratio is usually high enough to neglect the shear deformation and rotary inertia effects based on the Euler–Bernoulli beam assumptions [1–4]. The clamped–free Euler–Bernoulli beam shown in Figure 2.1 is subjected to the translation $g(t)$ of its base in the transverse direction with superimposed small rotation $h(t)$. Deformations of the geometrically uniform thin beam are assumed to be small and its composite structure (of perfectly bonded isotropic and/or transversely isotropic layers) is assumed to exhibit linear-elastic material behavior. For the purpose of demonstration, the cantilevered beam is depicted as a symmetric bimorph with three layers. Typically, the outer two layers are piezoceramics (poled in the thickness direction for series or parallel connection of the electrical outputs) and the layer bracketed in between is a metallic substructure (as covered in Chapter 3). The distributed-parameter formulations given in this chapter do *not* consider the piezoelectric coupling effect as the purpose here is to investigate the accuracy of the lumped-parameter harmonic base excitation relation in the absence of electromechanical coupling and to form the mechanical background for most vibration-based energy harvesting devices.

The transverse displacement of the beam at an arbitrary point x and time t along the neutral axis is denoted by $w(x, t)$. If the beam is assumed to be undamped, the equation of motion for free vibrations in the fixed frame of reference can be written as [1–4]

$$YI\frac{\partial^4 w(x, t)}{\partial x^4} + m\frac{\partial^2 w(x, t)}{\partial t^2} = 0 \tag{2.1}$$

where YI is the bending stiffness (given in terms of the elastic moduli and the cross-sectional geometric parameters of the layers) and m is the mass per unit length of the beam.[1]

Two types of damping mechanisms are included to the undamped beam: viscous air (or external) damping and Kelvin–Voigt (or strain-rate) damping.[2] Then the equation of motion

---

[1] Y is preferred to denote the elastic modulus (Young's modulus) in order to avoid confusion with the electric field term (E) in the following chapters. Note that the expression of the bending stiffness YI of a uniform bimorph cantilever is given in the electromechanical derivations of Chapter 3.

[2] Hence the material behavior is approximated as linear-viscoelastic hereafter.

of the damped beam becomes [5,6]

$$YI\frac{\partial^4 w(x,t)}{\partial x^4} + c_s I\frac{\partial^5 w(x,t)}{\partial x^4 \partial t} + c_a\frac{\partial w(x,t)}{\partial t} + m\frac{\partial^2 w(x,t)}{\partial t^2} = 0 \qquad (2.2)$$

where c_a is the viscous air damping coefficient and c_s is the strain-rate damping coefficient (it appears as an effective term $c_s I$ for the composite structure). Viscous air damping is a simple way of modeling the force acting on the beam due to the air particles displaced during the vibratory motion, while strain-rate damping accounts for the structural damping due to the energy dissipation internal to the beam. Both of these damping mechanisms satisfy the proportional damping criterion and they are mathematically convenient for the modal analysis solution procedure. Other beam damping mechanisms and the identification procedures of their respective damping parameters from experimental measurements are discussed by Banks and Inman [6]. In order to use analytical modal analysis techniques, one is usually restricted by the stiffness and mass proportional damping mechanisms (often referred to as *Rayleigh damping*, especially in dynamic analysis of discrete systems).

Following Timoshenko *et al.* [1], the absolute transverse displacement of the beam (i.e., the transverse displacement relative to the fixed reference frame) can be written as

$$w(x,t) = w_b(x,t) + w_{rel}(x,t) \qquad (2.3)$$

where $w_{rel}(x,t)$ is the transverse displacement relative to the clamped end of the beam and $w_b(x,t)$ is the base displacement given by

$$w_b(x,t) = \delta_1(x)g(t) + \delta_2(x)h(t) \qquad (2.4)$$

Here, $\delta_1(x)$ and $\delta_2(x)$ are the displacement influence functions for the transverse base displacement and small base rotation of the beam, respectively. For the cantilevered beam case, simply $\delta_1(x) = 1$ and $\delta_2(x) = x$ [1] (see Appendix B.1 for the modeling of the base excitation problem for beams with other boundary conditions). Using Equation (2.3) in Equation (2.2) leads to

$$YI\frac{\partial^4 w_{rel}(x,t)}{\partial x^4} + c_s I\frac{\partial^5 w_{rel}(x,t)}{\partial x^4 \partial t} + c_a\frac{\partial w_{rel}(x,t)}{\partial t} + m\frac{\partial^2 w_{rel}(x,t)}{\partial t^2}$$
$$= -m\frac{\partial^2 w_b(x,t)}{\partial t^2} - c_a\frac{\partial w_b(x,t)}{\partial t} \qquad (2.5)$$

Hence, after expressing the absolute transverse displacement $w(x,t)$ in terms of the base displacement $w_b(x,t)$ and the relative transverse displacement $w_{rel}(x,t)$, the free vibration equation for the absolute vibratory motion of the beam given by Equation (2.2) becomes a forced vibration equation for the relative vibratory motion of the beam. Note that, physically, the external damping acts on the absolute velocity field while the internal damping acts on the relative velocity field of the beam. Consequently, as far as the relative vibratory motion of the beam governed by Equation (2.5) is concerned, the excitation is not only due to the rigid body inertia of the beam, but also due to the effect of external damping on the rigid body motion.

The latter may or may not be negligible depending on the amount of external damping, which will be further discussed later in this chapter.

The boundary conditions for the relative vibratory motion of the beam (clamped at $x = 0$ and free at $x = L$ as depicted in Figure 2.1) can be written as

$$w_{rel}(0, t) = 0, \qquad \left.\frac{\partial w_{rel}(x, t)}{\partial x}\right|_{x=0} = 0 \tag{2.6}$$

$$\left[Y I \frac{\partial^2 w_{rel}(x, t)}{\partial x^2} + c_s I \frac{\partial^3 w_{rel}(x, t)}{\partial x^2 \partial t}\right]_{x=L} = 0, \quad \left[Y I \frac{\partial^3 w_{rel}(x, t)}{\partial x^3} + c_s I \frac{\partial^4 w_{rel}(x, t)}{\partial x^3 \partial t}\right]_{x=L} = 0 \tag{2.7}$$

Note that the strain-rate damping results in a moment as well as a transverse shear force term that appears in the natural boundary conditions written for the free end [6]. The solution of Equation (2.5) can be represented as a convergent series of eigenfunctions as [2]

$$w_{rel}(x, t) = \sum_{r=1}^{\infty} \phi_r(x) \eta_r(t) \tag{2.8}$$

where $\phi_r(x)$ and $\eta_r(t)$ are the mass normalized eigenfunction and the modal coordinate of a uniform clamped–free beam for the rth mode, respectively.[3] Since the system is proportionally damped, the eigenfunctions denoted by $\phi_r(x)$ are indeed the mass-normalized eigenfunctions of the corresponding *undamped* free vibration problem [7] along with the clamped–free boundary conditions[4]

$$w_{rel}(0, t) = 0, \qquad \left.\frac{\partial w_{rel}(x, t)}{\partial x}\right|_{x=0}, \qquad \left.Y I \frac{\partial^2 w_{rel}(x, t)}{\partial x^2}\right|_{x=L} = 0, \qquad \left.Y I \frac{\partial^3 w_{rel}(x, t)}{\partial x^3}\right|_{x=L} = 0 \tag{2.9}$$

Therefore, the resulting mass-normalized eigenfunction of the rth vibration mode is

$$\phi_r(x) = \sqrt{\frac{1}{mL}} \left[\cos \frac{\lambda_r}{L} x - \cosh \frac{\lambda_r}{L} x + \sigma_r \left(\sin \frac{\lambda_r}{L} x - \sinh \frac{\lambda_r}{L} x\right)\right] \tag{2.10}$$

where λ_r is the dimensionless frequency parameter (eigenvalue) of the rth mode obtained from the characteristic equation given by

$$1 + \cos \lambda \cosh \lambda = 0 \tag{2.11}$$

---

[3] Equation (2.8) follows from the solution of the partial differential equation using the method of separation of variables (Section C.1.2 in Appendix C).

[4] The expressions given in this section are the simplified forms of those given in Section C.1 in Appendix C, since the configuration discussed in this section does not have a tip mass attachment.

and σ_r is expressed as

$$\sigma_r = \frac{\sin \lambda_r - \sinh \lambda_r}{\cos \lambda_r + \cosh \lambda_r} \tag{2.12}$$

It should be noted that Equations (2.10)–(2.12) are valid for a clamped–free beam without a tip mass. The presence of a tip mass affects not only the eigenvalue problem but also the right hand side of Equation (2.5) since the inertia of a possible tip mass also contributes to the excitation amplitude of the beam in that case (see Section 2.2.2).

The mass-normalized form of the eigenfunctions given by Equation (2.10) satisfies the following orthogonality conditions:

$$\int_0^L \phi_s(x) m \phi_r(x)\, dx = \delta_{rs}, \quad \int_0^L \phi_s(x) Y I \frac{d^4 \phi_r(x)}{dx^4} dx = \omega_r^2 \delta_{rs} \tag{2.13}$$

where δ_{rs} is the Kronecker delta, defined as being equal to unity for $s = r$ and equal to zero for $s \neq r$, and ω_r is the undamped natural frequency of the rth vibration mode given by

$$\omega_r = \lambda_r^2 \sqrt{\frac{Y I}{m L^4}} \tag{2.14}$$

Using Equations (2.5)–(2.7) and (2.13), the partial differential equation of motion can be reduced to an infinite set of ordinary differential equations of the form[5]

$$\frac{d^2 \eta_r(t)}{dt^2} + 2 \zeta_r \omega_r \frac{d \eta_r(t)}{dt} + \omega_r^2 \eta_r(t) = f_r(t) \tag{2.15}$$

where $f_r(t)$ is the modal forcing function and

$$2 \zeta_r \omega_r = \frac{c_s I \omega_r^2}{Y I} + \frac{c_a}{m} \tag{2.16}$$

Therefore, the damping ratio ζ_r includes the effects of both strain-rate damping and viscous air damping and it can be expressed as $\zeta_r = \zeta_r^s + \zeta_r^a$ where the strain-rate and the air damping components of the damping ratio are $\zeta_r^s = c_s I \omega_r / 2 Y I$ and $\zeta_r^a = c_a / 2 m \omega_r$, respectively. It is clear from Equation (2.16) that the strain-rate damping coefficient is proportional to structural stiffness, whereas the viscous air damping coefficient is proportional to the mass per unit length. It is worth mentioning that identification of the proportional damping coefficients $c_s I$ and c_a (from experimental measurements) requires knowledge of the natural frequencies and damping ratios of two separate modes [5]. If one knows the undamped natural frequencies

---

[5] See Section C.1.7 in Appendix C for the steps to obtain the undamped form of Equation (2.15) from Equation (2.5). In obtaining the damped Equation (2.15) for the proportionally damped system described by Equation (2.5), one should use the boundary conditions given by Equations (2.6) and (2.7) along with the orthogonality conditions given by Equation (2.13).

(ω_j, ω_k) and the modal damping ratios[6] (ζ_j, ζ_k) of modes j and k, it is straightforward from Equation (2.16) to obtain the $c_s I$ and c_a values using[7]

$$\begin{bmatrix} c_s I \\ c_a \end{bmatrix} = \frac{2\omega_j\omega_k}{\omega_j^2 - \omega_k^2} \begin{bmatrix} \dfrac{YI}{\omega_k} & -\dfrac{YI}{\omega_j} \\ -m\omega_k & m\omega_j \end{bmatrix} \begin{bmatrix} \zeta_j \\ \zeta_k \end{bmatrix} \tag{2.17}$$

In Equation (2.15), the modal forcing function $f_r(t)$ can be expressed as

$$f_r(t) = f_r^m(t) + f_r^c(t) \tag{2.18}$$

Here, the inertia-related and the damping-related excitation terms are given by the following expressions, respectively:

$$f_r^m(t) = -m\left(\gamma_r^w \frac{d^2g(t)}{dt^2} + \gamma_r^\theta \frac{d^2h(t)}{dt^2}\right), \quad f_r^c(t) = -c_a\left(\gamma_r^w \frac{dg(t)}{dt} + \gamma_r^\theta \frac{dh(t)}{dt}\right) \tag{2.19}$$

where

$$\gamma_r^w = \int_0^L \phi_r(x)\,dx, \quad \gamma_r^\theta = \int_0^L x\phi_r(x)\,dx \tag{2.20}$$

Equations (2.18) and (2.19) show that the excitation function can be split into two separate components due to inertia and external damping (the latter term is often neglected). Knowing the modal forcing function, the modal response can be obtained by using the Duhamel integral with damping (assuming zero initial conditions) as

$$\eta_r(t) = \frac{1}{\omega_{rd}} \int_0^t f_r(\tau)e^{-\zeta_r\omega_r(t-\tau)} \sin\omega_{rd}(t - \tau)\,d\tau \tag{2.21}$$

where $\omega_{rd} = \omega_r\sqrt{1 - \zeta_r^2}$ is the damped natural frequency of the rth vibration mode. Eventually, the modal response obtained from Equation (2.21) can be used in Equation (2.8) along

---

[6] Concepts such as the quality factor, half-power points, or the Nyquist plot can be used for the identification of modal damping ratio in the frequency domain [8], or, alternatively, logarithmic decrement can be used for its identification in the time domain [5].

[7] This procedure of proportional damping assumes that the modal damping ratios of all the other modes can be determined if those of only two modes are known (which is not the case for most physical systems). The reader is referred to Banks and Inman [6] for a more realistic approach of identifying these damping coefficients. Often the damping ratios of the individual vibration modes are identified separately from experiments.

with the eigenfunction expression given by Equation (2.10) and the response $w_{rel}(x, t)$ at any point along the beam axis can be obtained as

$$w_{rel}(x, t) = \sum_{r=1}^{\infty} \frac{\phi_r(x)}{\omega_{rd}} \int_0^t f_r(\tau) e^{-\zeta_r \omega_r(t-\tau)} \sin \omega_{rd}(t - \tau) d\tau \qquad (2.22)$$

Note that the displacement at the tip of the beam (relative to the moving base) can be obtained by just setting $x = L$ in Equation (2.22). The response of the beam relative to the fixed reference frame can be obtained by just using the relative displacement expression and the base displacement input in Equation (2.3). However, the main concern in vibration-based energy harvesting is the response of the beam relative to its base. The expression obtained for the relative vibratory motion of the beam, Equation (2.22), is not restricted to harmonic base excitation and it can handle transient base histories as well as small base rotations.

2.1.2 Steady-State Response to Harmonic Base Excitation

In most of the theoretical and experimental work on piezoelectric energy harvesting, the base excitation is assumed to be a harmonic translation to simplify the problem and also to estimate the maximum power generation under resonance excitation. If the base translation is of the form $g(t) = W_0 e^{j\omega t}$ (where W_0 is the base displacement amplitude, ω is the excitation frequency, and $j = \sqrt{-1}$ is the unit imaginary number) and the base does not rotate (i.e., $h(t) = 0$), the steady-state modal response can be calculated to be

$$\eta_r(t) = \frac{m\omega^2 - j\omega c_a}{\omega_r^2 - \omega^2 + j2\zeta_r \omega_r \omega} \gamma_r^w W_0 e^{j\omega t} \qquad (2.23)$$

where

$$\gamma_r^w = \int_0^L \phi_r(x) \, dx = \frac{2\sigma_r}{\lambda_r} \sqrt{\frac{L}{m}} \qquad (2.24)$$

is obtained by integrating Equation (2.10) over the beam length.

Using Equations (2.8), (2.10), and (2.23), one can obtain the expression for the relative vibratory response at point x and time t as

$$w_{rel}(x, t) = 2W_0 e^{j\omega t} \sum_{r=1}^{\infty} \left[\cos\frac{\lambda_r}{L}x - \cosh\frac{\lambda_r}{L}x + \sigma_r \left(\sin\frac{\lambda_r}{L}x - \sinh\frac{\lambda_r}{L}x \right) \right]$$

$$\frac{\sigma_r \left(\omega^2 - j2\zeta_r^a \omega_r \omega \right)}{\lambda_r \left(\omega_r^2 - \omega^2 + j2\zeta_r \omega_r \omega \right)} \qquad (2.25)$$

Then, by setting $x = L$ one obtains

$$w_{rel}(L,t) = 2W_0 e^{j\omega t} \sum_{r=1}^{\infty} [\cos \lambda_r - \cosh \lambda_r + \sigma_r (\sin \lambda_r - \sinh \lambda_r)] \frac{\sigma_r \left(\omega^2 - j2\zeta_r^a \omega_r \omega \right)}{\lambda_r \left(\omega_r^2 - \omega^2 + j2\zeta_r \omega_r \omega \right)}$$

$$(2.26)$$

which is the distributed-parameter steady-state solution of the relative tip displacement due to harmonic base excitation (in the absence of a tip mass).

2.1.3 Lumped-Parameter Model of the Harmonic Base Excitation Problem

Early modeling efforts in vibration-based energy harvesting [9–15] used lumped-parameter modeling by considering the fundamental vibration mode (which is also known as single-degree-of-freedom modeling). In order to understand the limitations of such models, it is important to review the theory behind lumped-parameter modeling. This modeling approach requires a description of the dynamics of the point of interest (usually the free end of the beam) in terms of certain lumped parameters, which are the equivalent mass, stiffness, and the damping of the beam denoted by m_{eq}, k_{eq}, and c_{eq}, respectively (Figure 2.2a). The equivalent stiffness is obtained from the static deflection relation of a cantilevered beam due to a concentrated transverse load at the tip, while the equivalent mass is obtained by expressing the total kinetic energy of the beam in terms of the velocity at the tip through Rayleigh's quotient [16] for cantilevered end conditions where the base is not moving. It should be highlighted at this point that, in the base excitation problem (unlike the problem of a beam with a stationary base), the cantilevered beam is excited by its own inertia and there is an inertial contribution to the excitation from its distributed mass. The contribution of this distributed inertia to the excitation amplitude can be very significant as will be shown in the next section.

The commonly quoted [10–15] lumped-parameter representation of the base excitation problem is shown in Figure 2.2a and the lumped-parameter model with a more precise representation of the damping components is depicted in Figure 2.2b. The improvement made in Figure 2.2b is due to the modeling of the external viscous damping (air damping). It is analogous to the distributed-parameter solution in the sense that the structural (internal) damping acts on the relative velocity between the mass and the base, whereas the air (external) damping acts on the absolute velocity of the mass.

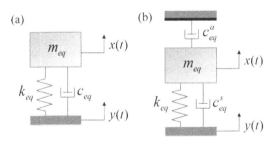

Figure 2.2 Lumped-parameter modeling of the base excitation problem: (a) the commonly used representation; and (b) a more precise representation of the damping components

In Figure 2.2, $y(t)$ is the harmonic base displacement ($y(t) = Y_0 e^{j\omega t}$) and $x(t)$ is the absolute displacement response of the mass (i.e., it is the absolute transverse displacement at the free end of the beam). If the displacement response of the mass relative to the base is $z(t) = x(t) - y(t)$, one can obtain the relative displacement of the mass for the system shown in Figure 2.2a as

$$z(t) = \frac{\omega^2 m_{eq}}{k_{eq} - \omega^2 m_{eq} + j\omega c_{eq}} Y_0 e^{j\omega t} \qquad (2.27)$$

which can be found in any elementary vibration text [2, 3] and which has also been frequently quoted in order to describe the dynamics of vibration-based energy harvesters [10–15]. In Equation (2.27), the equivalent flexural stiffness of the cantilever at the tip is

$$k_{eq} = \frac{3YI}{L^3} \qquad (2.28)$$

and the equivalent mass is due to Lord Rayleigh [16]:

$$m_{eq} = \frac{33}{140} mL + M_t \qquad (2.29)$$

where M_t is the tip mass (if one is present). The undamped natural frequency (the *fundamental* natural frequency of the structure) is[8]

$$\omega_n = \sqrt{\frac{k_{eq}}{m_{eq}}} = \sqrt{\frac{3YI/L^3}{(33/140)mL + M_t}} \qquad (2.30)$$

Then the equivalent damping coefficient is $c_{eq} = 2\zeta \omega_n m_{eq}$ where ζ is the equivalent damping ratio. According to Figure 2.2a, this model assumes a single damping coefficient which acts on the relative velocity of the tip mass.

Consider the lumped-parameter model presented in Figure 2.2b where air damping and structural damping are treated separately; the former acts on the absolute velocity of the mass, whereas the latter acts on the velocity of the mass relative to the base. The air damping coefficient c_{eq}^a is assumed to be proportional to m_{eq} ($c_{eq}^a = \tau m_{eq}$) whereas the structural damping coefficient c_{eq}^s is assumed to be proportional to k_{eq} ($c_{eq}^s = \upsilon k_{eq}$), where τ and υ are the constants of proportionality. Once again, obtaining the proportionality constants τ and υ (hence the damping coefficients c_{eq}^a and c_{eq}^s) from experimental measurements requires knowledge of the damping ratios and natural frequencies of two separate modes of the physical (multi-mode) system as in the case of the distributed-parameter model (see Equation (2.17)).

---

[8] It can be shown that the error due to using Equation (2.30) in predicting the fundamental natural frequency is about 1.5% in the absence of a tip mass (relative to the Euler–Bernoulli model fundamental natural frequency ω_1 obtained for $r = 1$ in Equation (2.14)). The prediction of Equation (2.30) is improved in the presence of a tip mass.

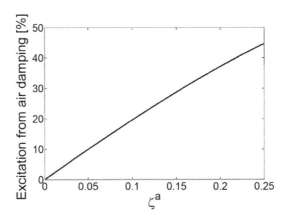

Figure 2.3 Contribution of excitation from air (or external) damping to the total excitation term as a function of air (or external) damping ratio (for excitation at $\omega = \omega_n$)

For the lumped-parameter model shown in Figure 2.2b, the relative displacement response of the mass can be expressed as

$$z(t) = \frac{\omega^2 m_{eq} - j\omega c_{eq}^a}{k_{eq} - \omega^2 m_{eq} + j\omega c_{eq}} Y_0 e^{j\omega t} \tag{2.31}$$

Here, $c_{eq}^a = 2\zeta^a \omega_n m_{eq}$ and $c_{eq} = c_{eq}^a + c_{eq}^s = 2\zeta \omega_n m_{eq}$, where ζ^a is the damping ratio due to the external viscous damping effect. Therefore, in the commonly quoted Equation (2.27), the forcing term due to air damping is missing. Consequently, representing all sources of mechanical damping by a single damping ratio acting on the relative velocity of the tip mass [10] is not necessarily the most general case. However, for the case where the damping ratio ζ^a due to the surrounding fluid (which is usually air) is very low, it is reasonable to expect the forcing term $-j\omega c_{eq}^a$ coming from the air damping to be much less than the inertial forcing term $\omega^2 m_{eq}$ and Equation (2.31) reduces to Equation (2.27). For a dimensionless comparison, dividing both of these forcing terms by m_{eq} gives the inertia contribution as ω^2 and the air damping contribution as $-j2\zeta^a \omega_n \omega$. For the case of excitation at the undamped natural frequency ($\omega = \omega_n$), the variation of the percentage forcing contribution of air damping with ζ^a is shown in Figure 2.3 (which is simply the plot of $2\zeta^a / \sqrt{1 + (2\zeta^a)^2} \times 100$). As can be seen from Figure 2.3, the excitation amplitude coming from the air damping is less than 5% of the total excitation (inertial and damping) if $\zeta^a < 0.025$. The conclusion of this brief discussion on damping is that the lumped-parameter model of the harmonic base excitation problem given by Figure 2.2a and Equation (2.27) implicitly assumes the excitation component due to the damping coming from the external fluid to be sufficiently low as compared to the inertial excitation component. For the base excitation problem of cantilevers operating in viscous fluids or for micro-scale cantilevers for which the external damping effect becomes more significant, the representation given in Figure 2.2b and Equation (2.31) is preferred.

2.1.4 Comparison of the Distributed-Parameter and the Lumped-Parameter Model Predictions

Consider the expressions of the relative tip displacement response obtained by using the distributed-parameter (Euler–Bernoulli) model and the lumped-parameter model, which are Equations (2.26) and (2.31), respectively. As mentioned previously, for the case of light external damping (i.e., for ζ^a, $\zeta_r^a \ll 1$), the excitation due to the inertia term dominates the numerators of Equations (2.26) and (2.31) and these equations can be reduced to

$$w_{rel}(L, t) = 2\omega^2 W_0 e^{j\omega t} \sum_{r=1}^{\infty} \frac{\sigma_r \left[\cos \lambda_r - \cosh \lambda_r + \sigma_r (\sin \lambda_r - \sinh \lambda_r)\right]}{\lambda_r \left(\omega_r^2 - \omega^2 + j2\zeta_r\omega_r\omega\right)} \tag{2.32}$$

$$z(t) = \frac{\omega^2}{\omega_n^2 - \omega^2 + j2\zeta\omega_n\omega} Y_0 e^{j\omega t} \tag{2.33}$$

respectively.

The ratio of the tip displacement to the base displacement gives the *relative displacement transmissibility function*, which forms an appropriate dimensionless basis for comparing the distributed-parameter and the lumped-parameter models. These relative displacement transmissibility functions can be extracted from Equations (2.32) and (2.33) as

$$T_{rel}^{DP}(\omega, \zeta_r) = 2\omega^2 \sum_{r=1}^{\infty} \frac{\sigma_r \left[\cos \lambda_r - \cosh \lambda_r + \sigma_r (\sin \lambda_r - \sinh \lambda_r)\right]}{\lambda_r \left(\omega_r^2 - \omega^2 + j2\zeta_r\omega_r\omega\right)} \tag{2.34}$$

$$T_{rel}^{LP}(\omega, \zeta) = \frac{\omega^2}{\omega_n^2 - \omega^2 + j2\zeta\omega_n\omega} \tag{2.35}$$

where T_{rel}^{DP} and T_{rel}^{LP} are the transmissibility functions obtained from the distributed-parameter and the lumped-parameter models, respectively. Note that the distributed-parameter and the lumped-parameter natural frequencies are not identical since the natural frequency prediction of the latter model (which is due to Equation (2.30)) is slightly different from that of the former model (which is due to Equation (2.14) for $r = 1$). It should be noted from Equations (2.34) and (2.35) that these are also functions of the damping ratios. Therefore, it is necessary to compare the results of these transmissibility functions for different values of the damping ratio. Here, three different values of the damping ratio ($\zeta = \zeta_1 = 0.01, 0.025, 0.05$) are used for comparison of the two models. The relative motion transmissibility functions given by Equations (2.34) and (2.35) are shown in Figures 2.4a and 2.4b, respectively. For convenience, the excitation frequency ω is normalized with respect to the fundamental natural frequency (of the distributed-parameter model as it is assumed to be the accurate one) and therefore the frequency axis is denoted by $\Omega = \omega/\omega_1$. As can be seen from Figure 2.4, the frequency of maximum relative displacement transmissibility corresponds to $\Omega \cong 1$ in both models since the lumped-parameter approach gives a good estimate of the fundamental natural frequency from Equation (2.30). However, it is not possible to draw the same conclusion from the amplitude results. It is clear from Figure 2.4 that the peak values for the same damping

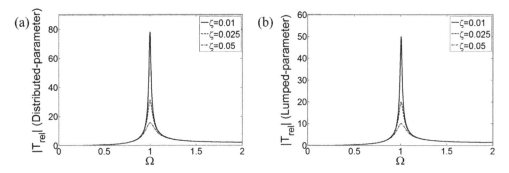

Figure 2.4 Relative motion transmissibility functions for the transverse vibrations of a cantilevered beam without a tip mass: (a) distributed-parameter model; and (b) lumped-parameter model

ratios are considerably different for the distributed-parameter and the lumped-parameter model predictions.

The percentage error in the lumped-parameter solution as a function of dimensionless frequency ratio is given by Figure 2.5 (relative to the distributed-parameter solution). As can be seen from Figure 2.5, the relative error due to using the lumped-parameter approach in predicting the relative motion at the tip of the beam is very large. In the vicinity of the first natural frequency, the error of the lumped-parameter model can be greater than 35% regardless of the damping ratio. The interesting behavior in the relative error plot around the resonance is due to the 1.5% error in the natural frequency predicted by the lumped-parameter approach. If the lumped-parameter natural frequency were taken to be identical to the first natural frequency of the distributed-parameter model, one would obtain a smooth behavior in the error. The significant error is in the prediction of the relative motion amplitude rather than the natural frequency. The error in the lumped-parameter model increases drastically at higher frequencies since higher vibration modes cannot be captured by the lumped-parameter approach with a single degree of freedom.

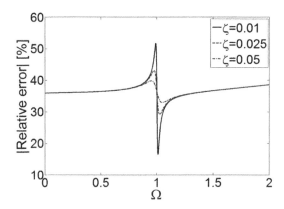

Figure 2.5 Error in the relative motion transmissibility due to using the lumped-parameter model for the transverse vibrations of a cantilevered beam without a tip mass

2.2 Correction of the Lumped-Parameter Base Excitation Model for Transverse Vibrations

2.2.1 Correction Factor for the Lumped-Parameter Model

Since much of the vibration-based energy harvesting literature uses lumped-parameter modeling for design and optimization [10–15], a correction factor is introduced for improving the predictions of the lumped-parameter model. Consider the relative motion transmissibility function of the distributed-parameter model given by Equation (2.34). If the beam is excited around its first natural frequency, taking only the first term in the summation sign (neglecting the terms for $r \geq 2$) gives a good approximation for the resulting motion transmissibility. This reduced form of the distributed-parameter model solution is denoted by $\hat{T}_{rel}^{DP}$:

$$\hat{T}_{rel}^{DP}(\omega, \zeta) = \frac{\mu_1 \omega^2}{\omega_1^2 - \omega^2 + j2\zeta\omega_1\omega} \tag{2.36}$$

which can be expressed as

$$\hat{T}_{rel}^{DP}(\Omega, \zeta) = \frac{\mu_1 \Omega^2}{1 - \Omega^2 + j2\zeta\Omega} \tag{2.37}$$

where $\Omega = \omega/\omega_1$ is used and μ_1 corrects the excitation amplitude for the first transverse vibration mode of a cantilevered Euler–Bernoulli beam without a tip mass (for predicting the vibration response at $x = L$). Using $\lambda_1 = 1.87510407$ and $\sigma_1 = 0.73409551$ obtained from Equations (2.11) and (2.12) gives the *correction factor* for the first mode as

$$\mu_1 = \frac{2\sigma_1 \left[\cos \lambda_1 - \cosh \lambda_1 + \sigma_1 \left(\sin \lambda_1 - \sinh \lambda_1\right)\right]}{\lambda_1} \cong 1.566 \tag{2.38}$$

It should be noted from Equations (2.35) and (2.36) that the reduced form of the distributed-parameter solution for the first mode is indeed the correction factor μ_1 multiplied by the lumped-parameter solution (assuming that the lumped-parameter natural frequency is sufficiently accurate so that $\Omega = \omega/\omega_1 \cong \omega/\omega_n$). Therefore, μ_1 corrects the amplitude of the relative motion obtained from the lumped-parameter solution. The comparison of the relative motion transmissibility functions obtained by using the distributed-parameter model, the lumped-parameter model, and the corrected lumped-parameter model is given in Figure 2.6 for $\zeta = 0.05$. The agreement between the distributed-parameter model (Equation (2.34)) and the corrected lumped-parameter model (Equation (2.36)) is very good over a wide frequency range around the fundamental resonance frequency and the corrected lumped-parameter relative motion transmissibility function starts deviating close to the region of the second natural frequency. The original lumped-parameter model prediction with Equation (2.35) underestimates considerably the relative motion transmissibility amplitude with an error as high as 35% (see Figure 2.5).

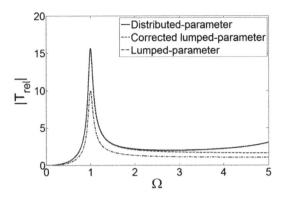

Figure 2.6 Relative motion transmissibility functions obtained from the distributed-parameter, the corrected lumped-parameter, and the original lumped-parameter models for $\zeta = 0.05$

If the beam is to be excited not at the first natural frequency but at one of the higher mode frequencies, one can obtain the correction factor of the mode of interest (rth mode) from the following relation:

$$\mu_r = \frac{2\sigma_r\,[\cos\lambda_r - \cosh\lambda_r + \sigma_r\,(\sin\lambda_r - \sinh\lambda_r)]}{\lambda_r} \tag{2.39}$$

and then use it in the following expression of reduced relative motion transmissibility:

$$\hat{T}_{rel}^{DP}(\Omega_r, \zeta_r) = \frac{\mu_r\Omega_r^2}{1 - \Omega_r^2 + j2\zeta_r\Omega_r} \tag{2.40}$$

where the dimensionless frequency ratio is now $\Omega_r = \omega / \omega_r$ and ω_r is the undamped natural frequency of the rth mode obtained from Equation (2.14) and ζ_r is the modal damping ratio of the rth mode. Note that, since the modal parameters λ_r and σ_r do not depend on the aspect ratio of the beam in the Euler–Bernoulli beam theory, the correction factor μ_r of the rth mode is unique in the absence of a tip mass. That is, the correction factor for the fundamental mode is $\mu_1 \cong 1.566$ for any uniform cantilevered Euler–Bernoulli beam without a tip mass in transverse vibrations (so long as the beam aspect ratio justifies the Euler–Bernoulli beam assumptions).[9] However, the presence of a tip mass affects the correction factor, which is discussed in the following section.

2.2.2 Effect of a Tip Mass on the Correction Factor

In some cases, it is necessary to attach a tip mass (proof mass) to the beam in order to tune its fundamental natural frequency to the excitation frequency and to improve its dynamic flexibility. If the differential eigenvalue problem is solved for a uniform cantilevered beam

---

[9] In the discussion given here, the correction factor is defined to predict the motion exactly at the tip. It should be noted that distributed-parameter modeling allows the prediction of the motion transmitted from the base to any arbitrary point of the beam. In such a case, the numerical value of the correction factor obviously changes (e.g., μ_1 takes a value lower that 1.566 for the fundamental mode) but it is still independent of the aspect ratio as long as the beam is sufficiently thin, geometrically and materially uniform, and linear vibrations are considered.

with a tip mass of M_t rigidly attached at $x = L$, the eigenfunctions can be obtained as (Appendix C.1)[10]

$$\phi_r(x) = A_r \left[\cos \frac{\lambda_r}{L} x - \cosh \frac{\lambda_r}{L} x + \varsigma_r \left(\sin \frac{\lambda_r}{L} x - \sinh \frac{\lambda_r}{L} x \right) \right] \qquad (2.41)$$

where ς_r is obtained from

$$\varsigma_r = \frac{\sin \lambda_r - \sinh \lambda_r + \lambda_r \dfrac{M_t}{mL} (\cos \lambda_r - \cosh \lambda_r)}{\cos \lambda_r + \cosh \lambda_r - \lambda_r \dfrac{M_t}{mL} (\sin \lambda_r - \sinh \lambda_r)} \qquad (2.42)$$

and A_r is a modal amplitude constant which should be evaluated by normalizing the eigenfunctions according to the following orthogonality conditions:

$$\int_0^L \phi_s(x) m \phi_r(x) \, dx + \phi_s(L) M_t \phi_r(L) = \delta_{rs}$$

$$\int_0^L \phi_s(x) YI \frac{d^4 \phi_r(x)}{dx^4} dx - \left[\phi_s(x) YI \frac{d^3 \phi_r(x)}{dx^3} \right]_{x=L} = \omega_r^2 \delta_{rs} \qquad (2.43)$$

where it is sufficient to use the first expression (then the second one is automatically satisfied and vice versa). The natural frequency expression given by Equation (2.14) still holds but the dimensionless eigenvalues (λ_r for the rth mode) should be obtained from

$$1 + \cos \lambda \cosh \lambda + \lambda \frac{M_t}{mL} (\cos \lambda \sinh \lambda - \sin \lambda \cosh \lambda) = 0 \qquad (2.44)$$

where M_t / mL is a dimensionless parameter as it is the ratio of the tip mass to the beam mass. In the above equations, the rotary inertia of the tip mass is neglected for convenience; that is, the tip mass is assumed to be a point mass.

In addition to the modification of the eigenvalue problem in the presence of a tip mass, the forcing term due to base excitation also changes since the tip mass also contributes to the inertia of the structure. Equation (2.5) becomes

$$YI \frac{\partial^4 w_{rel}(x, t)}{\partial x^4} + c_s I \frac{\partial^5 w_{rel}(x, t)}{\partial x^4 \partial t} + c_a \frac{\partial w_{rel}(x, t)}{\partial t} + m \frac{\partial^2 w_{rel}(x, t)}{\partial t^2}$$
$$= -[m + M_t \delta(x - L)] \frac{\partial^2 w_b(x, t)}{\partial t^2} \qquad (2.45)$$

---

[10] Here, it is assumed that the uniform thin beam with a tip mass can be approximated as a normal-mode system so that the eigenfunctions of the undamped problem (Appendix C.1) can be used for modal analysis of the damped problem [7]. Banks *et al.* [17] pointed out the mathematical limitations of the normal-mode assumption for combined dynamical systems, i.e., distributed-parameter systems with discrete elements.

where $\delta(x)$ is the Dirac delta function and the forcing term due to external damping is neglected. The modal forcing function corresponding to the right hand side of Equation (2.45) is then

$$f_r(t) = -m \left(\frac{d^2 g(t)}{dt^2} \int_0^L \phi_r(x)\,dx + \frac{d^2 h(t)}{dt^2} \int_0^L x\phi_r(x)\,dx \right) - M_t\phi_r(L) \left(\frac{d^2 g(t)}{dt^2} + L\frac{d^2 h(t)}{dt^2} \right)$$

(2.46)

As expected, the foregoing modification results in variation of the correction factor defined in the previous section. Since the base is assumed to be not rotating (i.e., $h(t) = 0$) in deriving the correction factor, one can extract the expression of the correction factor μ_1 in the presence of a tip mass as

$$\mu_1 = \phi_1(L) \left(M_t\phi_1(L) + m \int_0^L \phi_1(x)\,dx \right) \qquad (2.47)$$

The variation of the correction factor μ_1 of the fundamental transverse vibration mode given by Equation (2.47) with the ratio of tip mass (M_t) to beam mass (mL) is shown in Figure 2.7. Figure 2.7 illustrates that when there is no tip mass ($M_t/mL = 0$), $\mu_1 \cong 1.566$ as previously obtained, whereas when M_t/mL becomes larger ($M_t/mL \rightarrow \infty$), μ_1 approaches unity. The important conclusion drawn from Figure 2.7 is that the *uncorrected* lumped-parameter model can be used safely only when the tip mass is sufficiently larger than the beam mass. From a physical point of view, if the tip mass is sufficiently large, the inertia of the tip mass dominates in the forcing function and the distributed inertia of the beam (as a component of excitation) becomes negligible. Table 2.1 shows the values that μ_1 takes for different M_t/mL ratios. It should be noted that, for the uncorrected lumped-parameter formulation, $\mu_1 = 1$, and therefore the relative error in the motion at the tip of the beam predicted by the uncorrected lumped-parameter model is estimated from $(1 - \mu_1)/\mu_1 \times 100$.

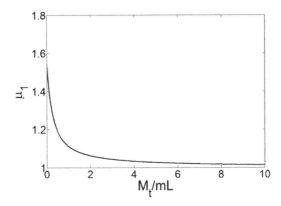

Figure 2.7 Variation of the correction factor for the fundamental transverse vibration mode with the ratio of tip mass to beam mass

Table 2.1 Correction factor for the fundamental transverse vibration mode and the error in the uncorrected lumped-parameter model for different values of tip mass - to - beam mass ratio

M_t/mL	μ_1	Error in the uncorrected lumped-parameter model (%)
0	1.56598351	-36.14
0.1	1.40764886	-28.96
0.5	1.18922917	-15.91
1	1.11285529	-10.14
5	1.02662125	-2.59
10	1.01361300	-1.34

The following quadratic polynomial ratio (obtained by using the Curve Fitting Toolbox of MATLAB) gives an estimate of μ_1 with an error less than $9 \times 10^{-3}\%$ for all values of M_t/mL:

$$\mu_1 = \frac{(M_t/mL)^2 + 0.603\,(M_t/mL) + 0.08955}{(M_t/mL)^2 + 0.4637\,(M_t/mL) + 0.05718} \tag{2.48}$$

2.3 Experimental Case Studies for Validation of the Correction Factor

This section provides experimental demonstrations for the use of the correction factor in order to improve the predictions of the lumped-parameter base excitation model in the absence and presence of a tip mass. The bimorph cantilever discussed in the following is investigated extensively in Chapter 4 for validation of the distributed-parameter electromechanical models developed in Chapter 3. Detailed information regarding the cantilever and the experimental setup can therefore be found in Chapter 4. Here, only the motion transmissibility FRFs are considered in order to validate the amplitude correction factor derived in this chapter. Each one of the two cases presented here uses the measurement taken for an external electrical load close to short-circuit conditions of the electrodes (very low external load impedance). Therefore, the following cases can be considered as very close to being electromechanically uncoupled (i.e., having negligible piezoelectric shunt damping effect), in agreement with the formulation given in this chapter (where the purpose is to improve the mechanical domain of the existing lumped-parameter electromechanical model). The electromechanically coupled distributed-parameter dynamics of these configurations are modeled and validated in Chapters 3 and 4, respectively.

2.3.1 Cantilevered Beam without a Tip Mass under Base Excitation

The cantilever considered here is the T226-H4-203X bimorph manufactured by Piezo Systems, Inc. Detailed electromechanical analysis of this sample can be found in Section 4.1. Here, the experimental data for a 470 Ω load resistance (close to short-circuit conditions) is considered to validate the corrected lumped-parameter base excitation relation given by Equation (2.36) and to demonstrate the failure of the original relation given by Equation (2.35) in predicting

(1) Shaker with a small
accelerometer and the cantilever
(2) Laser vibrometer
(3) Fixed gain amplifier (power
supply)
(4) Charge amplifier
(5) Data acquisition system
(6) Frequency response analyzer
(software)

Figure 2.8 Experimental setup used for the frequency response measurements of a uniform bimorph cantilever

the vibration response. It should be noted from Table 4.1 that the overhang length, width, and thickness of the cantilever are 24.53 mm, 6.4 mm, and 0.670 mm, respectively. Therefore the ratio of overhang length to total thickness of the bimorph cantilever is about 37.7 and the Euler–Bernoulli formulation can safely be used as far as the fundamental vibration mode is concerned.

The experimental setup used for the frequency response measurement is shown in Figure 2.8. The bimorph cantilever is clamped onto a small electromagnetic shaker (TMC Model TJ-2). A small accelerometer (PCB Piezotronics Model U352C67) is attached with wax close to the root of the cantilever on the clamp. The tip velocity response of the cantilever is measured using a laser vibrometer (Polytec PDV100) by attaching a small piece of reflector tape at the tip of the cantilever (see Figure 2.9a for an enlarged view). Chirp excitation (burst type with five averages) is used for the frequency response measurement through the data acquisition

(a)    (b)

Figure 2.9 Close views of the cantilever tested under base excitation (a) without and (b) with a tip mass attachment

system (SigLab Model 20-42). The ratio of the vibrometer measurement to the acceleration measurement in the frequency domain defines the FRF of tip velocity to base acceleration.

It is important to note that the laser vibrometer measures the velocity response at the center of the reflector tape that is attached near the tip of the cantilever (i.e., not exactly at $x = L$). If the point of the velocity measurement is $x = L_v$, Equation (2.47) should be modified to

$$\mu_1^* = \phi_1(L_v) \left(M_t \phi_1(L) + m \int_0^L \phi_1(x)\,dx \right) \tag{2.49}$$

where μ_1^* is the modified form of the correction factor that accounts for the small distance of the point of velocity measurement from the tip. In the experiments, the position of the velocity measurement on the cantilever is $L_v = 23$ mm from the root (i.e., approximately 1.5 mm from the tip). For this point of velocity measurement, the variation of the correction factor with the ratio of tip mass to beam mass is given by Figure 2.10. In the absence of a tip mass, the modified correction factor is $\mu_1^* = 1.431$ and it can be used in Equation (2.36) to predict the vibratory motion at $x = L_v$.

It should also be noted at this stage that the theoretical discussion given so far is for the FRF of tip displacement to base displacement. However, the measurements taken with this experimental setup are the tip velocity and the base acceleration of the cantilever. Moreover, the experimental velocity measurement is relative to the fixed frame; that is, it is not the tip velocity of the cantilever relative to its moving base. One option for comparing the experimental data and the analytical predictions is to process the experimentally measured FRF to put it into the form of Equations (2.35) and (2.36). The second option is to put the analytical relative displacement transmissibility functions given by these equations into the form of the experimental measurement. The second option is preferred in order not to create possible noise in the experimental data due to post-processing. The relative displacement transmissibility

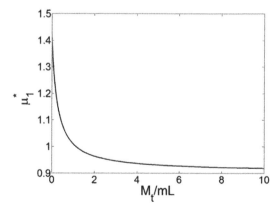

Figure 2.10 Variation of the modified correction factor for the fundamental transverse vibration mode with ratio of tip mass to beam mass

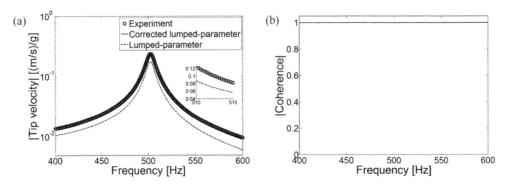

Figure 2.11   (a) Tip velocity - to - base acceleration FRFs of the cantilever without a tip mass: experimental measurement, corrected lumped-parameter, and uncorrected lumped-parameter model predictions; (b) coherence function of the experimental measurement

FRFs (Equations (2.35) and (2.36)) can be used in the following relation to give the absolute
tip velocity - to - base acceleration FRF as

$$\frac{\partial w(x, t)/\partial t|_{x=L_v}}{-\omega^2 W_0 e^{j\omega t}} = \frac{1 + T_{rel}(\omega, \zeta)}{j\omega} \tag{2.50}$$

In addition, the experimental FRFs are given per gravitational acceleration ($g =
9.81$ m/s^2); that is, Equation (2.50) should be multiplied by g in order to compare it to the
experimental data.

Figure 2.11a shows the experimental frequency response measurement and its prediction by
Equations (2.35) and (2.36) (when used in Equation (2.50)). The undamped natural frequency
of the cantilever (close to short-circuit conditions) is 502.5 Hz ($\omega_n = 3157.3$ rad/s) and the
mechanical damping ratio identified in Section 4.1 is 0.874% ($\zeta = 0.00874$). As can be seen
in Figure 2.11b, the coherence [18, 19] of the measurement is very good (unity). Therefore the
measurement is reliable for the frequency range of interest.

Since the cantilever does not have a tip mass ($M_t/mL = 0$), the modified amplitude correction factor is obtained from Equation (2.49) or from Figure 2.10 as $\mu_1^* \cong 1.431$. It is observed
from Figure 2.11a that the corrected lumped-parameter model (due to Equation (2.36)) is in
very good agreement with the experimental FRF, whereas the uncorrected lumped-parameter
model (due to Equation (2.35)) underestimates the vibration amplitude considerably (note that
the vertical axis in Figure 2.11a is given in log scale).[11] Therefore, in agreement with the theoretical discussion given with Figure 2.6, the commonly referred form of the lumped-parameter
base excitation model results in significantly inaccurate prediction of the vibration response
in the absence of a tip mass. The inaccurate prediction is in the form of *underestimation* of the
vibration response amplitude.

---

[11] The form of the uncorrected lumped-parameter expression given by Equation (2.35) is originally defined to
predict the tip motion but the experimental measurement belongs to the point $x = L_v$. Nevertheless, since this
lumped-parameter expression underestimates the motion amplitude at $x = L_v$ it would certainly underestimate the
motion amplitude at the exact point of $x = L$ where the vibration amplitude is larger for the fundamental mode.

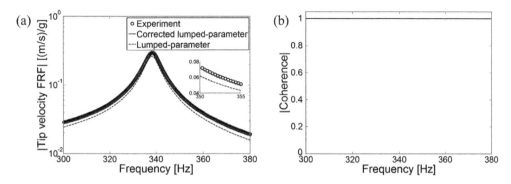

Figure 2.12 (a) Tip velocity - to - base acceleration FRFs of the cantilever with a tip mass: experimental measurement, corrected lumped-parameter, and uncorrected lumped-parameter model predictions; (b) coherence function of the experimental measurement

2.3.2 Cantilevered Beam with a Tip Mass under Base Excitation

The second case study is given for the same cantilever after a tip mass of 0.239 g is attached at the tip using a slight amount of wax (Figure 2.9b). Detailed electromechanical analysis of the cantilever with a tip mass is given in Section 4.2. The demonstration given here aims to estimate the correction factor and to verify the need for using it. For this tip mass ($M_t = 0.239 \times 10^{-3}$ kg) and for the numerical data of the cantilever given with Table 4.4, the ratio of the tip mass to beam mass is $M_t/mL = 0.291$. This ratio yields $\mu_1^* = 1.149$ using either Equation (2.49) or Figure 2.10. As discussed theoretically, in the presence of a tip mass, predictions of the original lumped-parameter equation are not expected to be as inaccurate as the case when there is no tip mass. Recall that, in the extreme case of having a very large tip mass, the correction factor is not required. However, in this experimental case study, the tip mass is not large enough to ignore the excitation coming from the distributed mass of the cantilever.

Figure 2.12a shows the experimental FRF of tip velocity to base acceleration (for a 470 Ω load resistance) along with the predictions of Equations (2.35) and (2.36) when used in Equation (2.50) (the coherence of the measurement is again very good in Figure 2.12b). As can be seen from the enlarged view in Figure 2.11a, the uncorrected lumped-parameter model underestimates the experimental measurement but the error is not as large as in the case of Figure 2.11a. The prediction of the corrected lumped-parameter equation (obtained using $\mu_1^* = 1.149$ in Equation (2.36)) agrees very well with the experimental data.

2.4 Base Excitation Problem for Longitudinal Vibrations and Correction of its Lumped-Parameter Model

So far the base excitation problem of a beam in transverse vibration has been discussed in detail and experimental validations have been provided for its corrected lumped-parameter model. This section summarizes the same problem for longitudinal vibrations of a uniform cantilevered bar under base excitation (Figure 2.13). As in the case of transverse vibrations,

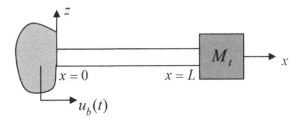

Figure 2.13 Cantilevered bar with a tip mass longitudinally excited by the translation of its base

the electromechanically uncoupled problem is considered in order to correct the excitation amplitude of its lumped-parameter model.

2.4.1 Analytical Modal Analysis and Steady-State Response to Harmonic Base Excitation

The uniform clamped–free bar shown in Figure 2.13 is subjected to the arbitrary translation of its base, which is denoted by $u_b(t)$. The equation of motion for the longitudinal free vibrations of a proportionally damped uniform bar can be written as

$$-YA\frac{\partial^2 u(x,t)}{\partial x^2} - c_s A\frac{\partial^3 u(x,t)}{\partial x^2 \partial t} + c_a \frac{\partial u(x,t)}{\partial t} + m\frac{\partial^2 u(x,t)}{\partial t^2} = 0 \qquad (2.51)$$

where YA is the axial stiffness (Y is the elastic modulus and A is the cross-sectional area) and m is the mass per unit length of the bar. The absolute longitudinal displacement at any point x and time t can be represented by $u(x,t) = u_b(t) + u_{rel}(x,t)$ (Appendix B.2). The damping mechanism is again represented by two terms: c_s is due to internal (strain-rate) damping and c_a accounts for external viscous (air) damping.[12] Proportional damping is assumed as it allows analytical modal analysis using the eigenfunctions and orthogonality conditions of the corresponding undamped system (Appendix C.2).

After following steps similar to the derivation given for the transverse vibrations case in Section 2.1, the longitudinal vibration response of the bar relative to its moving base can be obtained as

$$u_{rel}(x,t) = \sum_{r=1}^{\infty} \frac{\varphi_r(x)}{\omega_{rd}} \int_0^t f_r(\tau)e^{-\zeta_r \omega_r(t-\tau)} \sin \omega_{rd}(t-\tau)\,d\tau \qquad (2.52)$$

---

[12] Although the same notation is used for some of the terms (such as the damping coefficients) of the transverse vibrations case, they are not necessarily identical.

Here, the modal forcing function is

$$f_r(t) = -\frac{d^2 u_b(t)}{dt^2} \left[m \int_0^L \varphi_r(x)\,dx + M_t \varphi_r(L) \right] \qquad (2.53)$$

where the forcing term due to air damping is neglected. It should be noted from Equation (2.53) that the excitation coming from the tip mass is directly considered in the modal forcing. The undamped natural frequency ω_r of the rth vibration mode is given by[13]

$$\omega_r = \alpha_r \sqrt{\frac{YA}{mL^2}} \qquad (2.54)$$

and ω_{rd} in Equation (2.52) is the damped natural frequency given by

$$\omega_{rd} = \omega_r \sqrt{1 - \zeta_r^2}$$

where ζ_r is the modal damping ratio of the rth vibration mode. The mass-normalized eigen-function $\varphi_r(x)$ of the rth mode can be expressed as

$$\varphi_r(x) = \frac{1}{\sqrt{\dfrac{mL}{2}\left(1 - \dfrac{\sin 2\alpha_r}{2\alpha_r}\right) + M_t \sin^2 \alpha_r}} \sin \frac{\alpha_r}{L} x \qquad (2.55)$$

Along with the geometric and natural boundary conditions at $x = 0$ and $x = L$, respectively, the eigenfunctions given by Equation (2.55) satisfy the following orthogonality conditions:

$$\int_0^L \varphi_s(x) m \varphi_r(x)\,dx + \varphi_s(L) M_t \varphi_r(L) = \delta_{rs}$$

$$-\int_0^L \varphi_s(x) YA \frac{d^2 \varphi_r(x)}{dx^2} dx + \left[\varphi_s(x) YA \frac{d\varphi_r(x)}{dx} \right]_{x=L} = \omega_r^2 \delta_{rs} \qquad (2.56)$$

The eigenvalues (α_r for mode r) are the roots of the characteristic equation

$$\frac{M_t}{mL} \alpha \sin \alpha - \cos \alpha = 0 \qquad (2.57)$$

---

[13] The detailed modal analysis can be found in Appendix C.2.

For a harmonic base displacement input, $u_b(t) = U_0 e^{j\omega t}$, the steady-state displacement response relative to the moving base becomes

$$u_{rel}(L, t) = \sum_{r=1}^{\infty} \frac{\sin \alpha_r \left(\dfrac{1 - \cos \alpha_r}{\alpha_r} + \dfrac{M_t}{mL} \sin \alpha_r \right) \omega^2 U_0 e^{j\omega t}}{\left[\dfrac{2\alpha_r - \sin 2\alpha_r}{4\alpha_r} + \dfrac{M_t}{mL} \sin^2 \alpha_r \right] \left(\omega_r^2 - \omega^2 + j2\zeta_r \omega_r \omega \right)} \tag{2.58}$$

2.4.2 Correction Factor for Longitudinal Vibrations

The relative motion transmissibility between the tip of the bar and the moving base can be extracted from Equation (2.58) as

$$T_{rel}(\omega, \zeta_r) = \frac{u_{rel}(L, t)}{U_0 e^{j\omega t}} = \sum_{r=1}^{\infty} \frac{\kappa_r \omega^2}{\omega_r^2 - \omega^2 + j2\zeta_r \omega_r \omega} \tag{2.59}$$

where

$$\kappa_r = \frac{\sin \alpha_r \left(\dfrac{1 - \cos \alpha_r}{\alpha_r} + \dfrac{M_t}{mL} \sin \alpha_r \right)}{\dfrac{2\alpha_r - \sin 2\alpha_r}{4\alpha_r} + \dfrac{M_t}{mL} \sin^2 \alpha_r} \tag{2.60}$$

is the correction factor for the lumped-parameter model of the rth mode for longitudinal vibrations (to predict the vibratory motion at $x = L$). Note that κ_r is a function of M_t/mL and α_r due to Equation (2.60), and α_r is a function of M_t/mL from Equation (2.57). Therefore, for a given vibration mode, the correction factor κ_r is a function of M_t/mL only. In the absence of a tip mass ($M_t/mL = 0$), the correction factor for the fundamental mode can explicitly be obtained from Equations (2.57) and (2.60) as $\kappa_1 = 4/\pi \cong 1.273$. However, in the presence of a tip mass, the transcendental equation given by Equation (2.57) should be solved numerically to obtain the correction factor. The variation of the correction factor of the fundamental mode (κ_1) with M_t/mL is given in Figure 2.14.

As in the transverse vibrations case, the correction factor tends to unity as the ratio of tip mass to bar mass increases, meaning that the *uncorrected* lumped-parameter model can be used only for bars with a tip mass that is much larger than the bar mass. Table 2.2 shows the correction factor κ_1 for lumped-parameter modeling of the fundamental longitudinal vibration mode for different M_t/mL ratios. Note that the error in the relative motion u_{rel} at the tip of the bar predicted by using the uncorrected lumped-parameter model is simply obtained from $(1 - \kappa_1)/\kappa_1 \times 100$.

The following quadratic polynomial ratio (obtained by using the Curve Fitting Toolbox of MATLAB) represents the behavior of the correction factor shown in Figure 2.14 successfully with a maximum error less than $4.5 \times 10^{-2}\%$ for all values of M_t/mL:

$$\kappa_1 = \frac{(M_t/mL)^2 + 0.7664 \, (M_t/mL) + 0.2049}{(M_t/mL)^2 + 0.6005 \, (M_t/mL) + 0.161} \tag{2.61}$$

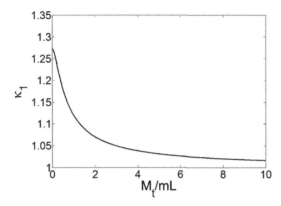

Figure 2.14   Variation of the correction factor for the fundamental longitudinal vibration mode with ratio of tip mass to bar mass

Table 2.2   Correction factor for the fundamental longitudinal vibration mode and the error in the uncorrected lumped-parameter model for different values of tip mass - to - bar mass ratio

M_t/mL	κ_1	Error in the uncorrected lumped-parameter model (%)
0	1.27323954	−21.46
0.1	1.26196259	−20.76
0.5	1.17845579	−15.14
1	1.11913201	−10.65
5	1.03108765	−3.02
10	1.01609422	−1.58

2.5   Correction Factor in the Electromechanically Coupled Lumped-Parameter Equations and a Theoretical Case Study

This section covers the insertion of the amplitude correction factor into the electromechanically coupled lumped-parameter piezoelectric energy harvester equations. The lumped-parameter electromechanical equations of a piezoelectric energy harvester are obtained by applying Newton's second law (or d'Alembert's principle) in the mechanical domain, Kirchhoff's laws in the electrical domain, and also including the electromechanical coupling effects coming from the piezoelectric constitutive equations [20].

2.5.1   An Electromechanically Coupled Lumped-Parameter Model for Piezoelectric Energy Harvesting

A lumped-parameter piezoelectric energy harvester model has been proposed by duToit *et al.* [10]. Figure 2.15 shows the schematic of their cantilevered energy harvester which is excited by the motion of its base in the longitudinal direction. Therefore, this model uses longitudinal

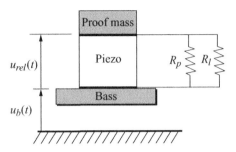

Figure 2.15   Schematic of a lumped-parameter piezoelectric energy harvester

vibrations of the piezoceramic for power generation. The design depicted in Figure 2.15 for longitudinal vibrations (almost like a uniaxial accelerometer) typically results in a very high fundamental natural frequency, which is not preferred for vibration-based energy harvesting as it will not match typical ambient vibration frequencies. Nevertheless, this lumped-parameter representation provides useful insight into the electromechanical problem for a simplified analysis of the coupled system dynamics.

As the longitudinal strain and the electric field directions are the same, the device shown in Figure 2.15 uses the 33-mode of piezoelectricity (and therefore the d_{33} constant of the piezoceramic) where the 3-direction (poling direction) is the longitudinal direction. The electrodes are connected to an equivalent resistive load R_{eq} (which is the parallel combination of the piezoelectric leakage resistance R_p and the load resistance R_l where $R_l \ll R_p$, hence $R_{eq} \approx R_l$). The coupled lumped-parameter equations are given by duToit *et al.* [10] as[14]

$$\frac{d^2 u_{rel}}{dt^2} + 2\zeta_m \omega_n \frac{d u_{rel}}{dt} + \omega_n^2 u_{rel} - \omega_n^2 d_{33} v = -\frac{d^2 u_b}{dt^2} \tag{2.62}$$

$$R_{eq} C_p \frac{dv}{dt} + v + m_{eq} R_{eq} d_{33} \omega_n^2 \frac{d u_{rel}}{dt} = 0 \tag{2.63}$$

where m_{eq} is the equivalent mass of the bar, ζ_m is the mechanical damping ratio, ω_n is the natural frequency, d_{33} is the piezoelectric constant, R_{eq} is the equivalent resistance, C_p is the capacitance, u_b is the harmonic base excitation, u_{rel} is the relative displacement of the proof mass, and v is the voltage output. One can then obtain the steady-state vibration response, voltage output, and power output FRFs (per base acceleration) by assuming $u_b(t) = U_0 e^{j\omega t}$ as

$$\left| \frac{u_{rel}}{\ddot{u}_b} \right| = \frac{1/\omega_n^2 \sqrt{1 + (r\Omega)^2}}{\sqrt{\left[1 - (1 + 2\zeta_m r)\,\Omega^2\right]^2 + \left[\left(1 + k_e^2\right) r\Omega + 2\zeta_m \Omega - r\Omega^3\right]^2}} \tag{2.64}$$

$$\left| \frac{v}{\ddot{u}_b} \right| = \frac{m_{eff} R_{eq} d_{33} \omega_n \Omega}{\sqrt{\left[1 - (1 + 2\zeta_m r)\,\Omega^2\right]^2 + \left[\left(1 + k_e^2\right) r\Omega + 2\zeta_m \Omega - r\Omega^3\right]^2}} \tag{2.65}$$

[14] Some of the variables have been adapted to the notation of this text.

$$\left|\frac{P}{(\ddot{u}_b)^2}\right| = \frac{\left(m_{eff}1/\omega_n r k_e^2 R_{eq}\right)/R_l\Omega^2}{\left[1 - (1 + 2\zeta_m r)\,\Omega^2\right]^2 + \left[\left(1 + k_e^2\right)r\Omega + 2\zeta_m\Omega - r\Omega^3\right]^2} \tag{2.66}$$

where $\Omega = \omega/\omega_n$ is the dimensionless frequency, k_e^2 is the lumped-parameter coupling coefficient, $r = \omega_n R_{eq} C_p$, and the over-dot represents differentiation with respect to time [10].

2.5.2 Correction Factor in the Electromechanically Coupled Lumped-Parameter Model and a Theoretical Case Study

From the theoretical discussion given in Section 2.4, it is known that the right hand side of the mechanical equilibrium equation given by Equation (2.62) should have a correction factor (to be valid for all values of the proof mass). Therefore the corrected form of Equation (2.62) is

$$\frac{d^2 u_{rel}}{dt^2} + 2\zeta_m\omega_n\frac{d u_{rel}}{dt} + \omega_n^2 u_{rel} - \omega_n^2 d_{33}v = -\kappa_1\frac{d^2 u_b}{dt^2} \tag{2.67}$$

Then the steady-state vibration response, voltage, and power output FRFs given by Equations (2.64)–(2.66) become

$$\left|\frac{u_{rel}}{\ddot{u}_b}\right| = \frac{\kappa_1/\omega_n^2\sqrt{1 + (r\Omega)^2}}{\sqrt{\left[1 - (1 + 2\zeta_m r)\,\Omega^2\right]^2 + \left[\left(1 + k_e^2\right)r\Omega + 2\zeta_m\Omega - r\Omega^3\right]^2}} \tag{2.68}$$

$$\left|\frac{v}{\ddot{u}_b}\right| = \frac{\kappa_1 m_{eff}R_{eq}d_{33}\omega_n\Omega}{\sqrt{\left[1 - (1 + 2\zeta_m r)\,\Omega^2\right]^2 + \left[\left(1 + k_e^2\right)r\Omega + 2\zeta_m\Omega - r\Omega^3\right]^2}} \tag{2.69}$$

$$\left|\frac{P}{(\ddot{u}_b)^2}\right| = \frac{\left(\kappa_1^2 m_{eff}1/\omega_n r k_e^2 R_{eq}\right)/R_l\Omega^2}{\left[1 - (1 + 2\zeta_m r)\,\Omega^2\right]^2 + \left[\left(1 + k_e^2\right)r\Omega + 2\zeta_m\Omega - r\Omega^3\right]^2} \tag{2.70}$$

Note that the steady-state vibration response and the voltage response are linearly proportional to κ_1. However, the power output is proportional to the square of κ_1 since it is proportional to the square of the voltage. Hence the inaccuracy of the predicted power output by using Equation (2.66) is expected to be more significant than that of the vibration amplitude or the voltage output.

Using the numerical values given in duToit et al. [10], the ratio of tip mass to bar mass for this sample device is obtained as $M_t/mL \cong 1.33$. From Equation (2.61), the correction factor for the fundamental mode for this ratio is $\kappa_1 \cong 1.0968$. If this factor is not used, the relative error in the predicted proof mass motion and the voltage output is about 8.83%, whereas the relative error in the estimated power output is about 16.9%. Figure 2.16 shows the vibration and the power FRFs obtained using the uncorrected and corrected equations for an arbitrary load resistance of $10\,k\Omega$ and a base acceleration of $9.81\,m/s^2$.

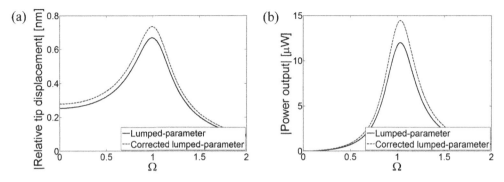

Figure 2.16 Corrected and uncorrected (a) relative tip displacement and (b) power output FRFs obtained using the lumped-parameter electromechanical model for a load resistance of $10\,k\Omega$ and a base acceleration of $9.81\,m/s^2$

2.6   Summary

Analytical distributed-parameter modeling of the transverse and longitudinal vibrations for cantilevered beams and bars under base excitation are reviewed. The general response to arbitrary base excitation is obtained based on the assumption of proportional damping. The total mechanical damping in the system is represented by internal and external damping mechanisms and it is shown that the external damping component contributes to the excitation amplitude in the base excitation problem. If the damping due to the surrounding fluid is relatively heavy, the contribution from the external damping related force component to the excitation amplitude may not be negligible. It is also shown that the frequently quoted lumped-parameter base excitation model might yield highly inaccurate results in predicting the vibration response (with an error as high as 35%) as it neglects the contribution of the distributed mass to the excitation amplitude. An amplitude correction factor is derived for improving the lumped-parameter model predictions. Variation of the correction factor with the ratio of tip mass to beam mass is also investigated and it is observed that the uncorrected lumped-parameter model can be accurate only when this ratio is sufficiently large. Two experimental case studies are presented and the corrected lumped-parameter model is validated for the transverse vibration problem. Analytical distributed-parameter modeling of the base excitation problem is summarized for the case of longitudinal vibrations and a correction factor is introduced for its lumped-parameter model as well. The amplitude correction factor is introduced to the lumped-parameter electromechanical equations of a piezo-stack energy harvester operating in the longitudinal vibration mode and a theoretical case study is given.

2.7   Chapter Notes

This chapter constitutes the analytical background for the mechanical domain of the electromechanical problem discussed in the next chapter (where piezoelectric coupling is introduced and distributed-parameter analytical solutions are obtained). It should be noted that cantilevers excited under base motion are employed in electromagnetic, electrostatic, and magnetostrictive energy harvesting as well. Therefore, the analytical distributed-parameter derivations given here can be coupled with the respective transduction mechanism for accurate prediction of the electromechanical response as is done for piezoelectric energy harvesting in the following chapter. While the cantilevered beam case is covered in detail throughout this chapter, the reader is referred to Chopra [21] in order to model the base excitation problem for relatively complicated structural configurations (such as open and closed frames).

Although the theoretical case study is given here for the electromechanically coupled longitudinal vibration problem, the inaccuracy of lumped-parameter modeling (or the conventional single-degree-of-freedom modeling) is the case for the transverse vibration problem as well. Moro *et al.* [22] and Benasciutti *et al.* [23] report a substantial difference between the predictions of the lumped-parameter model given by Roundy *et al.* [24, 25] (which ignores the distributed beam mass) and an electromechanically coupled finite-element model, while they observe that the finite-element simulations agree very well with the predictions of the distributed-parameter analytical model [26] covered in Chapter 3.

References

1. Timoshenko, S., Young, D.H., and Weaver, W. (1974) *Vibration Problems in Engineering*, John Wiley & Sons, Inc., New York.
2. Meirovitch, L. (2001) *Fundamentals of Vibrations*, McGraw-Hill, New York.
3. Inman, D.J. (2007) *Engineering Vibration*, Prentice Hall, Englewood Cliffs, NJ.
4. Rao, S.S. (2007) *Vibration of Continuous Systems*, John Wiley & Sons, Inc., Hoboken, NJ.
5. Clough, R.W. and Penzien, J. (1975) *Dynamics of Structures*, John Wiley & Sons, Inc., New York.
6. Banks, H.T. and Inman , D.J. (1991) On damping mechanisms in beams. *ASME Journal of Applied Mechanics*, **58**, 716–723.
7. Caughey, T.K. and O'Kelly, M.E.J. (1965) Classical normal modes in damped linear dynamic systems. *ASME Journal of Applied Mechanics*, **32**, 583–588.
8. Ewins, D.J. (2000) *Modal Testing: Theory, Practice and Application*, Research Studies Press, Baldock, Hertfordshire.
9. Roundy, S. and Wright, P.K. (2004) A piezoelectric vibration based generator for wireless electronics. *Smart Materials and Structures*, **13**, 1131–1144.
10. duToit, N.E., Wardle, B.L., and Kim, S. (2005) Design considerations for MEMS-scale piezoelectric mechanical vibration energy harvesters. *Journal of Integrated Ferroelectrics*, **71**, 121–160.
11. Stephen, N.G. (2006) On energy harvesting from ambient vibration. *Journal of Sound and Vibration*, **293**, 409–425.
12. Daqaq, F.M., Renno, J.M., Farmer, J.R., and Inman, D.J. (2007) Effects of system parameters and damping on an optimal vibration-based energy harvester. Proceedings of the 48th AIAA/ASME/ASCE/AHS/ASC Structures, Structural Dynamics, and Materials Conference, April 23–26, Honolulu, Hawaii.
13. Jeon, Y.B., Sood, R., Jeong, J.H., and Kim, S. (2005) MEMS power generator with transverse mode thin film PZT. *Sensors & Actuators A*, **122**, 16–22.
14. Fang, H.-B., Liu, J.-Q., Xu, Z.-Y., Dong, L., Chen, D., Cai, B.-C., and Liu, Y. (2006) A MEMS-based piezoelectric power generator for low frequency vibration energy harvesting. *Chinese Physics Letters*, **23**, 732–734.
15. Williams, C.B. and Yates, R.B. (1996) Analysis of a micro-electric generator for microsystems. *Sensors and Actuators A*, **52**, 8–11.
16. Strutt, J.W. (Lord Rayleigh) (1894) *The Theory of Sound*, Macmillan, London.
17. Banks, H.T., Luo, Z.H., Bergman, L.A., and Inman, D.J. (1998) On the existence of normal modes of damped discrete-continuous systems. *ASME Journal of Applied Mechanics*, **65**, 980–989.
18. Newland, D.E. (1993) *Random Vibrations, Spectral and Wavelet Analysis*, John Wiley & Sons, Inc., New York.
19. Bendat, J.S. and Piersol, A.G. (1986) *Random Data Analysis and Measurement Procedures*, John Wiley & Sons, Inc., New York.

20. Standards Committee of the IEEE Ultrasonics, Ferroelectrics, and Frequency Control Society (1987) *IEEE Standard on Piezoelectricity*, IEEE, New York.
21. Chopra, A.K. (2006) *Dynamics of Structures: Theory and Applications to Earthquake Engineering*, Prentice Hall, Englewood Cliffs, NJ.
22. Moro, L., Benasciutti, D., Brusa, E., and Zelenika, S. (2009) Caratterizzazione esperimentale di "Energy Scavengers" piezoelecttrici ottimizzati. AIAS XXXVIII Convengo Nazionale, Politecnico di Torino, 9–11 Septembre 2009.
23. Benasciutti, D., Moro, L., Zelenika, S., and Brusa, E. (2009) Vibration energy scavenging via piezoelectric bimorphs of optimized shapes. *Microsystem Technologies*, **16**, 657–668.
24. Roundy, S., Wright, P., and Rabaey, J. (2003) *Energy Scavenging for Wireless Sensor Networks*, Kluwer Academic, Boston, MA.
25. Roundy, S., Wright, P.K., and Rabaey, J.M. (2003) A study of low level vibrations as a power source for wireless sensor nodes. *Computer Communications*, **26**, 1131–1144.
26. Erturk, A. and Inman, D.J. (2009) An experimentally validated bimorph cantilever model for piezoelectric energy harvesting from base excitations. *Smart Materials and Structures*, **18**, 025009.

3

Analytical Distributed-Parameter Electromechanical Modeling of Cantilevered Piezoelectric Energy Harvesters

In this chapter, electromechanically coupled analytical solutions are presented for symmetric bimorph piezoelectric energy harvester configurations with series and parallel connections of the piezoceramic layers. Following the mechanical derivations given in Chapter 2, the base excitation is assumed to be translation in the transverse direction with superimposed small rotation. After describing the electromechanical modeling assumptions and the bimorph configurations, the distributed-parameter Euler–Bernoulli beam equation with piezoelectric coupling is derived and its modal analysis in short-circuit conditions is given. The electromechanically coupled circuit equation excited by infinitely many vibration modes is then derived based on the integral form of Gauss's law and the relevant piezoelectric constitutive equation. The governing electromechanical equations are then reduced to ordinary differential equations in modal coordinates and eventually an infinite set of algebraic equations is obtained for the complex modal vibration response and the complex voltage response of the energy harvester beam. For the series and parallel connections of the piezoceramic layers, the closed-form electromechanical expressions are first obtained for the steady-state response to harmonic excitation at arbitrary frequencies. The resulting multi-mode expressions are then simplified to single-mode expressions for modal excitations. A detailed theoretical case study is presented at the end of this chapter and experimental validations are given in Chapter 4.

3.1 Fundamentals of the Electromechanically Coupled Distributed-Parameter Model

3.1.1 Modeling Assumptions and Bimorph Configurations

The symmetric bimorph cantilever configurations shown in Figure 3.1 are modeled here as uniform composite beams based on the Euler–Bernoulli beam theory. This is a reasonable

Piezoelectric Energy Harvesting, First Edition. Alper Erturk and Daniel J. Inman.
© 2011 John Wiley & Sons, Ltd. Published 2011 by John Wiley & Sons, Ltd.

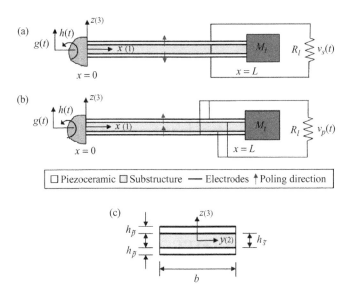

Figure 3.1   Geometrically uniform bimorph piezoelectric energy harvester configurations with (a) series and (b) parallel connections of the piezoceramic layers and (c) their cross-sectional view

assumption since typical cantilevered piezoelectric energy harvesters are designed and manufactured as fairly thin beams and most of the commercially available bimorphs are also thin structures (e.g., the samples used in Chapter 4 for model validation). Deformations are assumed to be small and the composite structure is assumed to exhibit linear material behavior (i.e., the excitation level is small enough to assume linear vibrations). The mechanical losses are represented by internal and external damping mechanisms as done in the electromechanically uncoupled discussion of Chapter 2. The internal damping mechanism is assumed to be in the form of strain-rate damping and the effect of external damping is considered with a separate damping coefficient. The homogeneous piezoceramic and substructure layers are assumed to be perfectly bonded to each other. The electrodes covering the opposite transverse faces of piezoceramic layers are assumed to be very thin as compared to the overall thicknesses of the harvester so that their contribution to the thickness dimension is negligible. In the following formulation, therefore, only the substructure and piezoceramic layers are assumed to be present. However, the presence of additional structural layers (such as epoxy and kapton layers) can be handled easily.[1] In Figure 3.1, the x-, y-, and z-directions, respectively, are coincident with the 1-, 2-, and 3-directions of piezoelectricity. These directional subscripts are used interchangeably as the former are preferred for mechanical derivations, whereas the latter are used in the piezoelectric constitutive relations (Appendix A).

The continuous electrode pairs covering the top and the bottom faces of the piezoceramic layers are assumed to be perfectly conductive so that a single electric potential difference can

---

[1] With the inclusion of additional inactive layers, the bending stiffness and the mass per length terms will be altered. In addition, the distance of the piezoceramic layers from the neutral axis will be modified, which changes the piezoelectric coupling terms in the derivation.

be defined across them. Therefore, the instantaneous electric fields induced in the piezoceramic layers are assumed to be uniform throughout the length of the beam. A resistive electrical load (denoted by R_l in Figure 3.1) is considered in the circuit along with the internal (or inherent) capacitances of the piezoceramic layers. Note that considering a resistive load in the electrical domain is common practice in the modeling of vibration-based energy harvesters [1–11].[2] As a consequence, it is assumed that the base motion input is persistent so that continuous electrical outputs can be extracted from the electromechanical system. The leakage resistance of the piezoceramic is ignored here as it is typically very large. It can easily be included as an additional resistance parallel to the load resistance (as in the lumped-parameter model of duToit *et al.* [5] discussed in Section 2.5).

It is known from the literature of static sensing and actuation that, depending on the voltage or current requirements, the piezoceramic layers of a symmetric bimorph can be combined in series or in parallel (see, for instance, Wang and Cross [12]). This common practice of static sensing and actuation problems is valid for the dynamic piezoelectric energy harvesting problem as well. Each of the two bimorph configurations displayed in Figures 3.1a and 3.1b undergoes bending vibrations due to the motion of its base. The piezoceramic layers are assumed to be identical and their conductive electrodes are assumed to be fully covering the respective surfaces (top and bottom) of these layers. The instantaneous average bending strains in the top and the bottom layers at an arbitrary position x over the beam length have the opposite sign (i.e., one is in tension while the other is in compression). As a consequence, since the piezoceramic layers of the bimorph shown in Figure 3.1a are poled oppositely in the thickness direction (i.e., z-direction), this configuration represents the *series connection* of the piezoceramic layers. Likewise, Figure 3.1b represents the *parallel connection* of the piezoceramic layers because the layers are poled in the same direction. The configuration in Figure 3.1a produces a larger voltage output whereas the one in Figure 3.1b produces a larger current output under optimal conditions.

3.1.2 Coupled Mechanical Equation and Modal Analysis of Bimorph Cantilevers

As far as the purely mechanical aspect of the problem is concerned, the bimorph configurations shown in Figures 3.1a and 3.1b are identical. That is, they have the same geometric and material properties for the extreme conditions of the external load resistance (short-circuit and open-circuit condition). However, the *backward* piezoelectric coupling effect[3] of electrical power generation in the beam equation for an arbitrary electrical load is different for series and parallel connections of the piezoceramic layers, which affects the vibration response of the cantilever. In the following, the beam equations are derived for these two configurations and the analytical modal analysis relations are presented.

The motion of the base for each of the cantilevers shown in Figures 3.1a and 3.1b is represented by the translational displacement $g(t)$ in the transverse direction with superimposed

---

[2] Typical nonlinear circuits used in piezoelectric energy harvesting are reviewed in Chapter 11.

[3] Throughout this text, the term *backward coupling* represents the electrical effect induced in the harvester structure due to the converse piezoelectric effect (the feedback sent from the electrical domain to the mechanical domain due to power generation). Hence the *forward coupling* term (to be defined later) is the term in the circuit equation due to the direct piezoelectric effect.

small rotational displacement $h(t)$. Therefore, the effective base displacement $w_b(x, t)$ in the transverse direction can be written as (Appendix B.1)

$$w_b(x, t) = g(t) + xh(t) \tag{3.1}$$

The partial differential equation governing the forced vibrations of a uniform cantilevered bimorph (with a tip mass) under base excitation is

$$-\frac{\partial^2 M(x, t)}{\partial x^2} + c_s I \frac{\partial^5 w_{rel}(x, t)}{\partial x^4 \partial t} + c_a \frac{\partial w_{rel}(x, t)}{\partial t} + m \frac{\partial^2 w_{rel}(x, t)}{\partial t^2}$$

$$= -[m + M_t \delta(x - L)] \frac{\partial^2 w_b(x, t)}{\partial t^2} \tag{3.2}$$

where $w_{rel}(x, t)$ is the transverse displacement of the beam (neutral axis) relative to its base at position x and time t, c_a is the viscous air damping coefficient, c_s is the strain-rate damping coefficient (it appears as an effective term $c_s I$ for the composite structure), m is the mass per unit length of the beam, M_t is its tip mass, $\delta(x)$ is the Dirac delta function, and $M(x, t)$ is the internal bending moment[4] (excluding the strain-rate damping effect). Recall from Equation (2.7) that the effect of strain-rate damping is an internal bending moment, which, in Equation (3.2), is directly written outside the undamped bending moment term $M(x, t)$.

The bimorph cantilevers shown in Figure 3.1 are assumed to be proportionally damped so that these configurations are normal-mode systems.[5] Hence the eigenfunctions of the respective undamped problem can be used for modal analysis. Indeed, instead of defining the damping coefficients in the physical equation of motion, one could consider the corresponding undamped equation (by setting $c_s I = c_a = 0$ in Equation (3.2)) and introduce modal damping into the equation of motion in modal coordinates as is common practice in structural dynamics. It is worth recalling that the foregoing consideration of the mechanical damping components results in an additional excitation term (peculiar to base excitation problems) due to external damping as shown in Section 2.1.1. For cantilevers operating in air, the excitation related to external damping is assumed to be negligible as compared to the inertial excitation term (Figure 2.3). Therefore the external damping-related excitation term is directly omitted in Equation (3.2) for simplicity.

The internal bending moment term in Equation (3.2) is the first moment of the axial stress field over the cross-section:

$$M(x, t) = b \left(\int_{-h_{\bar{p}} - h_{\bar{s}}/2}^{-h_{\bar{s}}/2} T_1^{\bar{p}} z \, dz + \int_{-h_{\bar{s}}/2}^{h_{\bar{s}}/2} T_1^{\bar{s}} z \, dz + \int_{h_{\bar{s}}/2}^{h_{\bar{p}} + h_{\bar{s}}/2} T_1^{\bar{p}} z \, dz \right) \tag{3.3}$$

where b is the width, $h_{\bar{p}}$ is the thickness of each piezoceramic layer, and $h_{\bar{s}}$ is the thickness of each substructure layer (Figure 3.1c). Furthermore, $T_1^{\bar{p}}$ and $T_1^{\bar{s}}$ are the stress components

---

[4] The convention for the bending moment is such that the *positive bending moment creates negative curvature* as in Sokolnikoff [13] and Dym and Shames [14], among others.

[5] See Banks *et al.* [15] for the limitations of the normal-mode assumption in combined dynamical systems.

(in the x-direction) in the piezoceramic and the substructure layers, respectively, and they are given by the following constitutive relations:

$$T_1^{\tilde{s}} = Y_{\tilde{s}} S_1^{\tilde{s}}, \quad T_1^{\tilde{p}} = \bar{c}_{11}^{E} S_1^{\tilde{p}} - \bar{e}_{31} E_3 \tag{3.4}$$

where Y_s is the elastic modulus (i.e., Young's modulus) of the substructure layer (which can be isotropic or orthotropic), $\bar{c}_{11}^{E}$ is the elastic modulus of the piezoceramic layer at constant electric field, $\bar{e}_{31}$ is the effective piezoelectric stress constant, and E_3 is the electric field component in the 3-direction (i.e., z-direction or the *poling* direction). The subscripts and superscripts $\tilde{p}$ and $\tilde{s}$ stand for the piezoceramic and the substructure layers, respectively. Based on the plane-stress assumption for a transversely isotropic thin piezoceramic beam (z-axis being the symmetry axis of transverse isotropy), the elastic modulus component of the piezoceramic can be expressed as $\bar{c}_{11}^{E} = 1/s_{11}^{E}$, where s_{11}^{E} is the elastic compliance at constant electric field. Furthermore, based on the same assumption, $\bar{e}_{31}$ can be given in terms of the more commonly used piezoelectric strain constant d_{31} as $\bar{e}_{31} = d_{31}/s_{11}^{E}$ (Appendix A.2). The axial strain components in the piezoceramic and the substructure layers are defined as $S_1^{\tilde{p}}$ and $S_1^{\tilde{s}}$, respectively, and they are due to bending only. Hence the axial strain at a certain level (z) from the neutral axis of the symmetric composite beam is simply proportional to the curvature of the beam at that position (x):[6]

$$S_1(x, z, t) = -z \frac{\partial^2 w_{rel}(x, t)}{\partial x^2} \tag{3.5}$$

The electric field component E_3 should be expressed in terms of the respective voltage term for each bimorph configuration (Figures 3.1a and 3.1b). This is the point where the resulting mechanical equations for the series and parallel connections of the piezoceramic layers differ from each other. Since the piezoceramic layers are assumed to be identical, the voltage across the electrodes of each piezoceramic layer is $v_s(t)/2$ in the series connection case (Figure 3.1a). As expected, for the parallel connection case (Figure 3.1b), the voltage across the electrodes of each piezoceramic layer is $v_p(t)$. It is worth adding that $\bar{e}_{31}$ has the opposite sign for the top and the bottom piezoceramic layers for the series connection case (due to opposite poling) so that the instantaneous electric fields are in the same direction (i.e., $E_3(t) = -v_s(t)/2h_{\tilde{p}}$ in both layers). For the configuration with the parallel connection, since $\bar{e}_{31}$ has the same sign in the top and bottom piezoceramic layers, the instantaneous electric fields are in the opposite directions (i.e., $E_3(t) = -v_p(t)/h_{\tilde{p}}$ in the top layer and $E_3(t) = v_p(t)/h_{\tilde{p}}$ in the bottom layer). Another important point is that, for both configurations, the piezoelectric coupling term from Equation (3.3) is a function of time only. Hence, before substituting Equation (3.3) into Equation (3.2), the electrical term must be multiplied by $[H(x) - H(x - L)]$ (so it survives after the spatial differentiation) where $H(x)$ is the Heaviside function. Since the voltage outputs of the series and parallel connection cases are different, the piezoelectric coupling effect in the mechanical equation (Equation (3.2)) is expected to be different. Thus, the mechanical response

---

[6] An independent axial displacement variable (in the x-direction) is not included in the strain expression as it decouples from the equation of motion for transverse vibrations due to the symmetric composite structure [16, 17] (see Section 6.2). Since there is no excitation in the axial direction, there is no strain contribution from the decoupled axial displacement.

expressions of the series and parallel connection configurations will be denoted by $w_{rel}^s(x, t)$ and $w_{rel}^p(x, t)$, respectively. Note that, here and hereafter, the subscripts and superscripts s and p, respectively, stand for the *series* and *parallel* connections of the piezoceramic layers (which should not be confused with $\tilde{s}$ and $\tilde{p}$ formerly introduced for the substructure and the piezoceramic layers, respectively).

The internal bending moment terms are then obtained from Equation (3.3) as

$$M^s(x, t) = -YI\frac{\partial^2 w_{rel}^s(x, t)}{\partial x^2} + \vartheta_s v_s(t)[H(x) - H(x - L)] \tag{3.6}$$

$$M^p(x, t) = -YI\frac{\partial^2 w_{rel}^p(x, t)}{\partial x^2} + \vartheta_p v_p(t)[H(x) - H(x - L)] \tag{3.7}$$

where the coefficients of the backward coupling terms (ϑ_s and ϑ_p) for the series and parallel connection cases can be expressed as

$$\vartheta_s = \frac{\bar{e}_{31}b}{2h_{\tilde{p}}}\left[\left(h_{\tilde{p}} + \frac{h_{\tilde{s}}}{2}\right)^2 - \frac{h_{\tilde{s}}^2}{4}\right] \tag{3.8}$$

$$\vartheta_p = 2\vartheta_s = \frac{\bar{e}_{31}b}{h_{\tilde{p}}}\left[\left(h_{\tilde{p}} + \frac{h_{\tilde{s}}}{2}\right)^2 - \frac{h_{\tilde{s}}^2}{4}\right] \tag{3.9}$$

The bending stiffness term YI of the composite cross-section for the constant electric field condition (short-circuit condition) of the piezoceramic is[7]

$$YI = \frac{2b}{3}\left\{Y_{\tilde{s}}\frac{h_{\tilde{s}}^3}{8} + \bar{c}_{11}^E\left[\left(h_{\tilde{p}} + \frac{h_{\tilde{s}}}{2}\right)^3 - \frac{h_{\tilde{s}}^3}{8}\right]\right\} \tag{3.10}$$

From Equation (3.2), the coupled beam equation can be obtained for the series connection case (Figure 3.1a) as follows:

$$YI\frac{\partial^4 w_{rel}^s(x, t)}{\partial x^4} + c_s I\frac{\partial^5 w_{rel}^s(x, t)}{\partial x^4 \partial t} + c_a\frac{\partial w_{rel}^s(x, t)}{\partial t} + m\frac{\partial^2 w_{rel}^s(x, t)}{\partial t^2}$$
$$-\vartheta_s v_s(t)\left[\frac{d\delta(x)}{dx} - \frac{d\delta(x - L)}{dx}\right] = -[m - M_t\delta(x - L)]\frac{\partial^2 w_b(x, t)}{\partial t^2} \tag{3.11}$$

Similarly, one can obtain the equation of motion for the case (Figure 3.1b) with the parallel connection of the piezoceramic layers as

$$YI\frac{\partial^4 w_{rel}^p(x, t)}{\partial x^4} + c_s I\frac{\partial^5 w_{rel}^p(x, t)}{\partial x^4 \partial t} + c_a\frac{\partial w_{rel}^p(x, t)}{\partial t} + m\frac{\partial^2 w_{rel}^p(x, t)}{\partial t^2}$$

---

[7] Here and in the following expressions, for a plate-like cantilever undergoing bending vibrations without torsion, one can replace the YI term with the effective plate bending stiffness (often denoted by D) to account for the Poisson effect.

$$-\vartheta_p v_p(t) \left[\frac{d\delta(x)}{dx} - \frac{d\delta(x-L)}{dx} \right] = -[m - M_t\delta(x-L)]\frac{\partial^2 w_b(x,t)}{\partial t^2} \qquad (3.12)$$

It is useful to note at this stage that the nth derivative of the Dirac delta function satisfies [18, 19]

$$\int_{-\infty}^{\infty} \frac{d^{(n)}\delta(x-x_0)}{dx^{(n)}} \gamma(x)\,dx = (-1)^n \frac{d^{(n)}\gamma(x)}{dx^{(n)}} \bigg|_{x=x_0} \qquad (3.13)$$

where $\gamma(x)$ is a smooth test function.

The mass per unit length term m is simply

$$m = b(\rho_{\bar{s}}h_{\bar{s}} + 2\rho_{\bar{p}}h_{\bar{p}}) \qquad (3.14)$$

where $\rho_{\bar{s}}$ and $\rho_{\bar{p}}$ are the mass densities of the substructure and the piezoceramic, respectively.

Based on the proportional damping (or modal damping) assumption, the vibration response relative to the base of the bimorph (Figures 3.1a and 3.1b) can be represented as a convergent series of the eigenfunctions as

$$w_{rel}^s(x,t) = \sum_{r=1}^{\infty} \phi_r(x)\eta_r^s(t), \quad w_{rel}^p(x,t) = \sum_{r=1}^{\infty} \phi_r(x)\eta_r^p(t) \qquad (3.15)$$

where $\phi_r(x)$ is the mass-normalized eigenfunction of the rth vibration mode, and $\eta_r^s(t)$ and $\eta_r^p(t)$ are the modal mechanical coordinate expressions of the series and parallel connection cases, respectively. The eigenfunctions denoted by $\phi_r(x)$ are the mass-normalized eigenfunctions of the corresponding undamped free vibration problem[8]

$$\phi_r(x) = A_r \left[\cos\frac{\lambda_r}{L}x - \cosh\frac{\lambda_r}{L}x + \varsigma_r \left(\sin\frac{\lambda_r}{L}x - \sinh\frac{\lambda_r}{L}x \right) \right] \qquad (3.16)$$

where ς_r is obtained from

$$\varsigma_r = \frac{\sin\lambda_r - \sinh\lambda_r + \lambda_r \dfrac{M_t}{mL}(\cos\lambda_r - \cosh\lambda_r)}{\cos\lambda_r + \cosh\lambda_r - \lambda_r \dfrac{M_t}{mL}(\sin\lambda_r - \sinh\lambda_r)} \qquad (3.17)$$

---

[8] See Appendix C.1 for the analytical modal analysis of the undamped and electromechanically uncoupled problem.

and A_r is a modal amplitude constant which should be evaluated by normalizing the eigenfunctions according to the following orthogonality conditions:[9]

$$
\int_0^L \phi_s(x) m \phi_r(x)\, dx + \phi_s(L) M_t \phi_r(L) + \left[\frac{d\phi_s(x)}{dx} I_t \frac{d\phi_r(x)}{dx} \right]_{x=L} = \delta_{rs}
$$

$$
\int_0^L \phi_s(x) Y I \frac{d^4\phi_r(x)}{dx^4} dx - \left[\phi_s(x) Y I \frac{d^3\phi_r(x)}{dx^3} \right]_{x=L} + \left[\frac{d\phi_s(x)}{dx} Y I \frac{d^2\phi_r(x)}{dx^2} \right]_{x=L} = \omega_r^2 \delta_{rs}
$$

$$(3.18)$$

Here, I_t is the mass moment of inertia of the tip mass M_t about the free end[10] $(x = L)$ and δ_{rs} is the Kronecker delta. Furthermore, ω_r is the undamped natural frequency of the rth vibration mode in short-circuit conditions (i.e., $R_l \rightarrow 0$) given by

$$
\omega_r = \lambda_r^2 \sqrt{\frac{YI}{mL^4}}
$$

$$(3.19)$$

where the eigenvalues of the system (λ_r for mode r) are obtained from

$$
1 + \cos\lambda \cosh\lambda + \lambda \frac{M_t}{mL} (\cos\lambda \sinh\lambda - \sin\lambda \cosh\lambda) - \frac{\lambda^3 I_t}{mL^3} (\cosh\lambda \sin\lambda + \sinh\lambda \cos\lambda)
$$

$$
+ \frac{\lambda^4 M_t I_t}{m^2 L^4} (1 - \cos\lambda \cosh\lambda) = 0
$$

$$(3.20)$$

It should be mentioned that the foregoing modal analysis is given for the short-circuit conditions (i.e., for $R_l \rightarrow 0$) so that the conventional form of the eigenfunctions given by Equation (3.16) is obtained since

$$
\lim_{R_l \rightarrow 0} v_s(t) = 0, \quad \lim_{R_l \rightarrow 0} v_p(t) = 0
$$

$$(3.21)$$

in Equations (3.11) and (3.12), respectively (i.e., the voltage output vanishes for zero load resistance). Note that the short-circuit condition is indeed a constant electric field condition ($E_3 \rightarrow 0$ as $R_l \rightarrow 0$), in agreement with the piezoelectric constitutive equation used here.

For a given bimorph, the form of the eigenfunctions given by $\phi_r(x)$ and their mass normalization conditions are the same regardless of the series or parallel connections of the piezoceramic layers. For non-zero values of load resistance, the voltage terms in the mechanical equations take finite values, generating *point moment excitations* at the boundaries of the piezoceramic layer according to Equations (3.11) and (3.12), and yielding two different modal mechanical response functions for these equations as $\eta_r^s(t)$ and $\eta_r^p(t)$, respectively. The feedback received from the voltage response for a given load resistance alters the mechanical response as well as

---

[9] It is sufficient to use the first expression in Equations (3.18) for normalizing the eigenfunctions (once the first one is satisfied, the second expression is automatically satisfied and vice versa).

[10] Knowing the mass moment of inertia of the tip mass about its centroid, the mass moment of inertia I_t about the end point of the elastic beam $(x = L)$ can be obtained by using the parallel axis theorem [20].

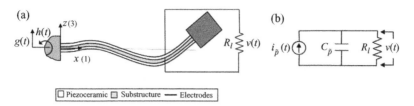

Figure 3.2 (a) Bimorph cantilever with a single layer connected to a resistive load and (b) the corresponding electrical circuit with a dependent current source

the resonance frequency of the energy harvester, as discussed theoretically here and validated experimentally in the next chapter.

3.1.3 Coupled Electrical Circuit Equation of a Thin Piezoceramic Layer under Dynamic Bending

In order to derive the governing electrical circuit equations of the bimorph configurations for the series and parallel connections of the piezoceramic layers, one should first examine the electroelastic dynamics of a single layer under bending vibrations. Consider Figure 3.2a, where the electrodes of a single layer are connected to a resistive electrical load (and the electrodes of the other layer are shorted). The deflections are exaggerated to highlight the space- and time-dependent radius of curvature at an arbitrary point x on the neutral axis at time t.

Since the only source of mechanical strain is assumed to be the axial strain due to bending, for the given electrode configuration, the tensorial representation of the relevant piezoelectric constitutive equation [21] that gives the vector of electric displacements can be reduced to the following scalar equation (Appendix A.2):

$$D_3 = \bar{e}_{31} S_1^{\bar{p}} + \bar{\varepsilon}_{33}^S E_3 \tag{3.22}$$

where D_3 is the electric displacement component and $\bar{\varepsilon}_{33}^S$ is the permittivity component at constant strain with the plane-stress assumption for a beam ($\bar{\varepsilon}_{33}^S = \varepsilon_{33}^T - d_{31}^2/s_{11}^E$ where ε_{33}^T is the permittivity component at constant stress as given in Appendix A.2) .

As the external circuit admittance across the electrodes is $1/R_l$, the electric current output can be obtained from the integral form of Gauss's law [21] as

$$\frac{d}{dt}\left(\int_A \mathbf{D} \cdot \mathbf{n}\, dA\right) = \frac{v(t)}{R_l} \tag{3.23}$$

where $\mathbf{D}$ is the vector of electric displacement components in the piezoceramic layer, $\mathbf{n}$ is the unit outward normal, and the integration is performed over the electrode area A. The only contribution to the inner product of the integrand in Equation (3.23) is from D_3, since the electrodes are perpendicular to the 3-axis (i.e., z-axis). After expressing the average bending strain in the piezoceramic layer in terms of the curvature (see Equation (3.5)) and the uniform electric field in terms of the electric potential difference ($E_3(t) = -v(t)/h_{\bar{p}}$), Equation (3.22) can be used in Equation (3.23) to obtain

$$\frac{\bar{\varepsilon}_{33}^S bL}{h_{\bar{p}}}\frac{dv(t)}{dt} + \frac{v(t)}{R_l} + \bar{e}_{31}h_{\bar{p}c}b\int_0^L \frac{\partial^3 w_{rel}(x,t)}{\partial x^2 \partial t}dx = 0 \qquad (3.24)$$

where b, $h_{\bar{p}}$, and L are the width, thickness, and length of the piezoceramic layer, respectively, and $h_{\bar{p}c}$ is the distance between the neutral axis and the center of each piezoceramic layer ($h_{\bar{p}c} = (h_{\bar{p}} + h_{\bar{s}})/2$). One can then substitute the modal expansion form of the transverse vibration response (relative to the base) given by

$$w_{rel}(x,t) = \sum_{r=1}^{\infty} \phi_r(x)\eta_r(t) \qquad (3.25)$$

into Equation (3.24) to obtain

$$\frac{\bar{\varepsilon}_{33}^S bL}{h_{\bar{p}}}\frac{dv(t)}{dt} + \frac{v(t)}{R_l} + \sum_{r=1}^{\infty}\kappa_r\frac{d\eta_r(t)}{dt} = 0 \qquad (3.26)$$

where κ_r is the modal coupling term in the electrical circuit equation:

$$\kappa_r = \bar{e}_{31}h_{\bar{p}c}b\int_0^L \frac{d^2\phi_r(x)}{dx^2}dx = \bar{e}_{31}h_{\bar{p}c}b\left.\frac{d\phi_r(x)}{dx}\right|_{x=L} \qquad (3.27)$$

The *forward* coupling term κ_r has important consequences that need to be addressed briefly. According to Equation (3.24), which originates from Equation (3.23), the excitation of the simple *RC* circuit considered here as well as that of more sophisticated energy harvesting circuit topologies [22–27] is proportional to the integral of the dynamic strain distribution over the electrode area. For vibration modes of a cantilevered beam other than the fundamental vibration mode, the dynamic strain distribution over the beam length changes sign at the *strain nodes*.[11] It is known from Equation (3.5) that the curvature at a point is a direct measure of the bending strain. Hence, for modal vibrations of a thin beam, strain nodes are the *inflection points* of the eigenfunctions and the integrand in Equation (3.27) is the curvature eigenfunction. If the electric charge developed at the opposite sides of a strain node is collected by continuous electrodes for vibrations with a certain mode shape, cancellation occurs due to the phase difference in the mechanical strain distribution. Mathematically, the partial areas under the integrand function of the integral in Equation (3.27) cancel each other over the domain of integration (Appendix D.1). As an undesired consequence, the excitation of the electrical circuit, and therefore the electrical outputs, might diminish drastically. In order to avoid electrical cancellations, segmented electrodes can be used in harvesting energy from the modes higher than the fundamental mode. The leads of the segmented electrodes can be combined

---

[11] See Appendix D for a detailed discussion of strain nodes in thin beams.

in the circuit in an appropriate manner. Note that the rth vibration mode of a clamped–free beam has $r - 1$ strain nodes, and, consequently, the first mode of a cantilevered beam has no cancellation problem. Some boundary conditions are more prone to strong cancellations [28]. For instance, a beam with clamped–clamped boundary conditions has $r + 1$ strain nodes for the rth vibration mode (see Appendix D.3).

Based on Equation (3.26), it is very useful to represent the electrical domain of the coupled system by the simple circuit shown in Figure 3.2b. It is known in the circuitry-based energy harvesting literature that a piezoelectric element can be represented as a current source in parallel with its internal capacitance [22,24].[12] Therefore, the simple circuit shown in Figure 3.2b is the complete circuit of the electrical domain for a single resistive load case. Note that this representation considers the electrical domain *only* and the electromechanical representation of the coupled system is actually a *transformer* because of the voltage feedback sent to the mechanical domain due to the piezoelectric coupling (which will be incorporated in the formulation here). Hence the current amplitude and phase are *not* constant in this representation (implying that the current source is a *dependent* variable). The components of the circuit are the internal capacitance $C_{\tilde{p}}$ of the piezoceramic layer, the resistive load R_l, and the dependent current source $i_{\tilde{p}}(t)$. In agreement with Figure 3.2a, the voltage across the resistive load is denoted by $v(t)$. Then, Kirchhoff's laws can be applied to the electrical circuit shown in Figure 3.2b to obtain

$$C_{\tilde{p}}\frac{dv(t)}{dt} + \frac{v(t)}{R_l} - i_{\tilde{p}}(t) = 0 \tag{3.28}$$

where the internal capacitance and the dependent current source terms can be extracted by matching Equations (3.26) and (3.28) as

$$C_{\tilde{p}} = \frac{\bar{\varepsilon}_{33}^S bL}{h_{\tilde{p}}}, \quad i_{\tilde{p}}(t) = -\sum_{r=1}^{\infty} \kappa_r \frac{d\eta_r(t)}{dt} \tag{3.29}$$

Identification of the above terms (especially the dependent current source term) has an important use for the modeling of multimorph piezoelectric energy harvesters. For a given number of piezoceramic layers, there is no need to derive the electrical circuit equation by using Equations (3.22) and (3.23). Each piezoceramic layer will have a capacitance and a dependent current source term and the layers can be combined to the resistive electrical load(s) in a desired way.

3.2 Series Connection of the Piezoceramic Layers

Based on the fundamentals given in Section 3.1, this section presents the derivation of the closed-form expressions for the coupled voltage response $v_s(t)$ and vibration response $w_{rel}^s(x, t)$ of the bimorph configuration shown in Figure 3.1a. First the coupled mechanical

---

[12] The alternative representation is a voltage source in series with its internal capacitance. See the discussion in Section 5.7.6 regarding a misuse of these representations to obtain the optimum load resistance.

equation is given in modal coordinates and then the coupled circuit equation is derived. The resulting electromechanical equations are then solved for the steady-state voltage response and vibration response for harmonic base motion inputs.

3.2.1   Coupled Beam Equation in Modal Coordinates

After substituting the first equation of Equations (3.15) into Equation (3.11) and applying the orthogonality conditions of the eigenfunctions for the undamped problem, the mechanical equation of motion in modal coordinates can be obtained as

$$\frac{d^2\eta_r^s(t)}{dt^2} + 2\zeta_r\omega_r\frac{d\eta_r^s(t)}{dt} + \omega_r^2\eta_r^s(t) - \chi_r^s v_s(t) = f_r(t) \tag{3.30}$$

where the modal electromechanical coupling term is[13]

$$\chi_r^s = \vartheta_s \left.\frac{d\phi_r(x)}{dx}\right|_{x=L} \tag{3.31}$$

The modal mechanical forcing function can be expressed as

$$f_r(t) = -m\left(\frac{d^2g(t)}{dt^2}\int_0^L \phi_r(x)\,dx + \frac{d^2h(t)}{dt^2}\int_0^L x\phi_r(x)\,dx\right) - M_t\phi_r(L)\left(\frac{d^2g(t)}{dt^2} + L\frac{d^2h(t)}{dt^2}\right) \tag{3.32}$$

In Equation (3.30), ζ_r is the modal mechanical damping ratio that includes the combined effects of strain-rate and air damping. In the absence of a tip mass, how to relate the modal damping ratio to the strain-rate and air damping terms $c_s I$ and c_a mathematically based on the assumption of proportional damping is described in Chapter 2. However, as in common experimental modal analysis practice, one can identify the modal damping ratio ζ_r of a desired mode directly from the frequency response or time-domain measurements (which avoids the requirement of defining and obtaining the physical damping terms $c_s I$ and c_a).[14]

3.2.2   Coupled Electrical Circuit Equation

As described in Section 3.1.1, the piezoceramic layers of the bimorph configuration shown in Figure 3.1a are connected in series. It is known from the discussion given in Section 3.1.3 that each piezoceramic layer can be represented as a dependent current source in parallel with its internal capacitance. Therefore, Figure 3.3 displays the series connection of the identical piezoceramic layers of the bimorph configuration shown in Figure 3.1a.

---

[13] Note that, here, Equation (3.13) is used to obtain the modal backward coupling term χ_r^s where the test function is the eigenfunction ($\gamma(x) = \phi_r(x)$) multiplying the partial differential equation in the modal analysis solution procedure.

[14] Chapter 5 provides closed-form relations to identify the modal mechanical damping ratio from electromechanical FRF measurements in the presence of a resistive electrical load.

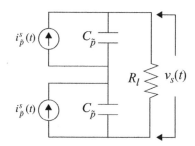

Figure 3.3 Electrical circuit representing the series connection of the piezoceramic layers with two dependent current sources

Kirchhoff's laws can be applied to the circuit depicted in Figure 3.3 to obtain

$$\frac{C_{\tilde{p}}}{2}\frac{dv_s(t)}{dt} + \frac{v_s(t)}{R_l} - i_{\tilde{p}}^s(t) = 0 \tag{3.33}$$

where the internal capacitance and the dependent current source terms of the bimorph (for each layer) are

$$C_{\tilde{p}} = \frac{\bar{\varepsilon}_{33}^S bL}{h_{\tilde{p}}}, \quad i_{\tilde{p}}^s(t) = -\sum_{r=1}^{\infty} \kappa_r \frac{d\eta_r^s(t)}{dt} \tag{3.34}$$

The modal electromechanical coupling term is then

$$\kappa_r = \bar{e}_{31} h_{\tilde{p}c} b \int_0^L \frac{d^2\phi_r(x)}{dx^2} dx = \frac{\bar{e}_{31}\left(h_{\tilde{p}} + h_{\tilde{s}}\right) b}{2} \left.\frac{d\phi_r(x)}{dx}\right|_{x=L} \tag{3.35}$$

where $h_{\tilde{p}c}$ (the distance between the neutral axis and the center of each piezoceramic layer) is expressed in terms of the piezoceramic and the substructure layer thicknesses $h_{\tilde{p}}$ and $h_{\tilde{s}}$ (Figure 3.1c). Hence, Equation (3.33) is the electrical circuit equation of the bimorph cantilever for the series connection of the piezoceramic layers.

3.2.3 Closed-Form Voltage Response and Vibration Response at Steady State

Equations (3.30) and (3.33) constitute the coupled equations for the modal mechanical response $\eta_r^s(t)$ of the bimorph and the voltage response $v_s(t)$ across the resistive load. In this section, the steady-state solution of these terms for harmonic motion inputs is derived. If the translational and rotational components of the base displacement given by Equation (3.1) are harmonic of the forms $g(t) = W_0 e^{j\omega t}$ and $h(t) = \theta_0 e^{j\omega t}$,[15] where W_0 and θ_0 are the translational and the small rotational displacement amplitudes of the base, ω is the excitation frequency, and j is the unit imaginary number, then the modal forcing function given by Equation (3.32) can be expressed as $f_r(t) = F_r e^{j\omega t}$ where the amplitude F_r is

$$
F_r = \omega^2 \left[m \left(W_0 \int_0^L \phi_r(x)\,dx + \theta_0 \int_0^L x\phi_r(x)\,dx \right) + M_t \phi_r(L)(W_0 + L\theta_0) \right] \tag{3.36}
$$

For the harmonic base motions at frequency ω, the steady-state modal mechanical response of the beam and the steady-state voltage response across the resistive load are assumed to be harmonic at the same frequency as $\eta_r^s(t) = H_r^s e^{j\omega t}$ and $v_s(t) = V_s e^{j\omega t}$ (linear system assumption), respectively, where H_r^s and V_s are complex valued. Therefore, Equations (3.30) and (3.33) yield the following two equations for H_r^s and V_s:[16]

$$
\left(\omega_r^2 - \omega^2 + j2\zeta_r\omega_r\omega\right) H_r^s - \chi_r^s V_s = F_r \tag{3.37}
$$

$$
\left(\frac{1}{R_l} + j\omega\frac{C_{\bar{p}}}{2}\right) V_s + j\omega \sum_{r=1}^{\infty} \kappa_r H_r^s = 0 \tag{3.38}
$$

The complex modal mechanical response H_r^s can be extracted from Equation (3.37) and substituted into Equation (3.38) to obtain the complex voltage V_s explicitly. The resulting complex voltage can then be used in $v_s(t) = V_s e^{j\omega t}$ to express the steady-state voltage response as

$$
v_s(t) = \frac{\displaystyle\sum_{r=1}^{\infty} \frac{-j\omega\kappa_r F_r}{\omega_r^2 - \omega^2 + j2\zeta_r\omega_r\omega}}{\displaystyle\frac{1}{R_l} + j\omega\frac{C_{\bar{p}}}{2} + \sum_{r=1}^{\infty} \frac{j\omega\kappa_r\chi_r^s}{\omega_r^2 - \omega^2 + j2\zeta_r\omega_r\omega}} e^{j\omega t} \tag{3.39}
$$

The complex voltage V_s can be substituted into Equation (3.37) to obtain the steady-state modal mechanical response of the bimorph as

$$
\eta_r^s(t) = \left(F_r - \chi_r^s \frac{\displaystyle\sum_{r=1}^{\infty} \frac{j\omega\kappa_r F_r}{\omega_r^2 - \omega^2 + j2\zeta_r\omega_r\omega}}{\displaystyle\frac{1}{R_l} + j\omega\frac{C_{\bar{p}}}{2} + \sum_{r=1}^{\infty} \frac{j\omega\kappa_r\chi_r^s}{\omega_r^2 - \omega^2 + j2\zeta_r\omega_r\omega}} \right) \frac{e^{j\omega t}}{\omega_r^2 - \omega^2 + j2\zeta_r\omega_r\omega} \tag{3.40}
$$

---

[15] If there is a relative phase difference between the small base rotation and the translation, it can be taken into account by representing either W_0 or θ_0 as a complex number.

[16] Although Equation (3.37) is nothing but a single linear algebraic equation, it represents an infinite number of equations since $r = 1, 2, \ldots$ for the mechanical vibration modes (hence Equations (3.37) and (3.38) represent "$\infty + 1$" equations).

The transverse displacement response (relative to the base) at point x on the bimorph can be obtained in physical coordinates by substituting Equation (3.40) into the first equation of Equations (3.15):

$$
w_{rel}^s(x,t) = \sum_{r=1}^{\infty} \left[\left(F_r - \chi_r^s \frac{\displaystyle\sum_{r=1}^{\infty} \frac{j\omega\kappa_r F_r}{\omega_r^2 - \omega^2 + j2\zeta_r\omega_r\omega}}{\dfrac{1}{R_l} + j\omega\dfrac{C_{\bar p}}{2} + \displaystyle\sum_{r=1}^{\infty} \frac{j\omega\kappa_r \chi_r^s}{\omega_r^2 - \omega^2 + j2\zeta_r\omega_r\omega}} \right) \frac{\phi_r(x)e^{j\omega t}}{\omega_r^2 - \omega^2 + j2\zeta_r\omega_r\omega} \right]
$$

$$(3.41)$$

Note that the vibration response given by Equation (3.41) is the displacement response of the beam relative to its moving base. If one is interested in the coupled beam displacement in absolute physical coordinates (relative to the fixed reference frame), it is the superposition of the base displacement and the vibratory displacement relative to the base:

$$
w^s(x,t) = w_b(x,t) + w_{rel}^s(x,t) \tag{3.42}
$$

where $w_b(x,t)$ is the effective base displacement given by Equation (3.1).

3.3 Parallel Connection of the Piezoceramic Layers

This section aims to derive the steady-state voltage response $v_p(t)$ and the vibration response $w_{rel}^p(x,t)$ of the bimorph configuration shown in Figure 3.1b to harmonic base motions. The coupled beam equation in modal coordinates and the electrical circuit equation are derived and the closed-form solutions are obtained in the following.

3.3.1 Coupled Beam Equation in Modal Coordinates

After substituting the second equation of Equations (3.15) into Equation (3.12), the partial differential equation given by Equation (3.12) can be reduced to an infinite set of ordinary differential equations in modal coordinates as follows:

$$
\frac{d^2\eta_r^p(t)}{dt^2} + 2\zeta_r\omega_r \frac{d\eta_r^p(t)}{dt} + \omega_r^2\eta_r^p(t) - \chi_r^p v_p(t) = f_r(t) \tag{3.43}
$$

where the modal electromechanical coupling term is

$$
\chi_r^p = \vartheta_p \frac{d\phi_r(x)}{dx}\bigg|_{x=L} \tag{3.44}
$$

and the modal mechanical forcing function is given by Equation (3.32). The discussion regarding the mechanical damping ratio ζ_r is the same as given in Section 3.2.1. Thus, Equation (3.43) is the coupled beam equation in modal coordinates for the bimorph configuration with the parallel connection of the piezoceramic layers.

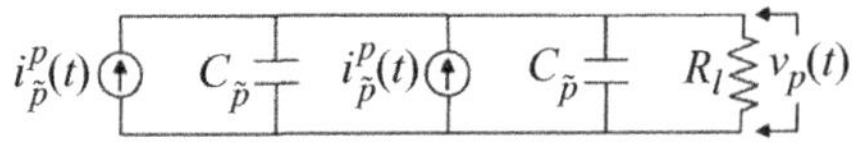

Figure 3.4 Electrical circuit representing the parallel connection of the piezoceramic layers with two dependent current sources

3.3.2 Coupled Electrical Circuit Equation

It was mentioned in Section 3.1.1 that the piezoceramic layers of the bimorph configuration shown in Figure 3.1b are connected in parallel. Since each of the piezoceramic layers can be represented as a dependent current source in parallel with its internal capacitance (Section 3.1.3), Figure 3.4 represents the parallel connection of the identical top and bottom piezoceramic layers of the bimorph configuration shown in Figure 3.1b.

One can then derive the governing electrical circuit equation based on Kirchhoff's laws as follows:

$$C_{\tilde{p}} \frac{dv_p(t)}{dt} + \frac{v_p(t)}{2R_l} - i_{\tilde{p}}^p(t) = 0 \tag{3.45}$$

where the internal capacitance and the dependent current source terms for each layer are

$$C_{\tilde{p}} = \frac{\bar{\varepsilon}_{33}^S bL}{h_{\tilde{p}}}, \quad i_{\tilde{p}}^p(t) = -\sum_{r=1}^{\infty} \kappa_r \frac{d\eta_r^p(t)}{dt} \tag{3.46}$$

The modal coupling term κ_r is given by Equation (3.35). Equation (3.45) is the electrical circuit equation of the bimorph cantilever for the parallel connection of the piezoceramic layers.

3.3.3 Closed-Form Voltage Response and Vibration Response at Steady State

In order to solve for $\eta_r^p(t)$ and $v_p(t)$ in Equations (3.43) and (3.45), the same procedure of Section 3.2.3 is followed by assuming the base excitation components in Figure 3.1b to be harmonic as $g(t) = W_0 e^{j\omega t}$ and $h(t) = \theta_0 e^{j\omega t}$. For these harmonic base motion components of the same frequency, the modal forcing is harmonic as $f_r(t) = F_r e^{j\omega t}$ where the amplitude F_r is given by Equation (3.36).

Based on the linear system assumption, the modal mechanical response $\eta_r^p(t)$ and the voltage response $v_p(t)$ are assumed to be harmonic at the frequency of excitation such that $\eta_r^p(t) = H_r^p e^{j\omega t}$ and $v_p(t) = V_p e^{j\omega t}$, where H_r^p and V_p are complex valued. Hence, Equations (3.43) and (3.45) yield the following equations for H_r^p and V_p:

$$\left(\omega_r^2 - \omega^2 + j2\zeta_r\omega_r\omega\right) H_r^p - \chi_r^p V_p = F_r \tag{3.47}$$

$$\left(\frac{1}{2R_l} + j\omega C_{\tilde{p}}\right) V_p + j\omega \sum_{r=1}^{\infty} \kappa_r H_r^p = 0 \tag{3.48}$$

where H_r^p and V_p can be obtained explicitly. Using the resulting complex voltage in $v_p(t) = V_p e^{j\omega t}$ gives the steady-state voltage response as

$$v_p(t) = \frac{\displaystyle\sum_{r=1}^{\infty} \frac{-j\omega\kappa_r F_r}{\omega_r^2 - \omega^2 + j2\zeta_r\omega_r\omega}}{\dfrac{1}{2R_l} + j\omega C_{\bar{p}} + \displaystyle\sum_{r=1}^{\infty} \frac{j\omega\kappa_r \chi_r^p}{\omega_r^2 - \omega^2 + j2\zeta_r\omega_r\omega}} e^{j\omega t} \tag{3.49}$$

Then the steady-state modal mechanical response of the bimorph can be obtained by using V_p in Equation (3.47) as

$$\eta_r^p(t) = \left(F_r - \chi_r^p \frac{\displaystyle\sum_{r=1}^{\infty} \frac{j\omega\kappa_r F_r}{\omega_r^2 - \omega^2 + j2\zeta_r\omega_r\omega}}{\dfrac{1}{2R_l} + j\omega C_{\bar{p}} + \displaystyle\sum_{r=1}^{\infty} \frac{j\omega\kappa_r \chi_r^p}{\omega_r^2 - \omega^2 + j2\zeta_r\omega_r\omega}} \right) \frac{e^{j\omega t}}{\omega_r^2 - \omega^2 + j2\zeta_r\omega_r\omega} \tag{3.50}$$

The modal mechanical response expression can then be used in the second equation of Equations (3.15) to obtain the transverse displacement response (relative to the base) at point x on the bimorph:

$$w_{rel}^p(x, t) = \sum_{r=1}^{\infty} \left[\left(F_r - \chi_r^p \frac{\displaystyle\sum_{r=1}^{\infty} \frac{j\omega\kappa_r F_r}{\omega_r^2 - \omega^2 + j2\zeta_r\omega_r\omega}}{\dfrac{1}{2R_l} + j\omega C_{\bar{p}} + \displaystyle\sum_{r=1}^{\infty} \frac{j\omega\kappa_r \chi_r^p}{\omega_r^2 - \omega^2 + j2\zeta_r\omega_r\omega}} \right) \frac{\phi_r(x)e^{j\omega t}}{\omega_r^2 - \omega^2 + j2\zeta_r\omega_r\omega} \right] \tag{3.51}$$

Having obtained the vibration response relative to the moving base, one can easily superimpose the base motion and the relative response to obtain the transverse displacement response at point x relative to the fixed frame as follows:

$$w^p(x, t) = w_b(x, t) + w_{rel}^p(x, t) \tag{3.52}$$

where the base displacement $w_b(x, t)$ is given by Equation (3.1).

3.4 Equivalent Representation of the Series and the Parallel Connection Cases

This section presents an equivalent (or a unified) representation of the distributed-parameter analytical solutions derived for the series and the parallel connection cases. For this purpose, the equivalent electromechanical coupling and the capacitance terms are obtained first and then the resulting equivalent representation is given.

3.4.1   Modal Electromechanical Coupling Terms

A careful investigation of the coefficient of the backward coupling term for the series connection case (Equation (3.8)) yields

$$\vartheta_s = \frac{\bar{e}_{31}b}{2h_{\bar{p}}}\left[\left(h_{\bar{p}} + \frac{h_{\bar{s}}}{2}\right)^2 - \frac{h_{\bar{s}}^2}{4}\right] = \frac{\bar{e}_{31}b}{2h_{\bar{p}}}\left[(h_{\bar{p}} + h_{\bar{s}})h_{\bar{p}}\right] = \bar{e}_{31}b\frac{h_{\bar{p}} + h_{\bar{s}}}{2} = \bar{e}_{31}bh_{\bar{p}c} \quad (3.53)$$

Then, Equation (3.9) for the parallel connection case becomes

$$\vartheta_p = 2\vartheta_s = 2\bar{e}_{31}bh_{\bar{p}c} \quad (3.54)$$

From Equations (3.31) and (3.44), the backward modal coupling terms lead to

$$\chi_r^s = \bar{e}_{31}bh_{\bar{p}c}\left.\frac{d\phi_r(x)}{dx}\right|_{x=L} \quad (3.55)$$

$$\chi_r^p = 2\bar{e}_{31}bh_{\bar{p}c}\left.\frac{d\phi_r(x)}{dx}\right|_{x=L} \quad (3.56)$$

while the forward modal coupling term given by Equation (3.35) is

$$\kappa_r = \bar{e}_{31}bh_{\bar{p}c}\left.\frac{d\phi_r(x)}{dx}\right|_{x=L} \quad (3.57)$$

3.4.2   Equivalent Capacitance for Series and Parallel Connections

The circuit equation for the series connection case given by Equation (3.33) can be rewritten as

$$C_{\bar{p}}^{eq,s}\frac{dv_s(t)}{dt} + \frac{v_s(t)}{R_l} + \sum_{r=1}^{\infty}\kappa_r\frac{d\eta_r^s(t)}{dt} = 0 \quad (3.58)$$

where

$$C_{\bar{p}}^{eq,s} = \frac{C_{\bar{p}}}{2} = \frac{\bar{\varepsilon}_{33}^S bL}{2h_{\bar{p}}} \quad (3.59)$$

is the equivalent capacitance of two identical capacitors ($C_{\bar{p}}$) connected in series.

Likewise, the circuit equation for the parallel connection case given by Equation (3.45) can be rearranged to give

$$C_{\bar{p}}^{eq,p}\frac{dv_p(t)}{dt} + \frac{v_p(t)}{R_l} + \sum_{r=1}^{\infty}2\kappa_r\frac{d\eta_r^p(t)}{dt} = 0 \quad (3.60)$$

Table 3.1 Modal electromechanical coupling and equivalent capacitance of a bimorph energy harvester for the series and the parallel connections of the piezocermaic layers

	Series connection	Parallel connection
$\tilde{\theta}_r$	$\bar{e}_{31}bh_{\tilde{p}c}\left.\dfrac{d\phi_r(x)}{dx}\right|_{x=L}$	$2\bar{e}_{31}bh_{\tilde{p}c}\left.\dfrac{d\phi_r(x)}{dx}\right|_{x=L}$
$C_{\tilde{p}}^{eq}$	$\dfrac{\bar{\varepsilon}_{33}^S bL}{2h_{\tilde{p}}}$	$\dfrac{2\bar{\varepsilon}_{33}^S bL}{h_{\tilde{p}}}$

where

$$C_{\tilde{p}}^{eq,p} = 2C_{\tilde{p}} = \frac{2\bar{\varepsilon}_{33}^S bL}{h_{\tilde{p}}} \tag{3.61}$$

is the equivalent capacitance of two identical capacitors ($C_{\tilde{p}}$) connected in parallel. Note that the difference between the representations of Equations (3.58) and (3.60) is a factor of two in the coupling terms, as in the backward coupling terms given by Equations (3.55) and (3.56). Indeed, from Equations (3.55)–(3.57), $\kappa_r = \chi_r^s$ and $2\kappa_r = \chi_r^p$.

3.4.3 Equivalent Representation of the Electromechanical Equations

The equivalent electromechanical equations governing the modal mechanical coordinate and the voltage response of a bimorph can be given by

$$\frac{d^2\eta_r(t)}{dt^2} + 2\zeta_r\omega_r\frac{d\eta_r(t)}{dt} + \omega_r^2\eta_r(t) - \tilde{\theta}_r v(t) = f_r(t) \tag{3.62}$$

$$C_{\tilde{p}}^{eq}\frac{dv(t)}{dt} + \frac{v(t)}{R_l} + \sum_{r=1}^{\infty}\tilde{\theta}_r\frac{d\eta_r(t)}{dt} = 0 \tag{3.63}$$

where the modal electromechanical coupling term $\tilde{\theta}_r$ and the equivalent capacitance $C_{\tilde{p}}^{eq}$ depend on the way piezoceramic layers are connected as summarized in Table 3.1. In Equations (3.62) and (3.63), $\eta_r(t)$ is the modal mechanical coordinate that gives the transverse displacement response of the cantilever relative to its moving base when used in Equation (3.25), whereas $v(t)$ is the voltage across the load resistance (i.e., $v(t) = v_s(t)$ in Figure 3.1a and $v(t) = v_p(t)$ in Figure 3.1b).

As mentioned previously, for the harmonic base displacement inputs of $g(t) = W_0 e^{j\omega t}$ and $h(t) = \theta_0 e^{j\omega t}$, the modal forcing function is harmonic of the form $f_r(t) = F_r e^{j\omega t}$ where the amplitude F_r is given by Equation (3.36). Substituting the steady-state response expressions

$\eta_r(t) = H_r e^{j\omega t}$ and $v(t) = V e^{j\omega t}$ into Equations (3.62) and (3.63) gives the coupled linear algebraic equations for the complex terms H_r and V:

$$\left(\omega_r^2 - \omega^2 + j2\zeta_r\omega_r\omega\right)H_r - \tilde{\theta}_r V = F_r \qquad (3.64)$$

$$\left(\frac{1}{R_l} + j\omega C_{\tilde{p}}^{eq}\right)V + j\omega\sum_{r=1}^{\infty}\tilde{\theta}_r H_r = 0 \qquad (3.65)$$

Eventually one obtains the steady-state voltage response and the vibration response expressions as follows:

$$v(t) = \frac{\displaystyle\sum_{r=1}^{\infty}\frac{-j\omega\tilde{\theta}_r F_r}{\omega_r^2 - \omega^2 + j2\zeta_r\omega_r\omega}}{\displaystyle\frac{1}{R_l} + j\omega C_{\tilde{p}}^{eq} + \sum_{r=1}^{\infty}\frac{j\omega\tilde{\theta}_r^2}{\omega_r^2 - \omega^2 + j2\zeta_r\omega_r\omega}}e^{j\omega t} \qquad (3.66)$$

$$w_{rel}(x,t) = \sum_{r=1}^{\infty}\left[\left(F_r - \tilde{\theta}_r\frac{\displaystyle\sum_{r=1}^{\infty}\frac{j\omega\tilde{\theta}_r F_r}{\omega_r^2 - \omega^2 + j2\zeta_r\omega_r\omega}}{\displaystyle\frac{1}{R_l} + j\omega C_{\tilde{p}}^{eq} + \sum_{r=1}^{\infty}\frac{j\omega\tilde{\theta}_r^2}{\omega_r^2 - \omega^2 + j2\zeta_r\omega_r\omega}}\right)\frac{\phi_r(x)e^{j\omega t}}{\omega_r^2 - \omega^2 + j2\zeta_r\omega_r\omega}\right] \qquad (3.67)$$

3.5 Single-Mode Electromechanical Equations for Modal Excitations

The steady-state voltage response and vibration response expressions derived in the previous sections are valid for harmonic excitations at any arbitrary excitation frequency ω (as long as the excitation frequency is not too high so the Euler–Bernoulli beam assumptions still hold, i.e., the shear deformation and rotary inertia effects remain to be negligible). That is, Equations (3.66) and (3.67) are the *multi-mode* solutions as they include all vibration modes of the bimorph energy harvester. Hence, these equations can predict the coupled system dynamics at steady state not only for resonance excitation, but also for excitations at the off-resonance frequencies of the energy harvester.

In order to obtain the maximum electrical response, it is preferable to excite a given energy harvester at its fundamental resonance frequency (or at one of the higher resonance frequencies). Most of the papers in the literature have focused on the problem of excitation at the fundamental resonance frequency in order to investigate the maximum power generation performance. Consequently, excitation of a bimorph at or very close to one of its natural frequencies is a very useful problem to investigate through the resulting equations derived here. This is the *modal excitation* condition and mathematically it implies $\omega \approx \omega_r$. With this assumption for the excitation frequency, the major contributions in the summation terms of Equations (3.66) and (3.67) are from the rth vibration mode. This approximation allows simplifications in the coupled voltage response and vibration response expressions.

For excitation at $\omega \approx \omega_r$, Equations (3.66) and (3.67), respectively, reduce to the following single-mode expressions:

$$\hat{v}(t) = \frac{-j\omega R_l \tilde{\theta}_r F_r e^{j\omega t}}{\left(1 + j\omega R_l C_p^{eq}\right)\left(\omega_r^2 - \omega^2 + j2\zeta_r \omega_r \omega\right) + j\omega R_l \tilde{\theta}_r^2} \tag{3.68}$$

$$\hat{w}_{rel}(x, t) = \frac{\left(1 + j\omega R_l C_p^{eq}\right) F_r \phi_r(x) e^{j\omega t}}{\left(1 + j\omega R_l C_p^{eq}\right)\left(\omega_r^2 - \omega^2 + j2\zeta_r \omega_r \omega\right) + j\omega R_l \tilde{\theta}_r^2} \tag{3.69}$$

where the circumflex or "hat" ($\hat{\ }$) denotes that the respective expression is given for a single vibration mode. Although the foregoing single-mode expressions are given for excitations at or very close to the rth natural frequency, it should be noted that the fundamental mode is the main concern in the vibration-based energy harvesting problem (which corresponds to $r = 1$).

3.6 Multi-mode and Single-Mode Electromechanical FRFs

In the electromechanical model proposed here, the two excitation inputs to the system are the translation of the base in the transverse direction and its small rotation (Figures 3.1a and 3.1b), while the resulting electromechanical outputs are the voltage response and the vibration response. Therefore, for harmonic base excitation, one can define four electromechanical FRFs between these two outputs and two inputs: voltage output to translational base acceleration, voltage output to rotational base acceleration, vibration response to translational base acceleration, and vibration response to rotational base acceleration. This section extracts these FRFs from the multi-mode (for arbitrary frequency excitations) and single-mode (for modal excitations) steady-state solutions derived in the previous sections.

Since the translational and small rotational displacements of the base are given by $g(t) = W_0 e^{j\omega t}$ and $h(t) = \theta_0 e^{j\omega t}$, the modal forcing function has the form of $f_r(t) = F_r e^{j\omega t}$ where F_r is given by Equation (3.36). Before identifying the aforementioned FRFs, the complex modal forcing given by Equation (3.36) should first be rearranged as follows:

$$F_r = -\sigma_r \omega^2 W_0 - \tau_r \omega^2 \theta_0 \tag{3.70}$$

where

$$\sigma_r = -m \int_0^L \phi_r(x)\,dx - M_t \phi_r(L) \tag{3.71}$$

$$\tau_r = -m \int_0^L x\phi_r(x)\,dx - M_t L \phi_r(L) \tag{3.72}$$

3.6.1   Multi-mode Electromechanical FRFs

Equation (3.66) can be re-expressed as

$$v(t) = \alpha\,(\omega)\left(-\omega^2 W_0 e^{j\omega t}\right) + \mu\,(\omega)\left(-\omega^2 \theta_0 e^{j\omega t}\right) \tag{3.73}$$

where the voltage output - to - translational base acceleration FRF is

$$\alpha\,(\omega) = \frac{\displaystyle\sum_{r=1}^{\infty}\frac{-j\omega\tilde{\theta}_r\sigma_r}{\omega_r^2 - \omega^2 + j2\zeta_r\omega_r\omega}}{\dfrac{1}{R_l} + j\omega C_{\tilde{p}}^{eq} + \displaystyle\sum_{r=1}^{\infty}\frac{j\omega\tilde{\theta}_r^2}{\omega_r^2 - \omega^2 + j2\zeta_r\omega_r\omega}} \tag{3.74}$$

and the voltage output - to - rotational base acceleration FRF can be given by

$$\mu\,(\omega) = \frac{\displaystyle\sum_{r=1}^{\infty}\frac{-j\omega\tilde{\theta}_r\tau_r}{\omega_r^2 - \omega^2 + j2\zeta_r\omega_r\omega}}{\dfrac{1}{R_l} + j\omega C_{\tilde{p}}^{eq} + \displaystyle\sum_{r=1}^{\infty}\frac{j\omega\tilde{\theta}_r^2}{\omega_r^2 - \omega^2 + j2\zeta_r\omega_r\omega}} \tag{3.75}$$

Similarly, Equation (3.67) can be written as

$$w_{rel}(x, t) = \beta\,(\omega, x)\left(-\omega^2 W_0 e^{j\omega t}\right) + \psi\,(\omega, x)\left(-\omega^2 \theta_0 e^{j\omega t}\right) \tag{3.76}$$

where the relative transverse displacement - to - translational base acceleration FRF is

$$\beta(\omega, x) = \sum_{r=1}^{\infty}\left[\left(\sigma_r - \tilde{\theta}_r\frac{\displaystyle\sum_{r=1}^{\infty}\frac{j\omega\tilde{\theta}_r\sigma_r}{\omega_r^2 - \omega^2 + j2\zeta_r\omega_r\omega}}{\dfrac{1}{R_l} + j\omega C_{\tilde{p}}^{eq} + \displaystyle\sum_{r=1}^{\infty}\frac{j\omega\tilde{\theta}_r^2}{\omega_r^2 - \omega^2 + j2\zeta_r\omega_r\omega}}\right)\frac{\phi_r(x)}{\omega_r^2 - \omega^2 + j2\zeta_r\omega_r\omega}\right] \tag{3.77}$$

and the relative transverse displacement - to - rotational base acceleration FRF is

$$\psi(\omega, x) = \sum_{r=1}^{\infty}\left[\left(\tau_r - \tilde{\theta}_r\frac{\displaystyle\sum_{r=1}^{\infty}\frac{j\omega\tilde{\theta}_r\tau_r}{\omega_r^2 - \omega^2 + j2\zeta_r\omega_r\omega}}{\dfrac{1}{R_l} + j\omega C_{\tilde{p}}^{eq} + \displaystyle\sum_{r=1}^{\infty}\frac{j\omega\tilde{\theta}_r^2}{\omega_r^2 - \omega^2 + j2\zeta_r\omega_r\omega}}\right)\frac{\phi_r(x)}{\omega_r^2 - \omega^2 + j2\zeta_r\omega_r\omega}\right] \tag{3.78}$$

3.6.2 Single-Mode Electromechanical FRFs

One can re-express Equation (3.68) to give

$$\hat{v}(t) = \hat{\alpha}(\omega)\left(-\omega^2 W_0 e^{j\omega t}\right) + \hat{\mu}(\omega)\left(-\omega^2 \theta_0 e^{j\omega t}\right) \tag{3.79}$$

where the single-mode voltage output - to - translational base acceleration FRF is

$$\hat{\alpha}(\omega) = \frac{-j\omega R_l \tilde{\theta}_r \sigma_r}{\left(1 + j\omega R_l C_{\tilde{p}}^{eq}\right)\left(\omega_r^2 - \omega^2 + j2\zeta_r\omega_r\omega\right) + j\omega R_l \tilde{\theta}_r^2} \tag{3.80}$$

and the single-mode voltage output - to - rotational base acceleration FRF is

$$\hat{\mu}(\omega) = \frac{-j\omega R_l \tilde{\theta}_r \tau_r}{\left(1 + j\omega R_l C_{\tilde{p}}^{eq}\right)\left(\omega_r^2 - \omega^2 + j2\zeta_r\omega_r\omega\right) + j\omega R_l \tilde{\theta}_r^2} \tag{3.81}$$

Likewise, from Equation (3.69),

$$\hat{w}_{rel}(x, t) = \hat{\beta}(\omega, x)\left(-\omega^2 W_0 e^{j\omega t}\right) + \hat{\psi}(\omega, x)\left(-\omega^2 \theta_0 e^{j\omega t}\right) \tag{3.82}$$

where the single-mode relative transverse displacement - to - translational base acceleration FRF is

$$\hat{\beta}(\omega, x) = \frac{\left(1 + j\omega R_l C_{\tilde{p}}^{eq}\right)\sigma_r \phi_r(x)}{\left(1 + j\omega R_l C_{\tilde{p}}^{eq}\right)\left(\omega_r^2 - \omega^2 + j2\zeta_r\omega_r\omega\right) + j\omega R_l \tilde{\theta}_r^2} \tag{3.83}$$

and the single-mode relative transverse displacement - to - rotational base acceleration FRF can be given by

$$\hat{\psi}(\omega, x) = \frac{\left(1 + j\omega R_l C_{\tilde{p}}^{eq}\right)\tau_r \phi_r(x)}{\left(1 + j\omega R_l C_{\tilde{p}}^{eq}\right)\left(\omega_r^2 - \omega^2 + j2\zeta_r\omega_r\omega\right) + j\omega R_l \tilde{\theta}_r^2} \tag{3.84}$$

3.7 Theoretical Case Study

This section presents theoretical demonstrations of the analytical model developed in this chapter. An extensive electromechanical analysis of a bimorph cantilever configuration is presented in the following subsections. First the series and the parallel connections of the piezoceramic layers are studied with a focus on multiple vibration modes. Then the electrical and the mechanical response simulation results of the multi-mode and the single-mode FRFs are compared. The shunt damping effect of piezoelectric power generation on the energy harvester beam is also explained.

Table 3.2 Geometric properties of the bimorph cantilever

	Piezoceramic	Substructure
Length (L) (mm)	30	30
Width (b) (mm)	5	5
Thickness ($h_{\bar{p}}$, $h_{\bar{s}}$) (mm)	0.15 (each)	0.05

3.7.1 Properties of the Bimorph Cantilever

The geometric properties of the bimorph cantilever investigated in this case study are given in Table 3.2. The ratio of overhang length to total thickness of the harvester is about 85.7, which makes it reasonable to neglect the shear deformation and the rotary inertia effects for the first few vibration modes (in agreement with the Euler–Bernoulli beam theory). The substructure layer is assumed to be aluminum, the piezoceramic layer is taken to be PZT-5A, and the configuration does not have a tip mass (i.e., $M_t = I_t = 0$). The material properties of the substructure and the piezoceramic layers are shown in Table 3.3. The elastic, piezoelectric, and permittivity constants of PZT-5A are obtained from Table E.2 in Appendix E (reduced constants for the Euler–Bernoulli beam theory). The analysis given here considers the frequency range of 0–5000 Hz. It can easily be shown that this cantilever has three vibration modes in this frequency range. The first three mechanically and electrically *undamped* natural frequencies of the bimorph cantilever (for $R_l \to 0$, i.e., in short-circuit conditions) are $f_1 = 185.1$ Hz, $f_2 = 1159.8$ Hz, and $f_3 = 3247.6$ Hz (where $f_r = \omega_r/2\pi$). For the purpose of simulation, if one takes $\zeta_1 = 0.010$ and $\zeta_2 = 0.012$ as the mechanical damping ratios of the first two modes and assumes proportional damping based on the discussion of Chapter 2, the proportionality constants $c_s I/YI$ and c_a/m are obtained from Equation (2.17) (by using ζ_1, ζ_2, ω_1, and ω_2) as $c_s I/YI = 2.93 \times 10^{-6}$ s/rad and $c_a/m = 19.295$ rad/s. Closed-form identification of mechanical damping in the presence of an arbitrary load resistance based on the single-mode approximation is presented in Chapter 5. Here, it can be assumed that the damping ratios are identified by setting the electrical boundary condition to short circuit ($R_l \to 0$) and using conventional techniques (e.g., half-power points of the mechanical FRF [29]).

Before the simulation results are presented and discussed, it is worth saying a few words regarding the assumption of proportional damping. The restriction about the proportional damping assumption is such that, once the proportionality constants are identified using the modal properties of two vibration modes, the rest of the modal mechanical damping ratios are not arbitrary and they are automatically set equal to the following numbers due to Equation (2.17):

Table 3.3 Material properties of the bimorph cantilever

	Piezoceramic	Substructure
Material	PZT-5A	Aluminum
Elastic modulus ($\bar{c}_{11}^E$, $Y_{\bar{s}}$) (GPa)	61	70
Mass density ($\rho_{\bar{p}}$, $\rho_{\bar{s}}$) (kg/m^3)	7750	2700
Piezoelectric constant ($\bar{e}_{31}$) (C/m^2)	−10.4	—
Permittivity constant ($\bar{\varepsilon}_{33}^S$) (nF/m)	13.3	—

$\zeta_3 = 0.030$, $\zeta_4 = 0.059$, $\zeta_5 = 0.097$, and so on. It should be noted that the concept of pro-portional damping is a convenient *mathematical* modeling assumption (to force the system to be a normal-mode system) and the *physical* system may not agree with this assumption [30]. In other words, the damping ratios of higher modes identified experimentally from the physical system may not converge to the aforementioned values. Therefore, as a relaxation in the proportional damping assumption, one might prefer to use the identified damping ratios of the vibration modes of interest (to be used in the modal expressions) directly without obtain-ing $c_s I/YI$ and c_a/m since the resulting electromechanical expressions developed here need only the ζ_r values. This is the modal damping assumption and it is identical to starting with the respective undamped equation (the normal-mode problem by definition) and introducing damping in the modal domain. The proportional damping (in the form of Rayleigh damping) and modal damping assumptions are often used in vibration engineering regardless of their practical limitations. As far as the problem of vibration energy harvesting is concerned, it is the fundamental vibration mode that has the highest importance and proportional damping (or modal damping) is a reasonable assumption to establish the bridge between the partial differential equation and its closed-form analytical solution given in this chapter.

3.7.2 Frequency Response of the Voltage Output

In the simulations given here, the base of the cantilever is assumed to be not rotating (i.e., $h(t) = 0$ in Figure 3.1) and the series connection case is considered first. The multi-mode voltage FRFs (per base acceleration) shown in Figure 3.5 are obtained from

$$\frac{v(t)}{-\omega^2 W_0 e^{j\omega t}} = \frac{\sum\limits_{r=1}^{\infty} \dfrac{-j\omega \tilde{\theta}_r \sigma_r}{\omega_r^2 - \omega^2 + j2\zeta_r \omega_r \omega}}{\dfrac{1}{R_l} + j\omega C_{\tilde{p}}^{eq} + \sum\limits_{r=1}^{\infty} \dfrac{j\omega \tilde{\theta}_r^2}{\omega_r^2 - \omega^2 + j2\zeta_r \omega_r \omega}} \tag{3.85}$$

where $\tilde{\theta}_r$ and $C_{\tilde{p}}^{eq}$ are as given in the first column of Table 3.1 for the series connection of the piezoceramic layers. Note that, here and hereafter, the electromechanical FRFs are given in the modulus form and the base acceleration in the frequency response graphs is normalized with respect to the gravitational acceleration for a convenient representation (i.e., the voltage FRF given by the foregoing equation is multiplied by the gravitational acceleration, $g = 9.81$ m/s^2). The set of electrical load resistance considered here ranges from 100 to 10 MΩ. As far as the fundamental vibration mode of this particular bimorph is concerned, the lowest resistance ($R_l = 100\,\Omega$) used here is very close to short-circuit conditions, whereas the largest load ($R_l = 10$ MΩ) is very close to open-circuit conditions.

As the load resistance is increased from the short-circuit to open-circuit conditions, the volt-age output at every frequency increases monotonically. To be precise, the voltage output for the exact short-circuit condition with zero external resistance ($R_l = 0$) will be zero, which would not allow a voltage FRF to be defined. Consequently, throughout this text, the short-circuit condition is defined as $R_l \to 0$. At the other extreme, the open-circuit condition ($R_l \to \infty$), the voltage output at every frequency converges to its maximum value. Another important aspect of the voltage FRFs plotted in Figure 3.5 is that, with increasing load resistance, the

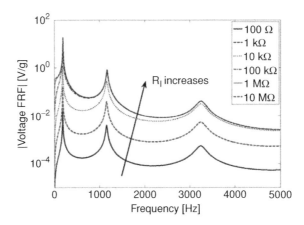

Figure 3.5 Voltage FRFs of the bimorph for a broad range of load resistance (series connection of the piezoceramic layers)

resonance frequency of each vibration mode moves from the short-circuit resonance frequency (ω_r^{sc} for $R_l \to 0$) to the open-circuit resonance frequency (ω_r^{oc} for $R_l \to \infty$). The short- and open-circuit resonance frequencies of the first three modes read from Figure 3.5 are listed in Table 3.4 (where $f_r^{sc} = \omega_r^{sc}/2\pi$ and $f_r^{oc} = \omega_r^{oc}/2\pi$). The direct conclusion based on this observation is that the resonance frequency of a given piezoelectric energy harvester depends on the external load resistance. Moreover, depending on the external load resistance, the resonance frequency of each mode can take a value only between the short- and open-circuit resonance frequencies f_r^{sc} and f_r^{oc}. Closed-form identification of the frequency shift ($\Delta f_r = f_r^{oc} - f_r^{sc}$) based on the single-mode approximation is given in Chapter 5. Here, the data in Table 3.4 is read from the resulting frequency response graph given by Figure 3.5.

Two enlarged views of the voltage FRFs with a focus on the first two vibration modes are shown in Figure 3.6 in order to display clearly the resonance frequency shift from the short-circuit to the open-circuit conditions. Note that the voltage FRFs of the largest two values of load resistance are almost indifferent, especially for the second vibration mode, implying a convergence of the curves to the open-circuit voltage FRF. That is, if the voltage FRF for the 100 MΩ case was also plotted, it would not be any different from that of the 10 MΩ case. Again, these numbers are for this particular cantilever. For a different configuration, it might be the case that even a load of 100 kΩ could be sufficient to represent the open-circuit conditions.

Table 3.4 First three short-circuit and open-circuit resonance frequencies read from the voltage FRF of the bimorph piezoelectric energy harvester

Mode (r)	f_r^{sc} (Hz)	f_r^{oc} (Hz)
1	185.1	191.1
2	1159.7	1171.6
3	3245.3	3254.1

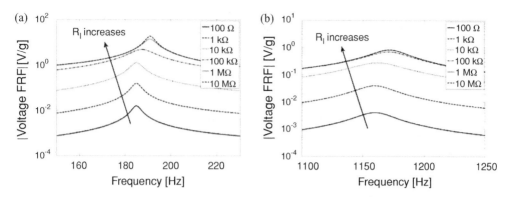

Figure 3.6 Voltage FRFs of the bimorph with a focus on the first two vibration modes: (a) mode 1; and (b) mode 2 (series connection)

As far as the fundamental vibration mode is concerned, the short- and open-circuit resonance frequencies are 185.1 Hz and 191.1 Hz, respectively. Excitation at these two frequencies will be of particular interest in this section as well as in the experimental validations to be discussed in the next chapter. Variation of the voltage output for excitations at the *fundamental short-circuit resonance frequency* and at the *fundamental open-circuit resonance frequency* are plotted in Figure 3.7. As can be seen from the figure, for low values of load resistance, the voltage output at the short-circuit resonance frequency is larger since the system is close to short-circuit conditions. With increasing load resistance, the curves intersect at a certain point (around 120 kΩ) and for the values of load resistance larger than the value at the intersection point, the voltage output at the open-circuit resonance frequency is larger. It is important at this stage to notice the linear asymptotic trends at the extrema of $R_l \to 0$ and $R_l \to \infty$ (which will be shown mathematically in Chapter 5). The graph given here in log–log scale shows a linear increase in the voltage output with increasing load resistance for low values of load resistance

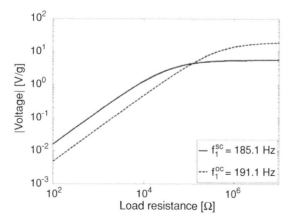

Figure 3.7 Variation of the voltage output with load resistance for excitations at the short-circuit and the open-circuit resonance frequencies of the first vibration mode (series connection)

(both at the short- and open-circuit resonance frequencies). The voltage output becomes less sensitive to variations in the load resistance for its large values due to the horizontal asymptotes of the $R_l \to \infty$ extremum.

3.7.3   Frequency Response of the Current Output

If the voltage FRF given by Equation (3.85) is divided by the load resistance, the multi-mode current FRF is obtained as

$$\frac{i(t)}{-\omega^2 W_0 e^{j\omega t}} = \frac{1}{R_l} \left(\frac{\displaystyle\sum_{r=1}^{\infty} \frac{-j\omega\tilde{\theta}_r \sigma_r}{\omega_r^2 - \omega^2 + j2\zeta_r\omega_r\omega}}{\displaystyle\frac{1}{R_l} + j\omega C_p^{eq} + \sum_{r=1}^{\infty} \frac{j\omega\tilde{\theta}_r^2}{\omega_r^2 - \omega^2 + j2\zeta_r\omega_r\omega}} \right) \tag{3.86}$$

The modulus of the current FRF is plotted versus the frequency in Figure 3.8. Unlike the voltage FRF shown in Figure 3.5, the amplitude of the current at every frequency decreases with increasing load resistance. Indeed this is the opposite of the voltage behavior shown in Figure 3.8, but the behavior at every frequency is still monotonic. For every excitation frequency, the maximum value of the current is obtained when the system is close to short-circuit conditions. The enlarged views of the current FRFs around the first two resonance frequencies are plotted in Figure 3.9, showing the change in the resonance frequency with increasing load resistance. Moreover, being analogous to the behavior of voltage output close to open-circuit conditions, the current FRFs become indistinguishable close to short-circuit conditions. That is, if one plotted the current FRF of the 10 Ω case, the resulting curve would not look any different than that of the 100 Ω case.

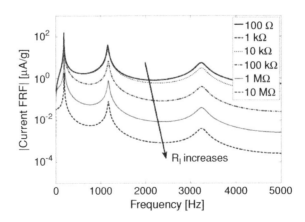

Figure 3.8   Current FRFs of the bimorph for a broad range of load resistance (series connection of the piezoceramic layers)

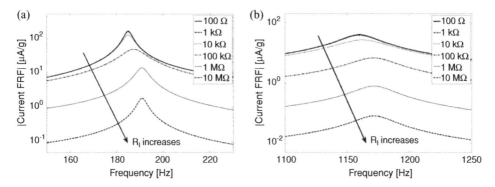

Figure 3.9 Current FRFs of the bimorph with a focus on the first two vibration modes: (a) mode 1; and (b) mode 2 (series connection)

Figure 3.10 shows the current output as a function of load resistance for excitations at the fundamental short-circuit and open-circuit resonance frequencies. It is clear from Figure 3.10 that the current output is very insensitive to the variations of the region of low load resistance (i.e., the slope is almost zero for $R_l \rightarrow 0$). In this region of relatively low load resistance, the current output is larger at the short-circuit resonance frequency, as in the case of the voltage output (in Figure 3.7), since the system is close to short-circuit conditions. Then, the current output starts decreasing with increasing load resistance and the curves intersect at a certain value of load resistance (around $120\,\mathrm{k\Omega}$). For the values of load resistance larger than the value at this intersection point, the current output at the open-circuit resonance frequency becomes larger since the system approaches the open-circuit conditions. As in the voltage versus load resistance graph, the asymptotic trends for $R_l \rightarrow 0$ and $R_l \rightarrow \infty$ appear to be linear.

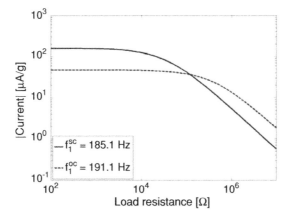

Figure 3.10 Variation of the current output with load resistance for excitations at the short-circuit and the open-circuit resonance frequencies of the first vibration mode (series connection)

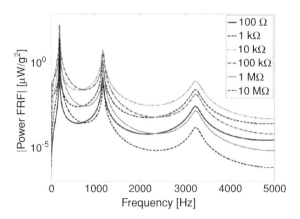

Figure 3.11 Power FRFs of the bimorph for a broad range of load resistance (series connection of the piezoceramic layers)

3.7.4 Frequency Response of the Power Output

The expression for the multi-mode power FRF is obtained from Equation (3.85) as

$$\frac{p(t)}{\left(-\omega^2 W_0 e^{j\omega t}\right)^2} = \frac{1}{R_l} \left(\frac{\displaystyle\sum_{r=1}^{\infty} \frac{j\omega\tilde{\theta}_r \sigma_r}{\omega_r^2 - \omega^2 + j2\zeta_r\omega_r\omega}}{\displaystyle\frac{1}{R_l} + j\omega C_p^{eq} + \sum_{r=1}^{\infty} \frac{j\omega\tilde{\theta}_r^2}{\omega_r^2 - \omega^2 + j2\zeta_r\omega_r\omega}} \right)^2 \qquad (3.87)$$

Therefore the power output is proportional to the square of the voltage output. As a result, the moduli of power FRFs plotted in Figure 3.11 are normalized with respect to the square of base acceleration (i.e., g^2). According to Equation (3.87), the power output is a product of two FRFs (current and voltage) with the opposite trends against the load resistance. It is clear from Figure 3.11 that the power output FRF does not necessarily exhibit monotonic behavior with increasing (or decreasing) load resistance for a given frequency. Among the sample values of load resistance considered in this work, the maximum power output for the first vibration mode corresponds to the load of $100\,\mathrm{k\Omega}$ (see the enlarged view in Figure 3.12a) at $187.5\,\mathrm{Hz}$ which is expectedly a frequency in between the fundamental short- and open-circuit resonance frequencies. Considering the second vibration mode (see the second enlarged view in Figure 3.12b), one can observe that the maximum power output is obtained for $10\,\mathrm{k\Omega}$ at $1161.9\,\mathrm{Hz}$. It should be noted that the values of the load resistance used in this analysis are taken arbitrarily to observe the general trends. Therefore, the maximum power outputs obtained from each vibration mode are for these sample values and they are *not* necessarily for the maximum possible (or the optimized) power outputs. Another interesting aspect of the power FRFs given in Figure 3.11 is that they intersect one another. These intersections are observed not only around the resonance frequencies (e.g., the curves of $10\,\mathrm{k\Omega}$ and $1\,\mathrm{M\Omega}$ intersect at 190.2 and $193.9\,\mathrm{Hz}$ in Figure 3.12a), but also at the *off-resonance* frequencies (e.g., the curves of $10\,\mathrm{k\Omega}$

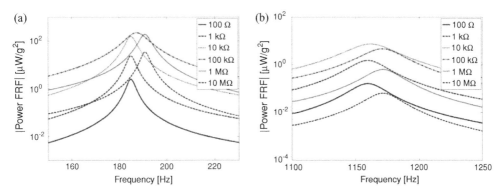

Figure 3.12 Power FRFs of the bimorph with a focus on the first two vibration modes: (a) mode 1; and (b) mode 2 (series connection)

and $100\,\text{k}\Omega$ intersect at $723.7\,\text{Hz}$ in Figure 3.11). At these intersection frequencies, the two respective load resistance values yield the same power output.

The behavior of power output with changing load resistance for excitations at the fundamental short- and open-circuit resonance frequencies is given in Figure 3.13. It can be recalled from Figures 3.7 and 3.10 that the voltage and the current outputs obtained at the short-circuit resonance frequency are larger than those obtained at the open-circuit resonance frequency up to a certain load resistance (approximately $120\,\text{k}\Omega$ in this case) after which the opposite is valid. Since the power output is simply the product of the voltage and current, this observation is valid for the power versus load resistance curves as well. As can be seen from Figure 3.13, the power output at the short-circuit resonance frequency is larger before the intersection point (at $120\,\text{k}\Omega$) whereas the power output at the open-circuit resonance frequency is larger after this point. The asymptotic trends for $R_l \to 0$ and $R_l \to \infty$ are again linear and they appear to have the same slope in log–log scale at these two frequencies.

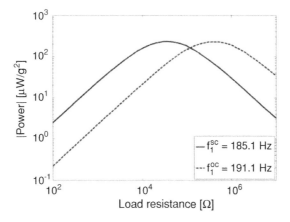

Figure 3.13 Variation of the power output with load resistance for excitations at the short-circuit and the open-circuit resonance frequencies of the first vibration mode (series connection)

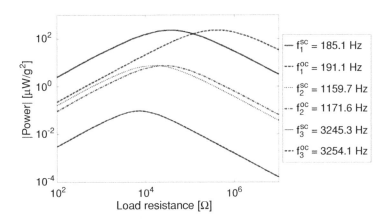

Figure 3.14   Variation of the power output with load resistance for excitations at the short-circuit and the open-circuit resonance frequencies of the first three vibration modes (series connection)

Since the behavior of power with changing load resistance is not monotonic, both of the power graphs shown in Figure 3.13 exhibit peak values which correspond to the optimum values of load resistance at the fundamental short- and open-circuit resonance frequencies of the voltage FRF. Considering Figures 3.7 and 3.10, it can be observed that neither the voltages nor the currents are identical at these optimum values of load resistance for excitations at the two special frequencies. Therefore, if one is flexible in terms of the excitation frequency within this narrow frequency band, the short-circuit resonance frequency is preferable for larger current, whereas the open-circuit resonance frequency is preferable for larger voltage. Another aspect of Figure 3.13 is that the optimum values of load resistance at these two frequencies are significantly different (around $36\,k\Omega$ for excitation at $185.1\,Hz$ and around $405\,k\Omega$ for excitation at $191.1\,Hz$). From another perspective, if the load resistance is a constraint, one can select the frequency of operation accordingly for the maximum power output.

It is worth adding a few words on the advantage of using the fundamental vibration mode for energy harvesting compared to higher vibration modes. Figure 3.14 shows the power versus load resistance curves obtained for excitation at the six frequencies listed in Table 3.4 (i.e., the short- and open-circuit resonance frequencies of the first three modes). The maximum power for mode 1 excitation is around $0.23\,mW/g^2$ at $185.1\,Hz$ or at $191.1\,Hz$ (for different optimum loads). For mode 2 excitation, the maximum power output is around $7.5\,\mu W/g^2$ (at $1159.7\,Hz$ or at $1171.6\,Hz$). The maximum power output for mode 3 excitation is as low as $98\,nW/g^2$ (at $3245.3\,Hz$ or at $3254.1\,Hz$).[17] As the moduli of the voltage and current FRFs decay by one order of magnitude with increasing mode number according to Figures 3.5 and 3.8, the modulus of the power FRF decays by around two orders of magnitude with increasing mode number. As expected, the maximum power output is obtained for the fundamental vibration mode. For this reason, the focus is usually placed on this vibration mode in practical considerations (as well as in the experimental validation cases presented in the next chapter).

---

[17] Recall from the modeling assumptions that these linear estimates strictly assume the base acceleration level to be small enough so that neither geometric nor material nonlinearities are pronounced.

While considering the substantial difference between the moduli of the power FRFs for different vibration modes, one should also note that these FRFs are given per base acceleration input. A base acceleration input of $0.2g$ at 185.1 Hz implies a base displacement amplitude of $1.45\,\mu m$. This base displacement amplitude, for instance, creates a base acceleration input of $7.9g$ at 1159.7 Hz (which may or may not be sustained by the brittle piezoceramic layers)[18]. As a result, one should define the electromechanical FRF according to the application and normalize with respect to the constant kinematic variable (displacement, velocity, or acceleration). In short, part of the reason for decreasing power moduli at higher vibration modes is that these FRFs are given here per base acceleration input (rather than displacement or velocity).

3.7.5   Frequency Response of the Relative Tip Displacement

The multi-mode expression for the tip displacement FRF (relative to the base) is

$$
\frac{w_{rel}(L,t)}{-\omega^2 W_0 e^{j\omega t}} = \sum_{r=1}^{\infty}\left[\left(\sigma_r - \tilde{\theta}_r \frac{\displaystyle\sum_{r=1}^{\infty}\frac{j\omega\tilde{\theta}_r\sigma_r}{\omega_r^2-\omega^2+j2\zeta_r\omega_r\omega}}{\dfrac{1}{R_l}+j\omega C_{\tilde{p}}^{eq}+\displaystyle\sum_{r=1}^{\infty}\frac{j\omega\tilde{\theta}_r^2}{\omega_r^2-\omega^2+j2\zeta_r\omega_r\omega}}\right)\frac{\phi_r(L)}{\omega_r^2-\omega^2+j2\zeta_r\omega_r\omega}\right]
$$

$$(3.88)$$

This FRF differs from the solution of the electromechanically uncoupled vibration FRF due to the voltage induced as a result of piezoelectric coupling. The solution of the uncoupled base excitation problem (addressed in Chapter 2) is obtained by neglecting the electrical term as

$$
\frac{w_{rel}(L,t)}{-\omega^2 W_0 e^{j\omega t}} = \sum_{r=1}^{\infty}\frac{\sigma_r\phi_r(L)}{\omega_r^2 - \omega^2 + j2\zeta_r\omega_r\omega}
$$

$$(3.89)$$

which is therefore identical to Equation (3.88) for $R_l \to 0$ (provided that the constant electric field elastic modulus is used in the uncoupled case as well). The feedback of piezoelectric power generation in the mechanical equation is associated with the $\tilde{\theta}_r$ term in Equation (3.88) and is basically a form of power dissipation due to Joule heating in the resistor. Therefore, what happens to the beam response due to power generation with a non-zero and finite load resistance is damping of the structure, which has a relatively sophisticated form compared to conventional viscous damping and structural damping mechanisms as demonstrated in the following.

Figure 3.15 shows the tip displacement FRFs[19] (per base acceleration) for the set of resistors and the frequency range of interest. One can see the three vibration modes in this figure but

---

[18] Given the bending strength of the piezoceramic and a safety factor, one can obtain a critical stress FRF as a function of load resistance by using the displacement and electric fields in the reduced piezoelectric constitutive equation. A linear estimate of the maximum acceleration level can be obtained for a given excitation frequency and load resistance.

[19] The vibration FRFs investigated here are for the response at the tip of the cantilever $(x = L)$. However, the distributed-parameter solution given here allows the coupled vibration response to be obtained at any point x_p on the beam by setting $x = x_p$ in the eigenfunction on the right hand side of Equation (3.88). The motion at the tip of the beam is of practical interest since it is the position of maximum deflection in the most flexible mode.

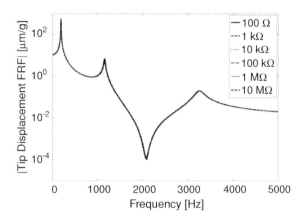

Figure 3.15 Tip displacement FRFs (relative to the vibrating base) of the bimorph for a broad range of load resistance (series connection of the piezoceramic layers)

it is not possible to distinguish between the curves of different load resistance. The enlarged views of the first two vibration modes are shown in Figure 3.16. The resistive shunt damping effect results in both frequency shift and vibration attenuation. With increasing load resistance, the electromechanical system moves from the short-circuit to the open-circuit conditions. In Figure 3.16a, the peak vibration amplitude of 566.4 μm/g for 100 Ω (at 185.1 Hz) is attenuated to a peak amplitude of 214.7 μm/g for 100 kΩ (at 187.4 Hz), which means an increased damping effect by a factor of about 2.6. In the second vibration mode, the peak amplitude of 6.67 μm/g for 100 Ω (at 1159 Hz) is attenuated to 4.97 μm/g for 10 kΩ (at 1160.8 Hz). One thing that is useful to mention is that, since the forms of the voltage and tip displacement FRFs are different, the resonance frequencies in these FRFs are slightly different (see Chapter 5). This is distinguishable only in the presence of relatively large mechanical damping. For instance, the short- and open-circuit resonance frequencies of the third vibration mode in Figure 3.15 are 3252.6 Hz and 3264.3 Hz, respectively. Compared to those of the voltage FRF given in Table 3.4 (3245.3 Hz and 3254.1 Hz) these numbers, respectively, are 0.2% and 0.3% larger. The difference is miniscule for the fundamental vibration mode. Since the main concern in energy harvesting is the electrical response, the short- and open-circuit resonance frequencies are defined here based on the voltage FRF (as listed in Table 3.4). It is worth adding that a mechanical *anti-resonance frequency* exists between modes 2 and 3 in Figure 3.25. Just like the resonance frequencies, this anti-resonance frequency also exhibits a shift (from 2078.8 Hz to 2090.2 Hz) as the electrical load is changed from the short-circuit to the open-circuit condition.

It is important to notice from the behavior around the modal frequencies of the vibration FRFs that the form of damping caused by piezoelectric power generation (or power dissipation in the resistor due to Joule heating) is more sophisticated than viscous damping (although oversimplification of the problem with the *electrically induced viscous damping* assumption has been made by several researchers [31–34] over the last decade). First, with increasing load resistance, the frequency of peak vibration amplitude changes considerably (and shifts to the right in the FRF, unlike the case of viscous damping). Secondly, with further increase in the load

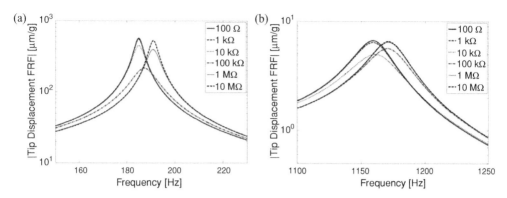

Figure 3.16 Tip displacement FRFs of the bimorph with a focus on the first two vibration modes: (a) mode 1; and (b) mode 2 (series connection)

resistance, although the vibration amplitude at the original (short-circuit) resonance frequency is attenuated to a certain amplitude, the vibration amplitude at the open-circuit resonance frequency is amplified considerably. The reason is the frequency shift due to changing electrical boundary conditions. Indeed, if one changes the load from one extremum ($R_l \rightarrow 0$) to the other ($R_l \rightarrow \infty$), the only modification in the energy harvester is a stiffness change. Physically, the elastic modulus of the piezoceramic increases from the constant electric field value to the constant electric displacement value and there is no overall energy dissipation. Only for non-zero and finite values of load resistance is there power dissipation in the mechanical domain (and hence power generation in the electrical domain). In short, the resonance frequency shift (from the mechanically damped natural frequency) associated with the presence of a finite electrical resistance *cannot* be represented by a real-valued viscous damping ratio or loss factor.

The behavior of the vibration response at the tip of the beam is further studied for excitations at the short- and open-circuit resonance frequencies of the fundamental vibration mode in Figure 3.17. It is interesting to note that the trends are not completely monotonic. For instance, the minimum vibration amplitude at 185.1 Hz is obtained as 156.8 μm/g for an electrical load of about 500 kΩ and a further increase of load resistance up to 10 MΩ slightly amplifies the vibration amplitude at this frequency (to 162 μm/g). Likewise, before it is amplified due to the shift in the resonance frequency, the vibration amplitude at 191.1 Hz is slightly attenuated until 40 kΩ (from 162 μm/g to 157.7 μm/g). Then it is strongly amplified as the load resistance is further increased. It is also useful to note that these electrical loads of maximum vibration attenuation are *not* those of maximum power generation. That is, the nature of electromechanical coupling is such that, for the electrical load of maximum power generation, the vibration amplitude at the tip of the beam does *not* necessarily take its minimum value.

3.7.6 Parallel Connection of the Piezoceramic Layers

In this section, the electromechanical coupling and the equivalent capacitance terms used in Equations (3.85)–(3.88) are replaced by those of the parallel connection case (from the second

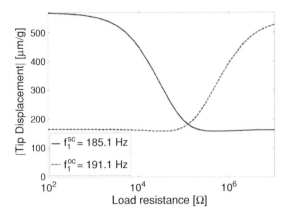

Figure 3.17 Variation of the tip displacement (relative to the vibrating base) with load resistance for excitations at the short-circuit and the open-circuit resonance frequencies of the first vibration mode (series connection)

column in Table 3.1) and sample results are presented. Figures 3.18–3.21 display the voltage, current, power, and tip displacement FRFs along with close-up views of the behavior around the fundamental vibration mode.

The short- and open-circuit resonance frequencies read from Figure 3.18 for the parallel connection case are identical to those listed in Table 3.4. The mathematical justification of this observation can be found in Chapter 5, where the single-mode approximation of the frequency shift ($\Delta f_r = f_r^{oc} - f_r^{sc}$) is shown to be proportional to $\tilde{\theta}_r^2 / C_p^{eq}$. As a result, from Table 3.1, the single-mode approximations of the resonance frequency shift for the series and parallel connection cases are identical (in agreement with Figures 3.18–3.21).

Comparisons of the series and parallel connection cases for excitation at the fundamental short-circuit resonance frequency are given in Figure 3.22. As anticipated, the maximum power outputs of the series and parallel connection cases are identical but they correspond

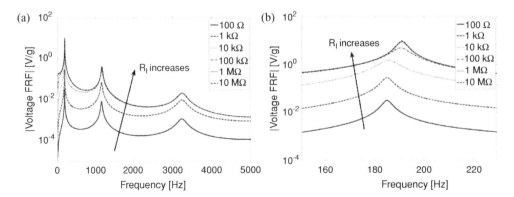

Figure 3.18 (a) Voltage FRFs for the parallel connection of the piezoceramic layers and (b) a close-up view around the first vibration mode

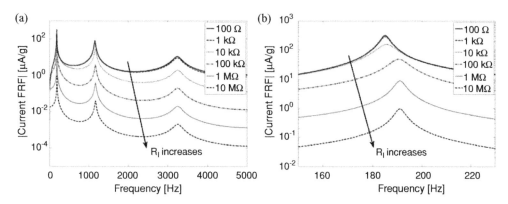

Figure 3.19 (a) Current FRFs for the parallel connection of the piezoceramic layers and (b) a close-up view around the first vibration mode

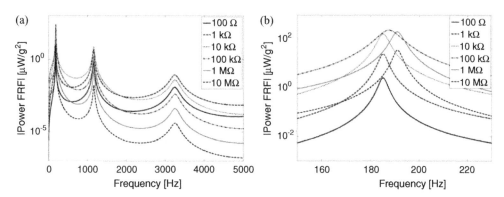

Figure 3.20 (a) Power FRFs for the parallel connection of the piezoceramic layers and (b) a close-up view around the first vibration mode

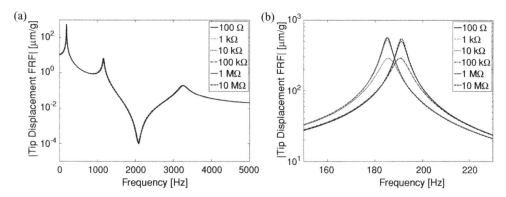

Figure 3.21 (a) Tip displacement FRFs for the parallel connection of the piezoceramic layers and (b) a close-up view around the first vibration mode

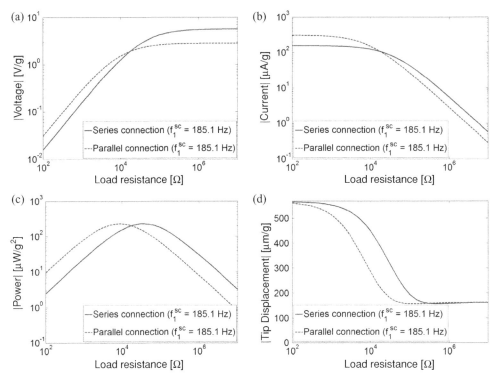

Figure 3.22 Comparison of the series and the parallel connection cases for excitations at the fundamental short-circuit resonance frequency: (a) voltage vs. load resistance; (b) current vs. load resistance; (c) power vs. load resistance; (d) tip displacement vs. load resistance

to different values of optimum load resistance. A maximum of $0.23\,\text{mW}/g^2$ is delivered to a resistive load of $36\,\text{k}\Omega$ in the series connection case, whereas the same maximum power output is delivered to a resistive load of $9\,\text{k}\Omega$ in the parallel connection case. The series connection case generates this power with a current amplitude of $0.08\,\text{mA}/g$ and a voltage amplitude of $2.8\,\text{V}/g$. In the parallel connection case, the same power output is obtained with a current amplitude of $0.16\,\text{mA}/g$ and a voltage amplitude of $1.4\,\text{V}/g$. Reasonably, series connection should be preferred for large voltage output, whereas parallel connection should be used for large current output. The last graph in Figure 3.22 shows that the resistive damping effect starts for lower values of load resistance in the parallel connection case, which agrees with the behavior of maximum power generation.

Piezoelectric energy harvesters, as observed in this case study, are poor current generators.[20] For instance, in the series connection case, for a base acceleration of $1g$ (achieved for the base displacement amplitude of $7.25\,\mu\text{m}$ at $185.1\,\text{Hz}$) the voltage output is around $2.8\,\text{V}$

---

[20] This fundamental fact about piezoelectric materials is indeed well known to vibration test engineers. This is the reason that piezoelectric-based transducers (such as accelerometers, force sensors, and impulse hammers) usually must have their signals run through a *charge amplifier* before they can be used as inputs to data acquisition hardware.

while the current associated with it is just 0.08 mA (linear estimates). While this voltage level is fairly good for charging a small battery, it is the current output that will make the duration of charging substantially long. A difference of several orders of magnitude between the voltage and the current outputs of the generator is a very typical case in piezoelectric energy harvesting. Therefore, if the purpose is to charge a battery using the piezoelectric power output, parallel connection of the piezoceramic layers can be preferred so long as the voltage limit for charging the battery is reached for a given excitation input.

3.7.7 Single-Mode FRFs

In the absence of base rotation, the single-mode voltage FRF can be expressed as

$$
\frac{\hat{v}(t)}{-\omega^2 W_0 e^{j\omega t}} = \frac{-j\omega R_l \tilde{\theta}_r \sigma_r}{\left(1 + j\omega R_l C_{\tilde{p}}^{eq}\right)\left(\omega_r^2 - \omega^2 + j2\zeta_r \omega_r \omega\right) + j\omega R_l \tilde{\theta}_r^2}
\tag{3.90}
$$

and the single-mode tip displacement FRF (relative to the base) is

$$
\frac{\hat{w}_{rel}(L, t)}{-\omega^2 W_0 e^{j\omega t}} = \frac{\left(1 + j\omega R_l C_{\tilde{p}}^{eq}\right)\sigma_r \phi_r(L)}{\left(1 + j\omega R_l C_{\tilde{p}}^{eq}\right)\left(\omega_r^2 - \omega^2 + j2\zeta_r \omega_r \omega\right) + j\omega R_l \tilde{\theta}_r^2}
\tag{3.91}
$$

In the following, the simulations of these single-mode expressions are compared to those of the multi-mode expressions for the first three vibration modes in the series connection case (i.e., $\tilde{\theta}_r$ and $C_{\tilde{p}}^{eq}$ are read from the first column of Table 3.1). Figure 3.23 shows the single-mode FRFs obtained from Equation (3.90) for $r = 1$, $r = 2$, and $r = 3$ along with the multi-mode solution for all values of load resistance considered here. As can be seen in this figure, the single-mode solutions agree with the multi-mode solution only around the resonance frequency of the respective mode of interest.

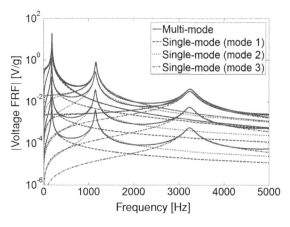

Figure 3.23 Comparison of the multi-mode and the single-mode voltage FRFs (series connection)

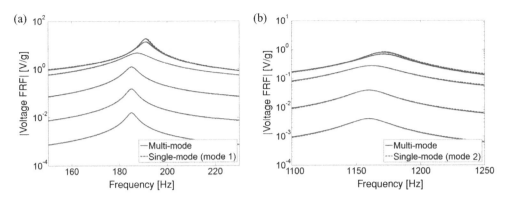

Figure 3.24 Comparison of the multi-mode and single-mode voltage FRFs with a focus on the first two vibration modes: (a) mode 1; and (b) mode 2 (series connection)

Figures 3.24a and 3.24b, respectively, show enlarged views of the single-mode and the multi-mode voltage FRF comparisons for the first two modes. The single-mode approximations work very well for all values of load resistance, yielding only slight overestimations of the resonance frequency and the voltage amplitude. The single-mode approximation for mode 1 shown in Figure 3.24a predicts the short-circuit resonance frequency accurately as 185.1 Hz and overestimates the open-circuit resonance frequency as 191.3 Hz (with an error of 0.1% compared to the multi-mode solution). In the second mode prediction illustrated by Figure 3.24b, the short-circuit resonance frequency is predicted as 1159.8 Hz (with an error of 0.009%) whereas the open-circuit resonance frequency is predicted as 1171.9 Hz (with an error of 0.03%). The single-mode voltage expression given by Equation (3.90) can therefore successfully represent the resonance behavior of the multi-mode voltage expression given by Equation (3.85) for a given mode of interest as a first approximation. The slight inaccuracy in the single-mode predictions is due to the residual effects of the neighboring modes, which are ignored in the single-mode approximation.

The frequency response predictions of the single-mode tip displacement FRFs obtained from Equation (3.91) (for $r = 1$, $r = 2$, and $r = 3$) are shown in Figure 3.25 along with the multi-mode tip displacement FRFs. Again, the single-mode FRFs exhibit agreement with the multi-mode FRFs around the modes of interest. The enlarged views for modes 1 and 2 are provided in Figure 3.26 for a clear picture of the quality of agreement. The slight overestimation of the resonance frequencies due to ignoring the neighboring vibration modes is the case here too, and the error in the single-mode resonance frequencies is less than 0.1% for these vibration modes (compared to the resonance frequencies of the multi-mode solution).

As a final comparison of the single-mode and multi-mode simulations, the focus is placed back on the fundamental vibration mode and variations of the electrical power and the vibration response with load resistance are plotted in Figures 3.27a and 3.27b, respectively, for excitations at the fundamental short- and open-circuit resonance frequencies. Note that the slightly overestimated open-circuit resonance frequency is used in the single-mode simulations. The predictions of the single-mode FRFs for this most important vibration mode are very accurate. Therefore, Equations (3.90) and (3.91) can comfortably be used as a first approximation in the modeling of a piezoelectric energy harvester beam for modal excitations.

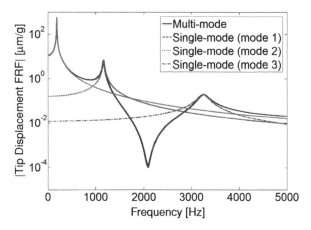

Figure 3.25 Comparison of the multi-mode and the single-mode tip displacement FRFs (series connection)

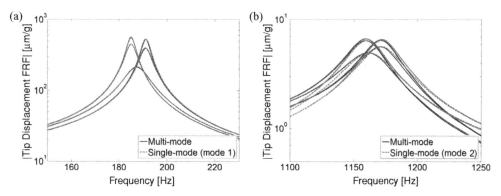

Figure 3.26 Comparison of the multi-mode and the single-mode tip displacement FRFs with a focus on the first two vibration modes: (a) mode 1; and (b) mode 2 (series connection)

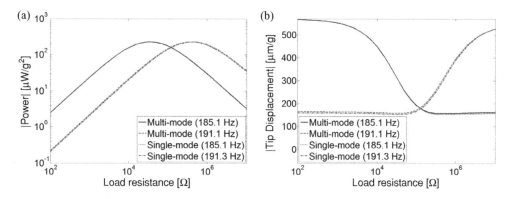

Figure 3.27 Variations of the (a) power output and (b) tip displacement with load resistance for the multi-mode and the single-mode solutions (excitations at the fundamental short-circuit and open-circuit resonance frequencies)

3.8   Summary

Analytical modeling of symmetric bimorph piezoelectric energy harvesters under base excitation is presented in this chapter. The distributed-parameter electromechanical formulation is based on the Euler–Bernoulli beam theory and is valid for thin piezoelectric energy harvesters for the typical vibration modes of interest in practice. The major steps in the analytical formulation for the series and the parallel connection cases of the piezoceramic layers are given independently. An equivalent representation of the series and the parallel connections is then given in a generic form. The distributed-parameter electromechanical equations are first obtained for excitation at any arbitrary frequency (including all vibration modes in the analytical solution). Then, for the problem of resonance excitation, the multi-mode solutions are reduced to the single-mode expressions (which are approximately valid only around the resonance frequencies). Electromechanical FRFs relating the steady-state electrical and mechanical response to translational and rotational base acceleration components are extracted from both the multi-mode and single-mode solutions. A detailed theoretical case study is presented where simulations for the series and the parallel connection cases are given using both the multi-mode and the single-mode electromechanical FRFs. The accuracy of the single-mode FRFs in predicting the multi-mode FRFs for modal excitations is also shown.

3.9   Chapter Notes

The analytical derivations given in this chapter can be used for various applications ranging from modeling, design, and performance prediction of cantilevered piezoelectric energy harvesters to possible combinations with advanced circuit topologies. The derivations given here should be used with caution due to the linear electromechanical system assumption. That is, the excitation level should be such that the geometric and piezoelastic nonlinearities are not pronounced. It is known that inherent piezoelastic nonlinearities [35–37] can become effective if the base acceleration level exceeds a few hundreds of milli-g (see Section 4.5). Fortunately, this range covers acceleration levels available in most ambient vibration energy sources [2]. However, for nonlinear distributed-parameter modeling of cantilevered piezoelectric energy harvesters excited under high acceleration levels, the reader is referred to Stanton and Mann [36] and Stanton et al. [37].

An alternative linear distributed-parameter formulation is the electromechanical version of the Rayleigh–Ritz method (which is an approximate analytical solution method [38]) based on the early derivation of Hagood et al. [39] for piezoelectric actuation. Following the formulation given in Hagood et al. [39], Rayleigh–Ritz type of cantilevered piezoelectric energy harvester models are presented by Sodano et al. [40] and duToit and Wardle [41]. Convergence of the Rayleigh–Ritz solution to the simplified unimorph version [42] of the analytical solutions covered here is shown by Elvin and Elvin [43], who combine the lumped parameters obtained from the Rayleigh–Ritz formulation with circuit simulation software to investigate more sophisticated circuits with storage components in the time domain. Finite-element simulations given by Rupp et al. [44], De Marqui et al. [45], Benasciutti et al. [46], Elvin and Elvin [47], and Yang and Tang [48] are also shown to agree with the analytical solutions covered here and they are introduced for various applications ranging from topology [44], added mass [45], and geometry [46] optimization to the analysis of nonlinear circuits with circuit simulators [47, 48].

The analytical solution given here can be extended to a great many other configurations such as symmetric multimorphs (cantilevers with multiple piezoceramic layers) as well as energy harvester configurations with multiple beams. Combining multiple cantilevers of different natural frequencies to obtain a broadband energy harvester (basically a mechanical band-pass filter) was proposed by Shahruz [49] without any electromechanical model. The analytical modeling approach given in this chapter can easily be extended to the problem of combining the electrical outputs of multiple bimorphs under the same mechanical excitation. As an example, for the series connection of N bimorphs, it can be shown that the governing electromechanical equations of the kth bimorph ($k = 1, \ldots, N$ are integers) in physical coordinates are

$$
YI_{(k)} \frac{\partial^4 w_{rel}^{(k)}(x, t)}{\partial x^4} + c_s I_{(k)} \frac{\partial^5 w_{rel}^{(k)}(x, t)}{\partial x^4 \partial t} + c_a^{(k)} \frac{\partial w_{rel}^{(k)}(x, t)}{\partial t} + m_{(k)} \frac{\partial^2 w_{rel}^{(k)}(x, t)}{\partial t^2}
$$

$$
- \vartheta_{(k)} v_{(k)}(t) \left[\frac{d\delta(x)}{dx} - \frac{d\delta(x - L_{(k)})}{dx} \right] = - \left[m_{(k)} + M_t^{(k)} \delta(x - L_{(k)}) \right] a(t) \qquad (3.92)
$$

$$
(C_p)_{(k)} \frac{dv_{(k)}(t)}{dt} + \frac{1}{R_l} \sum_{k=1}^{N} v_{(k)}(t) + \vartheta_{(k)} \int_0^{L_{(k)}} \frac{\partial^3 w_{rel}^{(k)}(x, t)}{\partial x^2 \partial t} dx = 0 \qquad (3.93)
$$

where $a(t)$ is the translational base acceleration input, $\vartheta_{(k)} = \bar{e}_{31}^{(k)} b_{(k)} h_{\bar{p}c}^{(k)}$ is the electromechanical coupling for the series connection of the layers in the kth bimorph, and x is the dummy axial position taken to be identical for all cantilevers (the remaining parameters are as defined previously but they are given for the kth cantilever: in particular, $v_{(k)}(t)$ is the voltage across the electrodes of the kth bimorph and $w_{rel}^{(k)}(x, t)$ is its vibration response relative to the base). The vibration response of the kth cantilever (coupled to the resistive load and therefore coupled to other cantilevers) is

$$
w_{rel}^{(k)}(x, t) = \sum_{r=1}^{\infty} \phi_r^{(k)}(x) \eta_r^{(k)}(t) \qquad (3.94)
$$

Following the standard modal analysis procedure, one can obtain an infinite set of ordinary differential equations

$$
\frac{d^2 \eta_r^{(k)}(t)}{dt^2} + 2\zeta_r^{(k)} \omega_r^{(k)} \frac{d\eta_r^{(k)}(t)}{dt} + \left(\omega_r^{(k)} \right)^2 \eta_r^{(k)}(t) - \tilde{\theta}_r^{(k)} v_{(k)}(t) = \sigma_r^{(k)} a(t) \qquad (3.95)
$$

$$
(C_p)_{(k)} \frac{dv_{(k)}(t)}{dt} + \frac{1}{R_l} \sum_{k=1}^{N} v_{(k)}(t) + \sum_{r=1}^{\infty} \tilde{\theta}_r^{(k)} \frac{d\eta_r(t)}{dt} = 0 \qquad (3.96)
$$

which can be reduced to an infinite set of algebraic equations for the steady-state response to harmonic excitation and closed-form solutions can be obtained as in the single bimorph

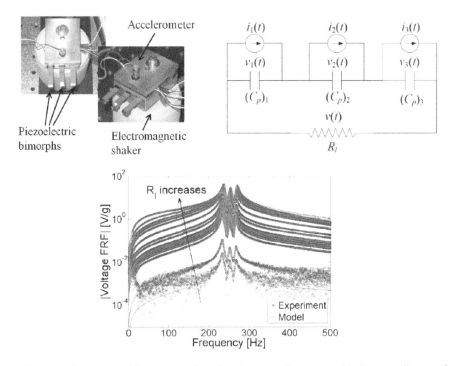

Figure 3.28 (a) Conventional broadband piezoelectric energy harvester with three cantilevers of different natural frequencies, (b) schematic representation of series connection with dependent current sources; and (c) voltage FRFs for a set of resistors

problem. Figure 3.28 provides an illustration for modeling and testing a conventional[21] broadband energy harvester made of three bimorph cantilevers (with different natural frequencies) combined in series.

If the resistive linear circuit is replaced by a nonlinear energy harvesting circuit with a storage component [22–27], the piecewise-defined circuit equations (see Section 11.1) can still be combined with the electromechanically coupled partial differential equation of motion and a distributed-parameter version of the analysis given by Shu *et al.* [26, 50] can be performed. The case of introducing additional linear circuit elements can easily be incorporated and closed-form solutions are available. One interesting example of theoretical value is to introduce an inductor to obtain a second-order circuit. Based on lumped-parameter modeling, Renno *et al.* [51] showed that a flat power FRF can be obtained if the inductance and the resistance are optimized as functions of frequency, mechanical damping, and electromechanical coupling.[22] For the bimorph configurations discussed in this chapter, if the external admittance seen across

---

[21] Unconventional broadband energy harvester configurations utilizing nonlinear phenomena are discussed in Chapter 8.

[22] Although the theoretical results obtained from a linear resistive–inductive circuit are remarkable, in practice, the matched inductance values are quite high as far as typical frequencies of interest in energy harvesting are concerned. This would require using a synthetic inductor or impedance [52], implying energy input to the system.

the terminals coming from the resultant of the piezoceramic layers is $Y(\omega)$, Equations (3.66) and (3.67) become

$$v(t) = \frac{\sum_{r=1}^{\infty} \dfrac{-j\omega\tilde{\theta}_r F_r}{\omega_r^2 - \omega^2 + j2\zeta_r\omega_r\omega}}{Y(\omega) + j\omega C_p^{eq} + \sum_{r=1}^{\infty} \dfrac{j\omega\tilde{\theta}_r^2}{\omega_r^2 - \omega^2 + j2\zeta_r\omega_r\omega}} e^{j\omega t} \tag{3.97}$$

$$w_{rel}(x,t) = \sum_{r=1}^{\infty}\left[\left(F_r - \tilde{\theta}_r \frac{\sum_{r=1}^{\infty}\dfrac{j\omega\tilde{\theta}_r F_r}{\omega_r^2-\omega^2+j2\zeta_r\omega_r\omega}}{Y(\omega)+j\omega C_p^{eq}+\sum_{r=1}^{\infty}\dfrac{j\omega\tilde{\theta}_r^2}{\omega_r^2-\omega^2+j2\zeta_r\omega_r\omega}}\right)\frac{\phi_r(x)e^{j\omega t}}{\omega_r^2-\omega^2+j2\zeta_r\omega_r\omega}\right] \tag{3.98}$$

where the admittance $1/R_l$ in Equations (3.66) and (3.67) is simply replaced by the generalized linear admittance $Y(\omega)$ and the voltage output is defined across the entire load (with real and/or imaginary parts). Table 3.5 lists the external admittance $Y(\omega)$ for various linear circuits (with capacitance C, resistance R, and inductance L). One can proceed with Equations (3.97) and (3.98) to obtain the steady-state voltage and vibration response to harmonic base excitation to perform a distributed-parameter version of the lumped-parameter analysis given by Renno et al. [51].

It is worth adding a few words regarding the use of the analytical model in the presence of a tip mass. Following Erturk and Inman [53], the derivations given here use the transcendental characteristic equation for a cantilevered beam with a tip mass (M_t) of non-zero mass moment of inertia (I_t) (Equation (3.20) based on the formulation given in Appendix C.1). The *point mass assumption* $(I_t = 0)$ can be used if the mass moment of inertia of the tip mass attachment is negligible. For a tip mass attachment with large dimensions, however, the point mass assumption fails and one should calculate I_t at the tip of the elastic beam to use in the relevant equations obtained from the differential eigenvalue problem. In calculating I_t,

Table 3.5 External admittance seen across the terminals coming from the resultant of the piezoceramic layers for various linear circuits (excluding the admittance due to the inherent piezoelectric capacitance)

External circuit	Admittance $Y(\omega)$
Resistive	$1/R$
Capacitive	$j\omega C$
Inductive	$1/j\omega L$
Resistive–inductive (in series)	$1/(R+j\omega L)$
Resistive–inductive (in parallel)	$1/R + 1/j\omega L$

the mass moment of inertia about the centroid of the tip mass should be shifted to the tip of the elastic beam (where the boundary condition is expressed) by using the parallel axis theorem.

Before ending this chapter, one final remark is regarding the validity of the analytical derivations for the base excitation problem of plate-like symmetric configurations under base excitation. As long as the bending stiffness of the physical system is correctly expressed (by accounting for the Poisson effect), the governing equations can represent the electromechanical behavior for bending vibrations of cantilevered piezoelectric energy harvesters with comparable length and width dimensions. Depending on the aspect ratio, the dynamic effect of Poisson contraction/expansion can be accounted for in the electric displacement expression as well.

References

1. Roundy, S., Wright, P., and Rabaey, J. (2003) *Energy Scavenging for Wireless Sensor Networks*, Kluwer Academic, Boston, MA.
2. Roundy, S., Wright, P.K., and Rabaey, J.M. (2003) A study of low level vibrations as a power source for wireless sensor nodes. *Computer Communications*, **26**, 1131–1144.
3. Jeon, Y.B., Sood, R., Jeong, J.H., and Kim, S. (2005) MEMS power generator with transverse mode thin film PZT. *Sensors & Actuators A*, **122**, 16–22.
4. Roundy, S. and Wright, P.K. (2004) A piezoelectric vibration based generator for wireless electronics. *Smart Materials and Structures*, **13**, 1131–1144.
5. duToit, N.E., Wardle, B.L., and Kim, S. (2005) Design considerations for MEMS-scale piezoelectric mechanical vibration energy harvesters. *Journal of Integrated Ferroelectrics*, **71**, 121–160.
6. Sodano, H.A., Park, G., and Inman, D.J. (2004) Estimation of electric charge output for piezoelectric energy harvesting. *Strain*, **40**, 49–58.
7. Lu, F., Lee, H.P., and Lim, S.P. (2004) Modeling and analysis of micro piezoelectric power generators for micro-electromechanical-systems applications. *Smart Materials and Structures*, **13**, 57–63.
8. Chen, S.-N., Wang, G.-J., and Chien, M.-C. (2006) Analytical modeling of piezoelectric vibration-induced micro power generator. *Mechatronics*, **16**, 397–387.
9. Lin, J.H., Wu, X.M., Ren, T.L., and Liu, L.T. (2007) Modeling and simulation of piezoelectric MEMS energy harvesting device. *Integrated Ferroelectrics*, **95**, 128–141.
10. duToit, N.E. and Wardle, B.L. (2007) Experimental verification of models for microfabricated piezoelectric vibration energy harvesters. *AIAA Journal*, **45**, 1126–1137.
11. Ajitsaria, J., Choe, S.Y., Shen, D., and Kim, D.J. (2007) Modeling and analysis of a bimorph piezoelectric cantilever beam for voltage generation. *Smart Materials and Structures*, **16**, 447–454.
12. Wang, Q.M. and Cross, L.E. (1999) Constitutive equations of symmetrical triple layer piezoelectric benders. *IEEE Transactions on Ultrasonics, Ferroelectrics, and Frequency Control*, **46**, 1343–1351.
13. Sokolnikoff, I.S. (1946) *Mathematical Theory of Elasticity*, McGraw-Hill, New York.
14. Dym, C.L. and Shames, I.H. (1973) *Solid Mechanics: A Variational Approach*, McGraw-Hill, New York.
15. Banks, H.T., Luo, Z.H., Bergman, L.A., and Inman, D.J. (1998) On the existence of normal modes of damped discrete-continuous systems. *ASME Journal of Applied Mechanics*, **65**, 980–989.
16. Hyer, M.W. (1998) *Stress Analysis of Fiber-Reinforced Composite Materials*, McGraw-Hill, New York.
17. Reddy, J.N. (2004) *Mechanics of Laminated Composite Plates and Shells: Theory and Analysis*, CRC Press, Boca Raton, FL.
18. Schwartz, L. (1978) *Théorie des Distributions*, Hermann, Paris.
19. Friedlander, F.G. (1998) *Introduction to the Theory of Distributions*, Cambridge University Press, Cambridge.
20. Meriam, J.L. and Kraige, L.G. (2001) *Engineering Mechanics: Dynamics*, John Wiley & Sons, Inc., New York.
21. Standards Committee of the IEEE Ultrasonics, Ferroelectrics, and Frequency Control Society (1987) *IEEE Standard on Piezoelectricity*, IEEE, New York.

22. Ottman, G.K., Hofmann, H.F., Bhatt, A.C., and Lesieutre, G.A. (2002) Adaptive piezoelectric energy harvesting circuit for wireless remote power supply. *IEEE Transactions on Power Electronics*, **17**, 669–676.

23. Guyomar, D., Badel, A., Lefeuvre, E., and Richard, C. (2005) Toward energy harvesting using active materials and conversion improvement by nonlinear processing. *IEEE Transactions on Ultrasonics, Ferroelectrics, and Frequency Control*, **52**, 584–595.

24. Guan, M.J. and Liao, W.H. (2007) On the efficiencies of piezoelectric energy harvesting circuits towards storage device voltages. *Smart Materials and Structures*, **16**, 498–505.

25. Lefeuvre, E., Audigier, D., Richard, C., and Guyomar, D. (2007) Buck-boost converter for sensorless power optimization of piezoelectric energy harvester. *IEEE Transactions on Power Electronics*, **22**, 2018–2025.

26. Shu, Y.C., Lien, I.C., and Wu, W.J. (2007) An improved analysis of the SSHI interface in piezoelectric energy harvesting. *Smart Materials and Structures*, **16**, 2253–2264.

27. Kong, N., Ha, D.S., Erturk, A., and Inman, D.J. (2010) Resistive impedance matching circuit for piezoelectric energy harvesting. *Journal of Intelligent Material Systems and Structures*, **21**, pp. 1293–1302.

28. Erturk, A., Tarazaga, P., Farmer, J.R., and Inman, D.J. (2009) Effect of strain nodes and electrode configuration on piezoelectric energy harvesting from cantilevered beams. *ASME Journal of Vibration and Acoustics*, **131**, 011010.

29. Ewins, D.J. (2000) *Modal Testing: Theory, Practice and Application*, Research Studies Press, Baldock, Hertfordshire.

30. Banks, H.T. and Inman, D.J. (1991) On damping mechanisms in beams. *ASME Journal of Applied Mechanics*, **58**, 716–723.

31. Lu, F., Lee, H.P., and Lim, S.P. (2004) Modeling and analysis of micro piezoelectric power generators for micro-electromechanical-systems applications. *Smart Materials and Structures*, **13**, 57–63.

32. Chen, S.-N., Wang, G.-J., and Chien, M.-C. (2006) Analytical modeling of piezoelectric vibration-induced micro power generator. *Mechatronics*, **16**, 397–387.

33. Lin, J.H., Wu, X.M., Ren, T.L., and Liu, L.T. (2007) Modeling and simulation of piezoelectric MEMS energy harvesting device. *Integrated Ferroelectrics*, **95**, 128–141.

34. Ajitsaria, J., Choe, S.Y., Shen, D., and Kim, D.J. (2007) Modeling and analysis of a bimorph piezoelectric cantilever beam for voltage generation. *Smart Materials and Structures*, **16**, 447–454.

35. von Wagner, U. and Hagedorn, P. (2002) Piezo-beam systems subjected to weak electric field: experiments and modeling of nonlinearities. *Journal of Sound and Vibration*, **256**, 861–872.

36. Stanton, S.C. and Mann, B.P. (2010) Nonlinear electromechanical dynamics of piezoelectric inertial generators: modeling, analysis, and experiment. *Nonlinear Dynamics* (in review).

37. Stanton, S.C., Erturk, A., Mann, B.P., and Inman, D.J. (2010) Nonlinear piezoelectricity in electroelastic energy harvesters: modeling and experimental identification. *Journal of Applied Physics*, **108**, 074903.

38. Meirovitch, L. (2001) *Fundamentals of Vibrations*, McGraw-Hill, New York.

39. Hagood, N.W., Chung, W.H., and Von Flotow, A. (1990) Modelling of piezoelectric actuator dynamics for active structural control. *Journal of Intelligent Material Systems and Structures*, **1**, 327–354.

40. Sodano, H.A., Park, G., and Inman, D.J. (2004) Estimation of electric charge output for piezoelectric energy harvesting. *Strain*, **40**, 49–58.

41. duToit, N.E. and Wardle, B.L. (2007) Experimental verification of models for microfabricated piezoelectric vibration energy harvesters. *AIAA Journal*, **45**, 1126–1137.

42. Erturk, A. and Inman, D.J. (2008) A distributed parameter electromechanical model for cantilevered piezoelectric energy harvesters. *ASME Journal of Vibration and Acoustics*, **130**, 041002.

43. Elvin, N.G. and Elvin, A.A. (2009) A general equivalent circuit model for piezoelectric generators. *Journal of Intelligent Material Systems and Structures*, **20**, 3–9.

44. Rupp, C.J., Evgrafov, A., Maute, K. and Dunn, M.L. (2009) Design of piezoelectric energy harvesting systems: a topology optimization approach based on multilayer plates and shells. *Journal of Intelligent Material Systems and Structures*, **20**, 1923–1939.

45. De Marqui, C. Jr., Erturk, A., and Inman, D.J. (2009) An electromechanical finite element model for piezoelectric energy harvester plates. *Journal of Sound and Vibration*, **327**, 9–25.

46. Benasciutti, D., Moro, L., Zelenika, S., and Brusa, E. (2009) Vibration energy scavenging via piezoelectric bimorphs of optimized shapes. *Microsystem Technologies*, **16**, 657–668.

47. Elvin, N.G. and Elvin, A.A. (2009) A coupled finite element–circuit simulation model for analyzing piezoelectric energy generators. *Journal of Intelligent Material Systems and Structures*, **20**, 587–595.

48. Yang, Y. and Tang, L. (2009) Equivalent circuit modeling of piezoelectric energy harvesters. *Journal of Intelligent Material Systems and Structures*, **20**, 2223–2235.
49. Shahruz, M.D. (2006) Design of mechanical band-pass filters for energy scavenging. *Journal of Sound and Vibration*, **292**, 987–998.
50. Shu, Y. C. and Lien, I. C. (2006) Analysis of power outputs for piezoelectric energy harvesting systems. *Smart Materials and Structures*, **15**, 1499–1502.
51. Renno, J.M., Daqaq, F.M., and Inman, D.J. (2009) On the optimal energy harvesting from a vibration source. *Journal of Sound and Vibration*, **320**, 386–405.
52. Fleming, J., Berhens, S., and Moheimani, S.O.R. (2000) Synthetic impedance for implementation of piezoelectric shunt-damping circuits. *Electronics Letters*, **36**, 1525–1526.
53. Erturk, A. and Inman, D.J. (2009) An experimentally validated bimorph cantilever model for piezoelectric energy harvesting from base excitations. *Smart Materials and Structures*, **18**, 025009.

4

Experimental Validation of the Analytical Solution for Bimorph Configurations

This chapter presents experimental validations of the analytical electromechanical equations derived in Chapter 3. The first two experimental cases consider a brass-reinforced PZT-5H bimorph cantilever in the absence and presence of a tip mass attachment in two separate sections. The voltage, current, power, and tip velocity FRFs of the bimorph (per base acceleration input) are analyzed extensively for both configurations and the focus is placed on the fundamental vibration mode in the frequency range of 0–1000 Hz. The general trends in the FRFs are addressed and model predictions in the absence and presence of a tip mass are presented. Excitations at the fundamental short-circuit and open-circuit resonance frequencies are investigated in detail and the electrical performance diagrams at these two frequencies are extracted. The shunt damping effect of piezoelectric power generation on the cantilever is also studied based on the experimental measurements and model predictions. The effect of rotary inertia of the tip mass is demonstrated by comparing the model predictions to the experimental results including and excluding the rotary inertia of the tip mass attachment. The electrical performance results of the same bimorph with and without the tip mass are compared. The last case study investigates a brass-reinforced PZT-5A bimorph cantilever with a focus on a wider frequency range (0–4000 Hz) covering the first two vibration modes. The model predictions are compared to the experimental results using both the multi-mode and the single-mode analytical FRFs.

4.1 PZT-5H Bimorph Cantilever without a Tip Mass

4.1.1 Experimental Setup and Guidelines for Testing an Energy Harvester

The first cantilever used for validation of the analytical model developed in Chapter 3 is a brass-reinforced bimorph (T226-H4-203X) manufactured by Piezo Systems Inc. The experimental

(1) Shaker with a small
accelerometer and the cantilever
(2) Laser vibrometer
(3) Fixed gain amplifier (power
supply)
(4) Charge amplifier
(5) Data acquisition system
(6) Frequency response analyzer
(software)

Figure 4.1   Experimental setup used for the electromechanical frequency response measurements

setup used for the electromechanical FRF measurements given in this chapter is shown in
Figure 4.1 and enlarged views of the equipment used in the experiments are displayed in
Figure 4.2. A small electromagnetic shaker (TMC model TJ-2) is used for excitation of the
bimorph cantilever and the acceleration at the base of the cantilever is measured by employing
a small accelerometer (PCB Piezotronics Model U352C67) attached to the aluminum clamp
of the cantilever using wax. The tip velocity of the cantilever in the transverse direction is
measured using a laser vibrometer (Polytec PDV100) by attaching a small piece of reflector
tape at the tip of the cantilever. The data acquisition box (SigLab Model 20-42) has four input

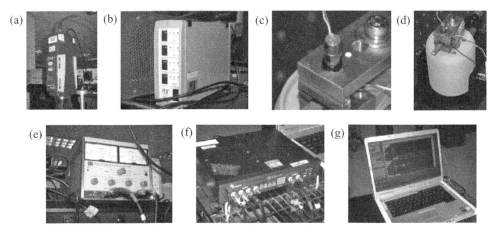

Figure 4.2   Equipment used in the experiments: (a) laser vibrometer; (b) charge amplifier; (c) accelerometer; (d) electromagnetic shaker; (e) fixed gain amplifier; (f) data acquisition system; (g) computer
with a frequency response analyzer

Table 4.1 Geometric and material properties of the PZT-5H bimorph cantilever without a tip mass

	Piezoceramic (PZT-5H)	Substructure (brass)
Length (L) (mm)	24.53	24.53
Width (b) (mm)	6.4	6.4
Thickness ($h_{\bar{p}}$, $h_{\bar{s}}$) (mm)	0.265 (each)	0.140
Mass density ($\rho_{\bar{p}}$, $\rho_{\bar{s}}$) (kg/m^3)	7500	9000
Elastic modulus ($\bar{c}_{11}^E$, $Y_{\bar{s}}$) (GPa)	60.6	105
Piezoelectric constant ($\bar{e}_{31}$) (C/m^2)	-16.6	—
Permittivity constant ($\bar{\varepsilon}_{33}^S$) (nF/m)	25.55	—

and two output channels. The base acceleration signal measured by the accelerometer is sent to the reference channel after being processed by a charge amplifier (PCB Piezotronics Model 482A16). The reference channel automatically becomes the denominator of the resulting FRFs in the frequency response analyzer. Two of the remaining input channels are used for the laser vibrometer and the piezoelectric voltage output signals. Therefore the two FRFs obtained using these three signals are for the tip velocity to base acceleration (or simply the tip velocity FRF) and the voltage output to base acceleration (or simply the voltage FRF). Therefore, in all frequency response measurements provided here, the reference input is the base acceleration in agreement with the derivations of Chapter 3. Chirp excitation (burst type with five averages) is provided to the shaker from the output channel of the data acquisition box (which is connected to a Hewlett-Packard 6826A fixed gain amplifier before the electromagnetic shaker). Since the purpose is to validate the linear electromechanical model, it is ensured that the base acceleration level in the FRF measurements is less than $0.1g$ so that the inherent nonlinearities are not pronounced.

The bimorph used in the first two experimental validation cases given here consists of two oppositely poled PZT-5H piezoelectric elements bracketing a brass substructure layer. The brass layer provides electrical conductivity between the bottom electrode of the top layer and the top electrode of the bottom layer. Therefore, collecting the charge output from the outermost electrodes becomes the series connection case described in Figure 3.1a (and that is what is used here). The geometric and material properties of the piezoceramic and the substructure layers in the cantilevered condition are given in Table 4.1. The data sheet of the manufacturer provides limited information regarding the elastic properties of the piezoceramic. Therefore, typical properties of PZT-5H referred from the literature (Appendix E) are used here. The reduced (plane-stress) properties of the piezoceramic for the Euler–Bernoulli beam theory are taken from Table E.2 in Appendix E. Note that, in agreement with the formulation given in the previous chapter, the length described by L is the overhang length of the harvester; that is, it is not the total free length (31.8 mm) of the bimorph as received from the manufacturer. The overhang length of the cantilever is measured as 24.53 mm. The reflector tape of negligible mass is attached close to the tip surface of the beam and the position of velocity measurement on the reflector is approximately 1.5 mm from the free end ($L_v = 23$ mm). A close-up view of the bimorph cantilever with the reflector tape is given in Figure 4.3 along with the resistive loads.

According to the geometric parameters in Table 4.1, the structure is assumed to be perfectly symmetric with respect to the neutral surface (xy-plane in Figure 3.1a). The nickel electrodes

Figure 4.3 A close-up view of the shaker, accelerometer, PZT-5H bimorph cantilever without a tip mass, its clamp, and a set of resistors

covering the surfaces of the piezoceramic layers are very thin and the thicknesses of the bonding layers are assumed to be negligible. The total thickness of the beam is 0.67 mm and therefore the ratio of overhang length to thickness is about 37.7. Since the focus in this case study (as well as in most energy harvesting applications) is placed on the fundamental vibration mode, shear deformation and rotary inertia effects are assumed to be negligible. This is in agreement with the model presented in the previous chapter based on the Euler–Bernoulli beam assumptions.

For a resistive load of 470 Ω, the experimentally measured tip velocity and voltage output FRFs are shown in Figure 4.4a. This resistive load is very close to the short-circuit conditions for this particular cantilever, therefore the resonance frequency of 502.5 Hz read from Figure 4.4a is the fundamental short-circuit resonance frequency of the PZT-5H bimorph cantilever shown in Figure 4.3 ($f_1^{sc} = 502.5$ Hz). It should be mentioned that, for the frequency range of measurement (0–1000 Hz), the data acquisition system automatically adjusts the frequency increment to 0.325 Hz (which is a device limitation). According to Figure 4.4b, the coherence [1, 2] of the tip velocity measurement is very good (unity) over the frequency range except for a slight reduction around 715 Hz. Around the resonance frequency, the voltage FRF also exhibits perfect coherence, which decays away from the resonance. The relatively poor

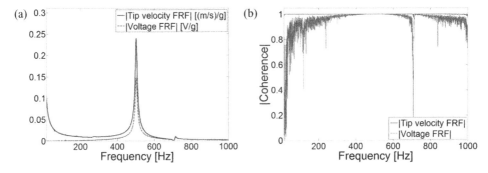

Figure 4.4 (a) Tip velocity and voltage output FRFs of the PZT-5H bimorph cantilever without a tip mass and (b) their coherence functions (for a load resistance of 470 Ω)

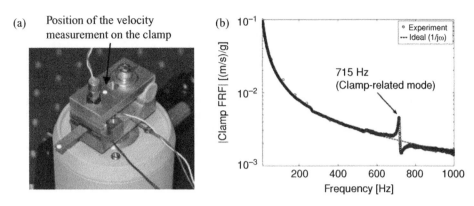

Figure 4.5 (a) A close view of the clamp showing the point of velocity measurement and (b) the FRF for clamp velocity to acceleration capturing the clamp-related imperfection

coherence of the measurement of voltage output to base acceleration is due to the low value of load resistance (deliberately chosen to realize a case close to the short-circuit conditions). Since the system is close to short-circuit conditions for a load resistance of 470 Ω, the signal from the piezoceramic layers is an acceptable measurement output (i.e., it is identified to be due to the input) only around the resonance frequency. Indeed, for a resistance of 10 Ω, the coherence of the voltage FRF becomes extremely low almost for the entire frequency range since the piezoelectric voltage output turns out to be at the noise level of the input channel for such a low resistance. As can be anticipated, with increasing load resistance, the coherence of the voltage measurement becomes as good as that of the tip velocity measurement due to the increased voltage signal.

The sudden reduction in the coherence around 715 Hz in Figure 4.4b is the case for both FRF measurements and it corresponds to a very small peak in the FRFs of Figure 4.4a. As will shortly be seen from the model predictions, the bimorph is not expected to have a mode around that frequency. The fundamental resonance of the shaker itself (without the attachment on it) is not expected for frequencies less than 15 kHz according to the manufacturer. Therefore, the only possible source of the low-amplitude peak around 715 Hz is the clamp of the bimorph that is attached to the shaker with a screw. This is checked very easily by pointing the laser vibrometer at the clamp after attaching the reflector tape to it. Figure 4.5a shows the location of the velocity measurement on the clamp and Figure 4.5b shows the FRF obtained using the laser vibrometer and the accelerometer measurements taken on the clamp. It can be seen from Figure 4.5b that the unexpected peak around 715 Hz in Figure 4.4a is indeed due to the clamp itself. If the clamp behaved ideally for the purpose intended here (i.e., if all the points on it moved identically in the vertical direction with a rigid body translation), the velocity-to-acceleration measurement would be a monotonically decaying function of frequency without any peaks. More precisely, this decaying function would be $1/j\omega$ (ratio of the velocity to acceleration in the frequency domain). This ideal behavior agrees very well with the measurement shown in Figure 4.5b for all frequencies away from the unexpected peak around 715 Hz. The source of the peak in the clamp FRF might be a possible rotation of the clamp as a result of the joint flexibility at the clamp–shaker interface or it might be due to the dynamic interaction of the clamp components. In either case the small peak appearing in the FRFs around this frequency is not a vibration mode of the bimorph cantilever. This imperfection will be ignored throughout this chapter since it is sufficiently far away from the fundamental vibration mode of the bimorph (which is the main concern here).

Before the model predictions are compared to the experimental measurements, it is important to note that the tip velocity measurement taken by the laser vibrometer is the velocity response of the cantilever relative to the *fixed* reference frame. In other words, it is not the velocity response *relative* to the base of the beam. Therefore, the relative tip displacement FRF given by Equation (3.77), $\beta(\omega, x)$, should be modified to express the absolute velocity response at the tip of the beam relative to the fixed frame of reference:

$$\beta^{\text{modified}}(\omega, x) = \frac{\dfrac{\partial w(x, t)}{\partial t}}{-\omega^2 W_0 e^{j\omega t}} = \frac{\dfrac{\partial}{\partial t}\left[W_0 e^{j\omega t} + w(x, t)\right]}{-\omega^2 W_0 e^{j\omega t}} = \frac{1}{j\omega} + j\omega\beta(\omega, x) \qquad (4.1)$$

where $x = L_v$ (the point of velocity measurement) should be used to predict the measurement taken by the laser vibrometer. Note that, instead of modifying the analytical FRF expression given by Equation (3.77), one could as well process the experimental tip velocity FRF to obtain a relative tip displacement FRF. However, post-processing of the analytical data is usually preferable in order not to amplify any noise by further processing the experimental data.

The last item to mention in the experimental setup concerns the resistance seen across the electrodes of the piezoceramic. A measurement taken on the input channel of the data acquisition system (when nothing is connected) shows a device resistance of 995 kΩ. Therefore, if the wires coming from the electrodes of the piezoceramic are directly connected to the data acquisition system in the experiments (without a probe), as done in the experiments of this chapter, the effective load resistance seen across the electrodes of the piezoceramic is the equivalent resistance of the resistive load used and 995 kΩ (which see each other in parallel): $R_l = 1/(1/R_l^{\text{used}} + 1/995 \times 10^3)$. This observation implies that the maximum resistance used in the experiments cannot exceed 995 kΩ.[1] The equivalent of the resistor used and the device resistance constitutes the R_l value to be used in the model and these effective values are listed in the second column of Table 4.2. Therefore the load resistance (R_l) ranges from 470 Ω (close to short-circuit conditions) to 995 kΩ (close to open-circuit conditions) in the experiments.

In summary, there are several important factors to consider when testing vibration-based energy harvesting devices. While the description and data presented here are specific to the hardware mentioned, the important issues can be generalized to any similar test:

1. Understanding the shaker dynamics to ensure that they do not interact with the beam dynamics.
2. Understanding and measuring the mounting bracket (clamp) dynamics and its interaction with the shaker.
3. Understanding the electrical resistance of the data acquisition system used as well as its limitations (such as the frequency resolution).

Very often these three steps are overlooked, leading to erroneous test results.

---

[1] One should use a probe if the measurements require load resistance values greater than 1 MΩ (commercially available 10:1 probes will allow the realization of resistance values safely up to several megohms below 10 MΩ).

Table 4.2   The resistors used in the experiment and their effective values
due to the internal resistance of the data acquisition system

Resistance of the resistor used (kΩ)	Effective load resistance (R_l) seen by the harvester (kΩ)
0.470	0.470
1.2	1.2
6.7	6.7
10	9.9
22	21.5
47	44.9
100	90.9
330	247.8
470	319.2
680	403.9
1000	498.7
Open	995.0

4.1.2   Validation of the Electromechanical FRFs for a Set of Resistors

In this first case study, the focus is placed on the fundamental vibration mode seen around 502.5 Hz (for a load resistance of 470 Ω). Since the performance of the energy harvester at resonance is the main concern, accurate identification of the modal mechanical damping ratio is very important. It is common practice to extract the mechanical damping ratio from the first measurement in experimental structural dynamics. The model developed here allows the identification of mechanical damping in the presence of an arbitrary load resistance. The FRFs of the resistance that are closest to $R_l \rightarrow 0$ are used here for damping identification (i.e., the FRFs given by Figure 4.4a for 470 Ω). The mechanical damping ratio identified graphically by matching the peaks of the experimental and analytical tip velocity FRFs in Figure 4.6a is $\zeta_1 = 0.00874$. For the numerical data given in Table 4.1, the analytical model predicts the fundamental short-circuit resonance frequency as 502.6 Hz with a relative error of 0.02% compared to the experimental short-circuit resonance frequency (502.5 Hz). If the electromechanical model is consistent, the analytical voltage FRF should accurately predict the peak amplitude of the experimental voltage FRF for the damping ratio of 0.874%. This mechanical damping ratio identified from the velocity FRF is substituted into the voltage FRF expressed by Equation (3.74) and the model prediction given in Figure 4.6b is obtained. The agreement is very good considering the fact that the two FRFs of Figures 4.6a and 4.6b are independent measurements (velocity and voltage) and the mechanical damping is identified from the former. One could as well identify the mechanical damping ratio from the voltage FRF (Figure 4.6b) since a piezoelectric energy harvester is itself a *transducer*. That is, the information of the velocity response of the cantilever is included in the voltage response due to piezoelectric coupling. As long as the linear electromechanical system assumption holds, analytical predictions of the voltage and the tip velocity have to be in agreement with each other for a given load resistance.

It is worth adding that, instead of identifying the modal mechanical damping ratio by matching the peak values of the experimental and analytical electromechanical FRFs

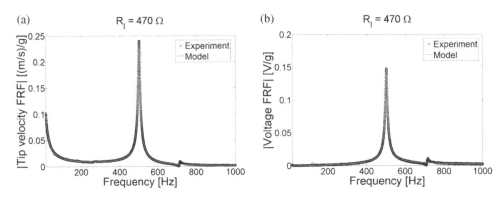

Figure 4.6 Measured and predicted (a) tip velocity and (b) voltage output FRFs of the PZT-5H bimorph cantilever without a tip mass for a load resistance of $470\,\Omega$

(or using the closed-form expressions of Chapter 5), one could as well follow a classic time-domain procedure to identify mechanical damping. For instance, without worrying about the voltage output, one could set the electrical boundary condition as close to $R_l \to 0$ as possible and perform a time-domain damping identification (e.g., logarithmic decrement). However, it should be mentioned that even a wire (that can be used to short the electrodes to realize $R_l \to 0$) has a certain resistance. Therefore the demonstration here is given for the presence of a finite load resistance (that is still very close to $R_l \to 0$), which is the general case. Indeed, using a resistive load that generates an acceptable voltage output (larger than the noise level) allows one to check whether or not the analytical model can predict the voltage output when the same damping ratio is substituted into the voltage FRF (as done here in Figure 4.6b). The upshot of this discussion is that the mechanical damping can be identified in the presence of an arbitrary load resistance using the electromechanical model. However, if one prefers to identify mechanical damping using conventional techniques (e.g., half-power points of the vibration FRF, logarithmic decrement in time domain, etc.), the electrical boundary condition should be set as close as possible to $R_l \to 0$ (which gives a noisy piezoelectric voltage output and does not allow the voltage prediction to be checked). The important point is that the conventional techniques of damping identification should *not* be used in the presence of *arbitrary* values of load resistance. In such a case, damping due to piezoelectric coupling contributes to the mechanical damping in the vibration response and the identified value will not be a pure mechanical damping ratio so it *cannot* be used as ζ_r in the equations derived in Chapter 3.

Having identified the mechanical damping ratio for a certain electrical load resistance, the next step is to keep the damping ratio fixed and predict the tip velocity and voltage FRFs for different values of load resistance. Figure 4.7 displays experimental measurements and model predictions for resistors of three different orders of magnitude: $1.2\,k\Omega$, $44.9\,k\Omega$, and $995\,k\Omega$. In all cases, the mechanical damping ratio is kept at $\zeta_1 = 0.00874$ and the model predictions are very good. Note that, among these three different values of load resistance, $44.9\,k\Omega$ results in the strongest attenuation of the peak vibration amplitude. The peak voltage amplitude, as expected from the theoretical case study of Section 3.7, increases monotonically with increasing load resistance. The peak voltage amplitude increases by two orders of magnitude as the load resistance is increased from $470\,\Omega$ (Figure 4.6b) to $995\,k\Omega$ (Figure 4.7c).

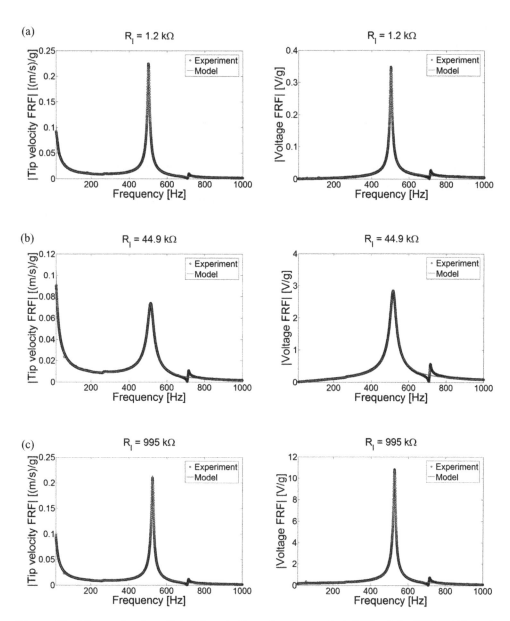

Figure 4.7 Measured and predicted tip velocity and voltage output FRFs of the PZT-5H bimorph cantilever without a tip mass for various resistors: (a) 1.2 kΩ; (b) 44.9 kΩ; and (c) 995 kΩ

As the load resistance is increased from 470 kΩ to 995 kΩ, the experimentally measured fundamental resonance frequency moves from the short-circuit value of 502.5 Hz to the open-circuit value of 524.7 Hz. The analytical model predicts these two resonance frequencies as 502.6 Hz and 524.5 Hz respectively. Table 4.3 summarizes these results along with the errors in the model prediction compared to the experimental frequencies. It will be seen from the modal

Table 4.3 Fundamental short-circuit and open-circuit resonance frequencies of the PZT-5H bimorph cantilever without a tip mass

Fundamental resonance frequency	Experiment	Model	% error
f_1^{sc}(short circuit) (Hz)	502.5	502.6	+0.02
f_1^{oc}(open circuit) (Hz)	524.7	524.5	−0.04

approximations given in Chapter 5 that the amount of resonance frequency shift from the short-circuit to the open-circuit value is directly proportional to the square of the modal electromechanical coupling term and inversely proportional to the capacitance and the respective undamped natural frequency. Although the errors in Table 4.3 are negligible for most practical purposes, possible inaccuracies in the piezoelectric and permittivity constants contribute to the error. In addition, the fact that the undamped natural frequency is slightly overestimated results in underestimation of the open-circuit resonance frequency (as the frequency shift is inversely proportional to the undamped natural frequency). The experimental source of error might be more significant than all these, since the frequency increment is automatically set equal to 0.325 Hz when the frequency range of interest is adjusted to 0–1000 Hz.

The piezoelectric voltage output and tip velocity FRFs measured experimentally and predicted analytically for all resistors used here are given by Figures 4.8 and 4.9, respectively. The gradual increase of the voltage output with increasing load resistance (Figure 4.8) is associated with a gradual attenuation of the vibration amplitude at the short-circuit resonance frequency (Figure 4.9). However, after a certain value of load resistance, the vibration amplitude at the short-circuit resonance frequency becomes saturated, while it becomes amplified at the open-circuit resonance frequency. Hence, if one focuses on the fundamental open-circuit resonance frequency, both the vibration amplitude and the voltage amplitude increase with increasing load resistance. Consequently, modeling the effect of piezoelectric coupling in the form of an electrically induced viscous damping term [3–6] fails in predicting what happens

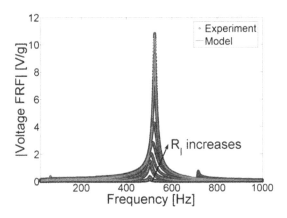

Figure 4.8 Voltage output FRFs of the PZT-5H bimorph cantilever without a tip mass for 12 different resistive loads (ranging from 470 Ω to 995 kΩ)

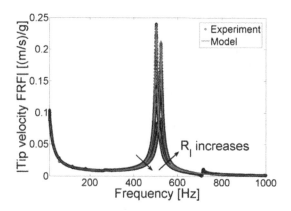

Figure 4.9 Tip velocity FRFs of the PZT-5H bimorph cantilever without a tip mass for 12 different resistive loads (ranging from 470 Ω to 995 kΩ)

to the piezoelectric energy harvester when it generates electricity (which dramatically affects the electrical predictions in the coupled system).

The close-up views of the voltage, current, power, and tip velocity FRFs are given in Figure 4.10 for all values of load resistance (in the 400–600 Hz region) in semi-log scale. It is clear from Figure 4.10a that the voltage FRFs monotonically converge to a single open-circuit voltage FRF as the load resistance increases toward the $R_l \rightarrow \infty$ extremum. The current FRFs in Figure 4.10b also behave monotonically and become similar to a single short-circuit current FRF for the $R_l \rightarrow 0$ extremum. The product of the voltage and the current FRFs gives the power FRF, which results in a relatively complicated picture when all 12 curves are plotted together in Figure 4.10c. For a given frequency and range of load resistance, the behavior is not necessarily monotonic, resulting in intersections of the curves. One can define different values of optimum load resistance at different frequencies. However, the short- and open-circuit resonance frequencies are of particular interest in a lightly damped, strongly coupled system. Variation of the vibration response is also non-monotonic and the trend of the peak vibration amplitude with load resistance is shown in Figure 4.10d. For a given excitation frequency around resonance, one can define an optimum load for the maximum vibration attenuation.

4.1.3 Electrical Performance Diagrams at the Fundamental Short-Circuit and Open-Circuit Resonance Frequencies

Focusing on the fundamental short- and open-circuit resonance frequencies, the electrical performance diagrams are extracted next. These diagrams can be useful for the electrical engineer to obtain an idea about the maximum voltage, current, and power output levels of the energy harvester as well as the optimum electrical load resistance to design a sophisticated energy harvesting circuit. Figure 4.11 shows the variation of the voltage amplitude at the fundamental short- and open-circuit resonance frequencies. The model predictions are in very good agreement with the experimental data points (for 12 resistors). The linear asymptotic trends can be noted from the experimental data as well. The model predicts the maximum

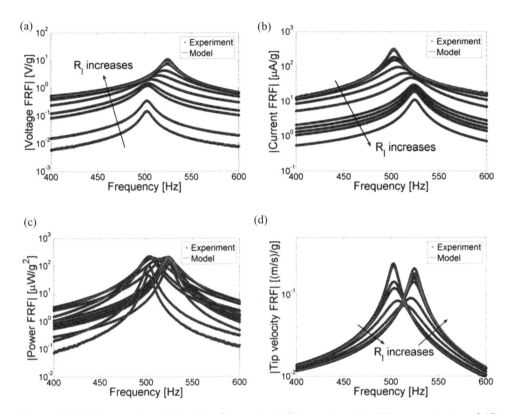

Figure 4.10 Close-up views of the (a) voltage output, (b) current output, (c) power output, and (d) tip velocity FRFs of the PZT-5H bimorph cantilever without a tip mass for 12 different resistive loads (ranging from 470 Ω to 995 kΩ)

voltage outputs ($R_l \rightarrow \infty$) as 2.6 V/g and 12.8 V/g for excitations at the fundamental short- and open-circuit resonance frequencies, respectively.

Variations of the electric current passing through the resistor for excitations at the fundamental short- and open-circuit resonance frequencies are plotted in Figure 4.12. For excitations at the fundamental short- and open-circuit resonance frequencies, the maximum current outputs ($R_l \rightarrow 0$) are predicted as 336 μA/g and 68 μA/g. Just like the asymptotic voltage behavior for $R_l \rightarrow \infty$, the asymptotic current trends are horizontal lines as $R_l \rightarrow 0$.

The final electrical performance diagram is the power versus load resistance diagram shown in Figure 4.13. According to the model, for excitations at the fundamental short- and open-circuit resonance frequencies, the maximum power output of about 0.22 mW/g^2 is delivered to the optimum electrical loads of 7.6 kΩ and 189 kΩ, respectively. Although these exact values of load resistance are not used in the experiments, the loads of 6.7 kΩ and 247.8 kΩ are relatively close to these optimum ones. The resistive load of 6.7 kΩ yields an experimental power output of 0.218 mW/g^2 at the fundamental short-circuit resonance frequency, whereas the resistive load of 247.8 kΩ yields 0.212 mW/g^2 at the fundamental open-circuit resonance frequency. The model predicts the power outputs for 6.7 kΩ and 247.8 kΩ as 0.217 mW/g^2

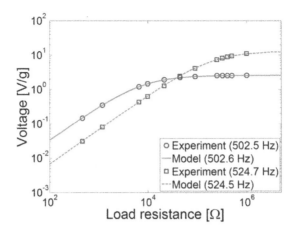

Figure 4.11 Variation of the voltage output with load resistance for excitations at the fundamental short-circuit and open-circuit resonance frequencies of the PZT-5H bimorph cantilever without a tip mass

and $0.215\,\mathrm{mW}/g^2$, respectively, with relative errors of -0.5% and $+1.4\%$ (compared to the experimental power amplitudes).

It is useful to report two normalized power measures: the *power density* and the *specific power* outputs. The former is defined as the power output divided by the overhang volume of the device and the latter is the power output divided by the total overhang mass. The overhang volume of the cantilever is obtained from the data in Table 4.1 as $0.105\,\mathrm{cm}^3$ and the overhang mass is $0.822\,\mathrm{g}$. Therefore the maximum power density of this configuration for

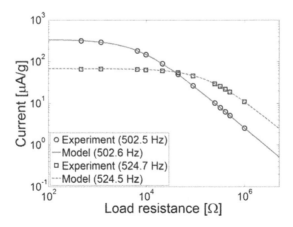

Figure 4.12 Variation of the current output with load resistance for excitations at the fundamental short-circuit and open-circuit resonance frequencies of the PZT-5H bimorph cantilever without a tip mass

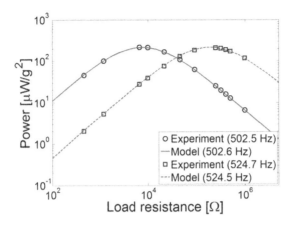

Figure 4.13 Variation of the power output with load resistance for excitations at the fundamental short-circuit and open-circuit resonance frequencies of the PZT-5H bimorph cantilever without a tip mass

resonance excitation is about 2.1 mW/(g^2 cm^3) while the maximum specific power output is about 0.27 mW/(g^2 g).[2]

4.1.4   *Vibration Response Diagrams at the Fundamental Short-Circuit and Open-Circuit Resonance Frequencies*

Although the main concern in energy harvesting is the electrical performance outputs, it is useful to summarize what happens to the piezoelectric energy harvester beam due to power generation. It was emphasized while discussing the tip velocity FRF that the form of piezoelectric coupling in the mechanical domain is substantially different than conventional damping mechanisms such as viscous or structural damping.

Figure 4.14 shows that the vibration amplitude at the fundamental short-circuit resonance frequency is attenuated as the load resistance is increased. At the fundamental short-circuit resonance frequency, the experimental vibration amplitude of 0.240 (m/s)/g for 470 Ω is attenuated to 0.050 (m/s)/g as the load resistance is increased to 995 kΩ. The model predicts these amplitudes as 0.240 (m/s)/g and 0.051 (m/s)/g,[3] respectively. As $R_l \rightarrow \infty$, the vibration amplitude at this frequency converges to 0.052 (m/s)/g after a slight increase (based on the model prediction). For excitation at the fundamental open-circuit resonance frequency, the experimental vibration amplitude of 0.047 (m/s)/g for 470 Ω increases to 0.210 (m/s)/g for 995 kΩ. The model predicts these two amplitudes as 0.050 (m/s)/g and 0.223 (m/s)/g, respectively. As $R_l \rightarrow \infty$, the vibration amplitude at the fundamental open-circuit resonance frequency settles to 0.265 (m/s)/g (based on the model prediction).

---

[2] Here, g stands for "grams" and should not be confused with italic g for gravitational acceleration.

[3] The mechanical damping ratio was identified for the data point 0.240 (m/s)/g, as can be recalled from Figure 4.6a.

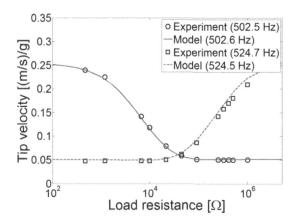

Figure 4.14   Variation of the tip velocity with load resistance for excitations at the fundamental short-circuit and open-circuit resonance frequencies of the PZT-5H bimorph cantilever without a tip mass

4.2   PZT-5H Bimorph Cantilever with a Tip Mass

4.2.1   *Experimental Setup*

In order to investigate the effect of a tip mass and to demonstrate the validity of the model in the presence of a tip mass, a cube-shaped rectangular mass of 0.239 g is attached to the tip of the PZT-5H bimorph cantilever investigated in the previous section (Figure 4.15). The experimental setup and the devices used for the measurements are as described in Section 4.1.1. The purpose here is to introduce the tip mass information directly to the model and check the accuracy of the model predictions without changing the overhang length of the beam (as well as the clamping condition).

A close-up view of the PZT-5H bimorph cantilever with the tip mass is shown in Figure 4.16a and a schematic showing the configuration with the tip mass is given by Figure 4.16b. The tip mass is attached to the tip of the cantilever by a slight amount of wax (with negligible mass) such that the center line of the cube is at the tip of the beam. Each edge of the tip mass (simply a rare earth magnet) is 0.125 inches long ($a = 3.2$ mm). As the center line of the tip

Figure 4.15   A view of the experimental setup for the PZT-5H bimorph cantilever with a tip mass

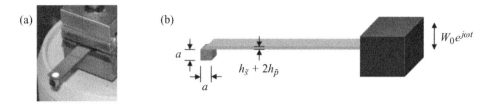

Figure 4.16 (a) A close view of the PZT-5H bimorph cantilever with a tip mass and (b) a schematic view showing the geometric detail of the cube-shaped tip mass

mass lies on the tip of the cantilever, the mass moment of inertia about the center axis of the bimorph is obtained from

$$I_t = M_t \left[\frac{a^2}{6} + \left(\frac{a + h_{\tilde{s}}}{2} + h_{\tilde{p}} \right)^2 \right] \tag{4.2}$$

where $M_t = 0.239 \times 10^{-3}$ kg (measured), the first term inside the parentheses is for the mass moment of inertia about the center axis of the cube and the second term is due to the parallel axis theorem [7] to account for the offset of the tip mass to one side.[4] Substituting the numerical data into Equation (4.2), the mass moment of inertia at the tip is calculated as 1.285×10^{-9} kg m^2. Therefore, the numerical values of the mass and the mass moment of inertia at the tip are the only additions to the data of the previous section as listed in Table 4.4. Following the same procedure, the first step is to obtain the experimental tip velocity and voltage output FRFs for the load resistance of 470 Ω.

Figure 4.17a shows the tip velocity and voltage FRFs of the PZT-5H bimorph cantilever with a tip mass for the same frequency range as before (0–1000 Hz) by connecting the outermost electrodes to a resistive load of 470 Ω. The resonance frequency of the fundamental mode measured for this low value of load resistance is 338.4 Hz. For this tip mass (which is about 29% of the overhang mass of the beam), the reduction in the fundamental short-circuit resonance frequency compared to that of the configuration without a tip mass is about 33%. The coherence of the velocity FRF is very good and the reduction of coherence in the voltage FRF is due to the low value of resistance as in the previous case. The clamp-related imperfection (recall Figure 4.5) appears in the tip velocity and the voltage FRFs and is to be ignored as the resonance frequency is sufficiently far away from it. The set of resistors used in the experiments is the same as the one given in Table 4.2.

---

[4] Although the offset of the tip mass is taken into account in calculating the mass moment of inertia to improve the model predictions, it can still be argued that the structural symmetry of the bimorph is distorted as the model in the previous chapter is given for a symmetrically located tip mass. However, mathematically, the tip mass is a singularity at $x = L$ (associated with the Dirac delta function in theory) and the analytical model takes it into account through the eigenvalues and normalization of the eigenfunctions (Section C.1.3 in Appendix C). The singularity assumption might fail if a large and/or elastic tip attachment covers a considerable region at the tip.

Table 4.4   Geometric and material properties of the PZT-5H bimorph cantilever with a tip mass

	Piezoceramic (PZT-5H)		Substructure (brass)
Length (L) (mm)	24.53		24.53
Width (b) (mm)	6.4		6.4
Thickness ($h_{\bar{p}}$, $h_{\bar{s}}$) (mm)	0.265 (each)		0.140
Tip mass (M_t) (kg)		0.239×10^{-3}	
Mass moment of inertia at the tip (I_t) (kg m^2)		1.285×10^{-9}	
Mass density ($\rho_{\bar{p}}$, $\rho_{\bar{s}}$) (kg/m^3)	7500		9000
Elastic modulus ($\bar{c}_{11}^E$, $Y_{\bar{s}}$) (GPa)	60.6		105
Piezoelectric constant ($\bar{e}_{31}$) (C/m^2)	-16.6		—
Permittivity constant ($\bar{\varepsilon}_{33}^S$) (nF/m)	25.55		—

4.2.2   Validation of the Electromechanical FRFs for a Set of Resistors

When the tip mass and the moment of inertia data are used in the model for a resistive load of 470 Ω (which is close to short-circuit conditions), the fundamental resonance frequency of the PZT-5H bimorph cantilever with a tip mass is obtained as 338.5 Hz (with an error of 0.03% compared to the experimental resonance frequency). Therefore, the analytical model predicts very accurately the resonance frequency shift due to the addition of the tip mass as shown in Figure 4.18a. Since this work does not aim to study the behavior of mechanical damping with changing structural configuration (due to the added tip mass in this case), the damping ratio of the original configuration without the tip mass is slightly tuned to $\zeta_1 = 0.00845$ in order to better match the peak amplitude in Figure 4.18a. This damping ratio is then used in the analytical voltage FRF expression and the experimental voltage FRF shown in Figure 4.18b is predicted with very good accuracy. Therefore, the model exhibits consistency between the vibration response and the electrical response in the presence of a tip mass as well.

To further study the performance of the electromechanical model, three particular resistive loads are investigated next: 1.2 kΩ, 44.9 kΩ, and 995 kΩ. Figure 4.19 shows the tip velocity and the voltage FRFs for these resistive loads (with fixed mechanical damping). The trend is very similar to the previous case (bimorph without a tip mass). Compared to Figure 4.7,

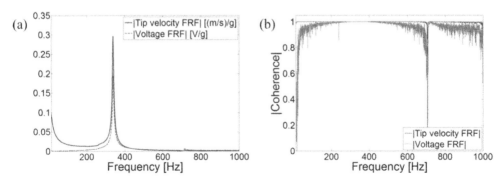

Figure 4.17   (a) Tip velocity and voltage output FRFs of the PZT-5H bimorph cantilever with a tip mass and (b) their coherence functions (for a load resistance of 470 Ω)

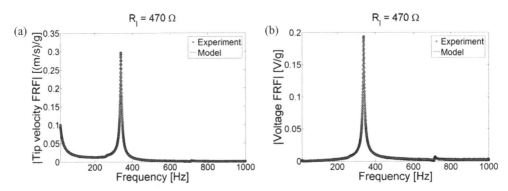

Figure 4.18 Measured and predicted (a) tip velocity and (b) voltage output FRFs of the PZT-5H bimorph cantilever with a tip mass for a load resistance of 470 Ω

both the vibration and the voltage amplitudes are larger in Figure 4.19. The main reason is the increased forcing amplitude for the same acceleration input since the force amplitude in the base excitation problem of a cantilever is proportional to the mass of the structure according to Equation (3.36). Once again, among the loads used here, 44.9 kΩ results in a very strong attenuation of the peak vibration amplitude. The electromechanical FRFs for these three resistive loads are successfully predicted in Figure 4.19. The experimentally measured fundamental short- and open-circuit resonance frequencies are 338.4 Hz and 356.3 Hz, respectively, and the model predicts these frequencies as 338.5 Hz and 355.4 Hz as listed in Table 4.5. Once again, although the error levels are very low for practical purposes, the primary theoretical sources of the slight inaccuracy in predicting the frequency shift are the possible inaccuracies of the piezoelectric and permittivity constants. Overestimation of the short-circuit resonance also results in underestimation of the frequency shift as mentioned earlier. As an experimental source of error, the frequency increment of the data acquisition system for the frequency range of interest is 0.325 Hz.

Figures 4.20 and 4.21, respectively, display the tip velocity and the voltage output FRFs of the PZT-5H bimorph cantilever with a tip mass for all values of load resistance used here. Variations of the peak amplitudes and the resonance frequencies are successfully predicted in both figures. The enlarged views of the voltage, current, power, and tip velocity FRFs for the frequency range of 200–500 Hz are given in Figure 4.22, where the variation of the resonance frequency with load resistance is better visualized. Based on these illustrations, one can conclude that the model can successfully predict the amplitude-wise and the frequency-wise coupled system dynamics of the bimorph cantilever in the presence of a tip mass as well.

4.2.3 Electrical Performance Diagrams at the Fundamental Short-Circuit and Open-Circuit Resonance Frequencies

In the electromechanical curves plotted in Figures 4.22a–4.22c, the focus is placed on the amplitude-wise results for excitations at the fundamental short- and open-circuit resonance frequencies to obtain the electrical performance diagrams.

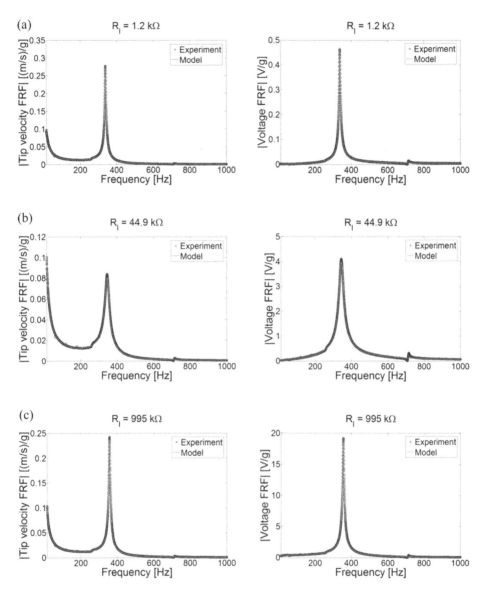

Figure 4.19 Measured and predicted tip velocity and voltage output FRFs of the PZT-5H bimorph cantilever with a tip mass for various resistors: (a) $1.2\,\mathrm{k\Omega}$; (b) $44.9\,\mathrm{k\Omega}$; and (c) $995\,\mathrm{k\Omega}$

The variations of the voltage amplitude at these two frequencies are plotted in Figure 4.23 along with the experimental data points for the 12 resistive loads. The analytical curve agrees very well with the experimental data points and the analytical asymptotes of $R_l \rightarrow \infty$ for excitations at fundamental short- and open-circuit resonance frequencies are $4.2\,\mathrm{V}/g$ and $24.7\,\mathrm{V}/g$, respectively.

Table 4.5 Fundamental short-circuit and open-circuit resonance frequencies of the PZT-5H bimorph
cantilever with a tip mass

Fundamental resonance frequency	Experiment	Model	% error
f_1^{sc}(short circuit) (Hz)	338.4	338.5	+0.03
f_1^{oc}(open circuit) (Hz)	356.3	355.4	−0.25

The current versus load resistance diagrams for excitations at the fundamental short- and
open-circuit resonance frequencies are plotted in Figure 2.24, where the asymptotes of $R_l \to 0$
for these frequencies are 435 μA/g and 75 μA/g, respectively.

The power versus load resistance diagrams are shown in Figure 4.25 for excitations at the
fundamental short- and open-circuit resonance frequencies of the PZT-5H bimorph with a tip

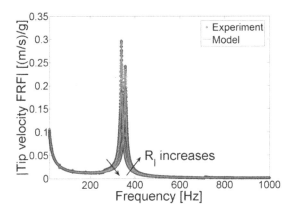

Figure 4.20 Tip velocity FRFs of the PZT-5H bimorph cantilever with a tip mass for 12 different
resistive loads (ranging from 470 Ω to 995 kΩ)

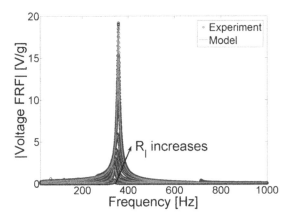

Figure 4.21 Voltage output FRFs of the PZT-5H bimorph cantilever with a tip mass for 12 different
resistive loads (ranging from 470 Ω to 995 kΩ)

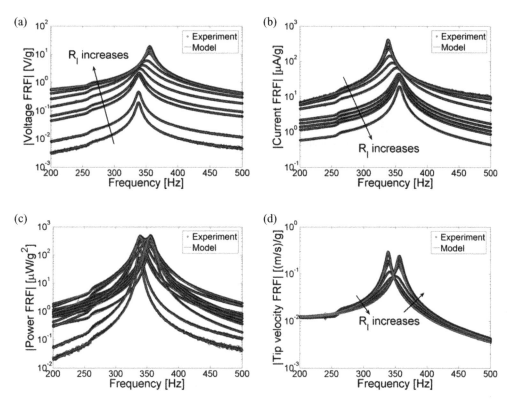

Figure 4.22 Enlarged views of the (a) voltage output, (b) current output, (c) power output, and (d) tip velocity FRFs of the PZT-5H bimorph cantilever with a tip mass for 12 different resistive loads (ranging from 470 Ω to 995 kΩ)

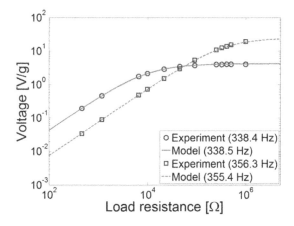

Figure 4.23 Variation of the voltage output with load resistance for excitations at the fundamental short-circuit and open-circuit resonance frequencies of the PZT-5H bimorph cantilever with a tip mass

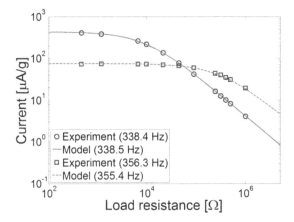

Figure 4.24 Variation of the current output with load resistance for excitations at the fundamental short-circuit and open-circuit resonance frequencies of the PZT-5H bimorph cantilever with a tip mass

mass. The analytical curves exhibit good agreement with the experimental data points and predict a peak power amplitude of $0.46\,mW/g^2$ for the optimum resistive loads of $9.7\,k\Omega$ and $331\,k\Omega$, respectively, at the fundamental short- and open-circuit resonance frequencies. These exact values of the predicted optimum loads are not available in the set of resistors used here. The experimental results for the nearest resistive loads are $0.460\,mW/g^2$ for $9.9\,k\Omega$ at the short-circuit resonance frequency and $0.463\,mW/g^2$ for $319.2\,k\Omega$ at the open-circuit resonance frequency. These resistors are very close to the optimum ones and their power outputs are in good agreement with the predicted maximum power output.

With the volume of the tip mass, the overhang volume of the cantilever becomes $0.137\,cm^3$ and the overhang mass becomes $1.061\,g$. Therefore the maximum power density of this configuration for resonance excitation is about $3.4\,mW/(g^2\,cm^3)$ and the maximum specific power output is about $0.43\,mW/(g^2\,g)$.

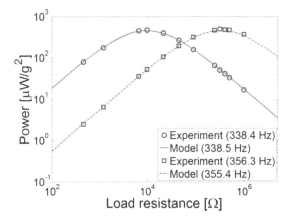

Figure 4.25 Variation of the power output with load resistance for excitations at the fundamental short-circuit and open-circuit resonance frequencies of the PZT-5H bimorph cantilever with a tip mass

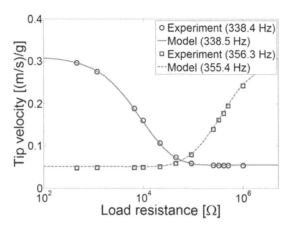

Figure 4.26 Variation of the tip velocity with load resistance for excitations at the fundamental short-circuit and open-circuit resonance frequencies of the PZT-5H bimorph cantilever with a tip mass

4.2.4 Vibration Response Diagrams at the Fundamental Short-Circuit and Open-Circuit Resonance Frequencies

Variations of the tip velocity amplitude with load resistance for excitations at the fundamental short- and open-circuit resonance frequencies are plotted in Figure 4.26. At the fundamental short-circuit resonance frequency, the experimental vibration amplitude of $0.296\,(\text{m/s})/g$ for $470\,\Omega$ is attenuated to $0.054\,(\text{m/s})/g$ as the load resistance is increased to $995\,\text{k}\Omega$. The model predicts these amplitudes as $0.296\,(\text{m/s})/g$ and $0.055\,(\text{m/s})/g$,[5] respectively. For $R_l \rightarrow \infty$ in the model, the vibration amplitude at the fundamental short-circuit resonance frequency converges approximately to the same amplitude: $0.055\,(\text{m/s})/g$. For excitation at the open-circuit resonance frequency, the experimental vibration amplitude of $0.048\,(\text{m/s})/g$ for $470\,\Omega$ increases to $0.242\,(\text{m/s})/g$ for $995\,\text{k}\Omega$. The model predicts these two amplitudes as $0.051\,(\text{m/s})/g$ and $0.246\,(\text{m/s})/g$, respectively. As $R_l \rightarrow \infty$ in the model, the vibration amplitude at the fundamental open-circuit resonance frequency reaches $0.327\,(\text{m/s})/g$.

4.2.5 Model Predictions with the Point Mass Assumption

In order to check the effect of the mass moment of inertia of the tip mass (therefore the rotary inertia of the tip mass), two simulations are performed using the model with the point mass assumption. The mass of the tip attachment is considered in the model ($M_t = 0.239 \times 10^{-3}\,\text{kg}$); however, its mass moment of inertia is ignored ($I_t = 0$).

Figures 4.27 and 4.28, respectively, display the voltage FRFs and the tip velocity FRFs of the experimental measurements along with the model predictions including and excluding the rotary inertia of the tip mass. As can be expected, when the rotary inertia of the tip mass is neglected, the fundamental short-circuit resonance frequency is overestimated (as $340.2\,\text{Hz}$ compared to the analytically obtained value of $338.5\,\text{Hz}$ in the presence of the rotary inertia

---

[5] The data point of $0.296\,(\text{m/s})/g$ was used for the identification of modal mechanical damping.

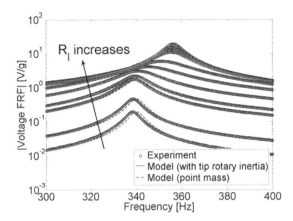

Figure 4.27 Comparison of the voltage FRFs predicted with and without considering the rotary inertia of the tip mass

and the experimental value of 338.4 Hz). Since the open-circuit resonance frequency was slightly underestimated for the case with the rotary inertia of the tip mass, it is now slightly overestimated with the point mass assumption. These data are given in Table 4.6. Once again, the frequency shift from the short-circuit to the open-circuit conditions depends on the piezoelectric and permittivity constants. Therefore the accuracy of the prediction with the point mass assumption should be checked considering the short-circuit resonance frequency. This frequency, obtained from the analytical model with the point mass assumption, overestimates the experimentally measured short-circuit resonance frequency by 0.5%. This error is reduced to 0.03% when the mass moment of inertia of the tip mass is included in the model. The effect of the mass moment of inertia can be much more important for tip mass attachments with larger dimensions.

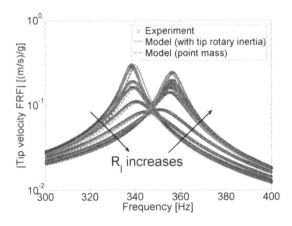

Figure 4.28 Comparison of the tip velocity FRFs predicted with and without considering the rotary inertia of the tip mass

Table 4.6 Fundamental short-circuit and open-circuit resonance frequencies of the PZT-5H bimorph cantilever with a tip mass in the presence and absence of the tip rotary inertia

Fundamental resonance frequency	Experiment	Model (with rotary inertia)	Model (with rotary inertia)
f_1^{sc}(short circuit) (Hz)	338.4	338.5	340.2
f_1^{oc}(open circuit) (Hz)	356.3	355.4	357.0

4.2.6 Performance Comparison of the PZT-5H Bimorph with and without the Tip Mass

The electrical performance comparisons of the PZT-5H bimorph cantilever with and without the tip mass are given in this section. Table 4.7 shows the detailed performance results for the maximum power output (along with the respective values of the optimum load resistance), the maximum voltage (for $R_l \to \infty$), and the maximum current (for $R_l \to 0$) outputs at the fundamental short- and open-circuit resonance frequencies. The maximum power density and specific power values are also given. It is observed that the maximum power output for resonance excitation increases by a factor of more than two with the tip mass attachment. For both resonance frequencies (short circuit and open circuit), the optimum load that gives the maximum power output increases considerably in the presence of the tip mass. As will be seen in Chapter 5, the optimum load is inversely proportional to the undamped natural frequency for resonance excitation. Therefore the increase in the optimum load due to the reduction in the resonance frequency is expected. It is worth noting that the increase in the maximum values of the current output is not as substantial as the increase in the maximum voltage. The power density calculation accounts for the additional volume of the tip mass and it exhibits an increase of about 61% (from 2.1 mW/(g^2 cm^3) to 3.4 mW/(g^2 cm^3)). The specific power calculation includes the additional mass of the tip attachment and it exhibits an increase of 59% (from 0.27 mW/μ(g^2 g) to 0.43 mW/μ(g^2 g)). For the purpose of comparing various

Table 4.7 Electrical performance comparisons of the PZT-5H bimorph cantilever with and without a tip mass of 0.239 g

	Without the tip mass	With the tip mass
f_1^{sc}(experimental) (Hz)	502.5	338.4
f_1^{oc}(experimental) (Hz)	524.7	356.3
Maximum power at f_1^{sc} (mW/g^2)	0.22	0.46
Optimum load at f_1^{sc} (kΩ)	7.6	9.7
Maximum power at f_1^{oc} (mW/g^2)	0.22	0.46
Optimum load at f_1^{oc} (kΩ)	189	331
Maximum voltage at f_1^{sc} (mV/g)	2.6	4.2
Maximum voltage at f_1^{oc} (mV/g)	12.8	24.7
Maximum current at f_1^{sc} (μA/g)	336	435
Maximum current at f_1^{oc} (μA/g)	68	75
Maximum power density (mW/(g^2 cm^3))	2.1	3.4
Maximum specific power (mW/(g^2 g))	0.27	0.43

energy harvester designs and materials, it is convenient to use these expressions normalized with respect to base acceleration (as well as the device volume and/or mass). However, it is extremely important to note that the amount of power output is completely dependent on the source of energy, that is, the frequency and the amplitude of the acceleration input. For instance, although the *resonance performance* of the configuration with the tip mass is better in Table 4.7, one should prefer the case without the tip mass for excitations around 500 Hz since the fundamental short-circuit resonance of the case without the tip mass is 502.5 Hz whereas that with the tip mass is 338.4 Hz.

It is important to recall that these FRFs are based on linear electromechanical modeling and the experiments are conducted under excitation levels less than $0.1g$. For high acceleration levels, piezoelastic nonlinearities as well as nonlinear dissipation become effective and the linear predictions tend to overestimate the experimental results. For the device tested here, the linear predictions overestimate the experimental results for excitation levels higher than a few hundreds of milli-g acceleration (see Section 4.5). For instance, the power density of $2.1 \, \text{mW}/(g^2 \, \text{cm}^3)$ in Table 4.7 accurately estimates the power output as $84 \, \mu\text{W/cm}^3$ for $0.2g$ acceleration input. However, the prediction of $8.4 \, \text{mW/cm}^3$ for $2g$ acceleration input will likely overestimate the experimental power output significantly (see Figure 4.35). Nevertheless, the valid range of the linear electromechanical model presented in Chapter 3 (up to a few hundreds of milli-g acceleration) covers the acceleration levels available in most ambient vibration energy sources [8].

4.3   PZT-5A Bimorph Cantilever

4.3.1   *Experimental Setup*

The last experimental case summarizes the analysis of a bimorph cantilever made of a different type of piezoceramic: PZT-5A. The brass-reinforced PZT-5A bimorph (T226-A4-203X) shown in Figure 4.29 is manufactured by the same company. It has the same overall geometric properties as given in Table 4.8 (except for the overhang length) and it is also manufactured for the series connection of the oppositely poled layers (Figure 3.1a). The main difference is due to the piezoceramic material. The reduced form of the typical PZT-5A properties is given in Table E.2 in Appendix E. The overhang length of the configuration shown in Figure 4.29 is 25.35 mm and the laser vibrometer measures the response at a point that is approximately

PZT-5H
bimorph
cantilever

Figure 4.29   A view of the experimental setup for the PZT-5A bimorph cantilever

Table 4.8 Geometric and material properties of the PZT-5A bimorph cantilever

	Piezoceramic (PZT-5A)	Substructure (brass)
Length (L) (mm)	25.35	25.35
Width (b) (mm)	6.4	6.4
Thickness $(h_{\bar{p}}, h_{\bar{s}})$ (mm)	0.265 (each)	0.140
Mass density $(\rho_{\bar{p}}, \rho_{\bar{s}})$ (kg/m^3)	7750	9000
Elastic modulus $(\bar{c}_{11}^E, Y_{\bar{s}})$ (GPa)	61	105
Piezoelectric constant $(\bar{e}_{31})$ (C/m^2)	-10.4	—
Permittivity constant $(\bar{\varepsilon}_{33}^S)$ (nF/m)	13.3	—

23.2 mm away from the root ($x = L_v = 23.2$ mm in Equation (4.1)). The experimental setup used and the procedure followed for the frequency response measurements are identical to those of Section 4.1.1. The same set of resistors is used (Table 4.2).

The two purposes of this section are to study a configuration with a different piezoelectric material (PZT-5A) and to investigate a wider frequency range to include a higher vibration mode (so that a demonstration comparing the single-mode and multi-mode FRFs can be presented). Therefore, the frequency bandwidth in the experiments is increased, which affects the frequency resolution (in an undesirable way) due to the limitations of the data acquisition system. For a frequency bandwidth of 0–5000 Hz, the frequency increment automatically becomes 1.5625 Hz in chirp excitation through the data acquisition system used here (which should be taken into account as a source of experimental error when comparing the model predictions).

For a resistive load of 470 Ω, the tip velocity and voltage FRFs of the bimorph cantilever are measured as shown in Figure 4.30a. The FRFs are given in semi-log scale to view the second vibration mode (which has an amplitude about an order of magnitude less compared to the first mode). The short-circuit resonance frequencies of the first and second vibration modes are read from the voltage FRF as 456.6 Hz and 2921.9 Hz, respectively. The clamp-related imperfection is still in the FRFs (around 715 Hz) but it is sufficiently far away from both vibration modes. The coherence functions of these measurements are given in Figure 4.30b. The coherence of

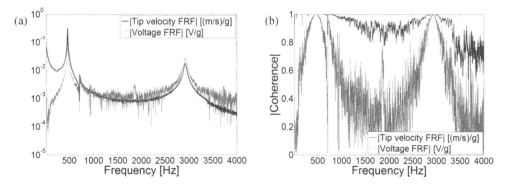

Figure 4.30 (a) Tip velocity and voltage output FRFs of the PZT-5A bimorph cantilever and (b) their coherence functions (for a load resistance of 470 Ω)

the tip velocity measurement drops around the minimum between these vibration modes and after the second mode due to low amplitude vibration response (low signal from the laser vibrometer). The poor behavior of the coherence of the voltage measurement away from the resonance frequencies is due to the low load resistance used deliberately close to the short-circuit conditions. However, the frequencies around the peak values have acceptable coherence (close to unity).

4.3.2 Validation of the Electromechanical FRFs for a Set of Resistors

When the numerical data given in Table 4.8 are used in the model, the tip velocity FRF of the 470 Ω case is predicted as shown in Figure 4.31a. The mechanical damping ratios of these two modes are identified using the model as $\zeta_1 = 0.00715$ and $\zeta_2 = 0.00740$. When these damping ratios are used in the voltage FRF expression, Figure 4.31b is obtained. The voltage amplitude of the fundamental mode is predicted with good accuracy (as 0.12 V/g compared to the experimental value of 0.13 V/g) and that of the second mode is overestimated (as 0.036 V/g compared to the experimental value of 0.028 V/g) for the mechanical damping ratios identified from the tip velocity FRF. The short-circuit resonance frequencies of the first and second vibration modes are obtained using from the model as 466.2 Hz and 2921.3 Hz, respectively. Note that the noisy behavior in Figures 4.31a and 4.31b away from the resonance frequencies corresponds to the low coherence regions in Figure 4.30.

For the set of resistors used in the experiments (Table 4.2), the voltage, current, power and tip velocity FRFs are obtained as shown in Figure 4.32. The overall analytical trends are in good agreement with the experimental measurements. The trends in Figure 4.32 (such as intersecting power curves) are similar to those of the theoretical case study with multiple vibration modes given in Section 3.7. The amplitude-wise predictions in the electrical FRFs are more accurate around the first vibration mode compared to those around the second vibration mode. The reason for the relative inaccuracy around the second mode in the voltage FRFs (and in the electrical FRFs derived from the voltage FRFs) might be possible inaccuracies in the

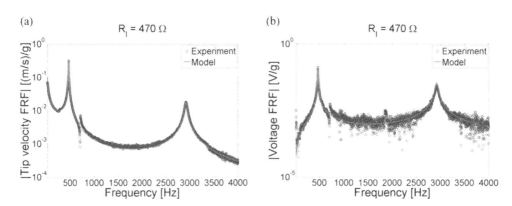

Figure 4.31 Measured and predicted (a) tip velocity and (b) voltage output FRFs of the PZT-5A bimorph cantilever for a load resistance of 470 Ω

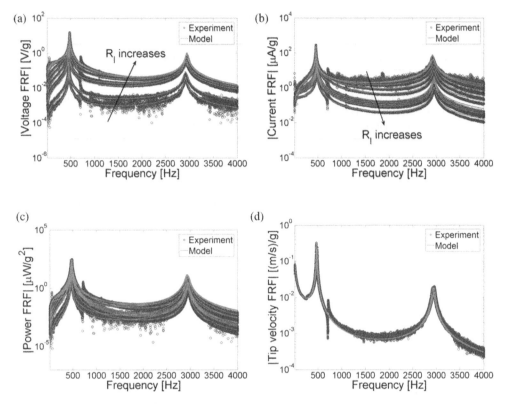

Figure 4.32 Electromechanical FRFs of the PZT-5A bimorph cantilever: (a) voltage output; (b) current output; (c) power output; and (d) tip velocity FRFs for 12 different resistive loads (ranging from 470 Ω to 995 kΩ)

piezoelectric and dielectric constants used in the model (based on the standard PZT-5A data of Appendix E.2) for representing the dynamics of this particular sample.

The frequency-wise predictions for the PZT-5A bimorph cantilever are listed in Table 4.9. The absolute value of the relative error in the model predictions is less than 0.6%, meaning that the Euler–Bernoulli theory works well even around the second natural frequency of this thin structure. It should also be noted that for the large frequency band considered here, the experimental frequency increment in the measured FRFs is relatively coarse (1.5625 Hz) for this data acquisition system, as mentioned previously.

4.3.3 Comparison of the Single-Mode and Multi-mode Electromechanical FRFs

It is useful practice to check briefly the performance of the single-mode electromechanical FRFs given by Equations (3.80) and (3.83) for modes 1 and 2. When $r = 1$ is used in the single-mode voltage FRF given by Equation (3.80), the experimental voltage behavior is predicted as shown in Figure 4.33a. Substituting $r = 2$ into the same expression gives the single-mode

Table 4.9   First two short-circuit and open-circuit resonance frequencies of the
PZT-5A bimorph cantilever

Resonance frequency	Experiment	Model	% error
f_1^{sc} (short circuit) (Hz)	465.6	466.2	+0.13
f_1^{oc} (open circuit) (Hz)	484.4	481.7	−0.56
f_2^{sc} (short circuit) (Hz)	2921.9	2921.3	−0.02
f_2^{oc} (open circuit) (Hz)	2954.7	2952.3	−0.08

curves shown in Figure 4.33b. Note that these two separate single-mode representations (for
mode 1 and mode 2 independently) are approximately valid around the respective resonance
frequencies only. The single-mode predictions shown in Figure 4.33a lack the information of
all vibration modes other than the fundamental vibration mode, whereas those illustrated in
Figure 4.33b have the information of the second vibration mode only.

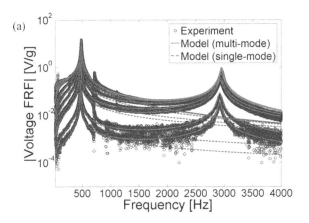

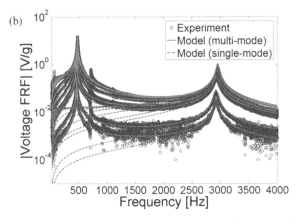

Figure 4.33   Prediction of the modal voltage frequency response using the single-mode FRFs for (a)
mode 1 and (b) mode 2 of the PZT-5A bimorph cantilever

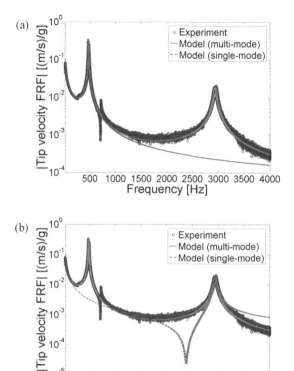

Figure 4.34 Prediction of the single-mode tip velocity frequency response using the single-mode FRFs for (a) mode 1 and (b) mode 2 of the PZT-5A bimorph cantilever

As discussed in the theoretical case study of Chapter 3, the effects of the neighboring modes are lost in the single-mode representations. The fundamental single-mode open-circuit resonance frequency is slightly overestimated as 484.3 Hz compared to the multi-mode case in Figure 4.33a. The single-mode open-circuit resonance frequency of the second mode in Figure 4.33b is also slightly overestimated compared to the multi-mode solution as 2953 Hz. In this case study, these open-circuit resonance frequencies of the single-mode prediction agree better with the experimental measurement compared to their multi-mode counterparts. However, one should note that the experimental results are not necessarily perfectly accurate. Secondly, it is very likely that the inaccuracy of the single-mode approximation (due to excluding the residuals of the neighboring modes) might be compensating for the possible inaccuracy of the frequency shift prediction (which also depends on the piezoelectric and the permittivity constants).

Single-mode predictions of the tip velocity FRFs are shown in Figure 4.34. As in the case of Figure 4.33, the single-mode FRFs given in Figure 4.34a are approximately valid around the fundamental resonance frequency only, whereas the single-mode FRFs shown in Figure 4.34b are approximations given for a narrow band around the resonance frequency of the second vibration mode only. Although the single-mode expressions are derived for $\omega \approx \omega_r$, it is clear from Figures 4.33 and 4.34 that the single-mode solution given for the fundamental vibration mode is valid over a wider frequency range around the respective mode of interest (compared to the valid range of the single-mode solution obtained for the second vibration mode). In Figures 4.33a and 4.34a, the single-mode solutions obtained for $\omega \approx \omega_1$ agree with the multi-mode solutions and the experimental results not only for $\omega \approx \omega_1$, but also for excitation frequencies over the wide range $0 < \omega < \omega_1$. The valid frequency range of the single-mode solution for mode 2 is narrower in Figures 4.33b and 4.34b because the effect of the mass-controlled region of the fundamental mode is lost in the $\omega \approx \omega_2$ approximation.

4.4 Summary

The analytical solutions derived in the previous chapter are validated for various experimental cases. The first experimental case is given for a brass-reinforced PZT-5H bimorph cantilever without a tip mass. After validating the analytical model predictions using the voltage, current, power, and tip velocity FRFs for this configuration and providing an extensive electromechanical analysis, a tip mass is attached to create a configuration for the second case study. The variations of the fundamental short-circuit and open-circuit resonance frequencies after the attachment of the tip mass are successfully predicted by the model. Performance diagrams for excitations at the fundamental short- and open-circuit resonance frequencies are extracted for both configurations (with and without the tip mass) and comparisons are made. Improvement of the overall power output (as well as the power density and the specific power) due to the addition of a tip mass is shown. The analytical model predicts the electrical and the mechanical response with very good accuracy for all values of load resistance. The effect of the rotary inertia of the tip mass is also studied by providing further analytical simulations with the point mass assumption. It is shown that the resonance frequencies can be overestimated if the rotary inertia of the tip mass is neglected. The damping effect of piezoelectric power generation is discussed in detail based on the experimental measurements and the model predictions. The final case study is given for a PZT-5A bimorph cantilever and the frequency range of interest is increased to cover the second vibration mode as well. Predictions of the multi-mode and the single-mode FRFs are provided (for the first two modes independently) and good agreement is observed with the experimental results. This chapter also provides detailed steps for conducting electromechanical modal analysis to test and compare vibration-based energy harvesters.

4.5 Chapter Notes

In order to characterize the electromechanical behavior of a piezoelectric energy harvester or to compare different energy harvester designs, performing reliable experiments is as crucial as using accurate models for design and analysis. Therefore, in addition to validation of the analytical derivations given in Chapter 3, the details given in this chapter aim to provide the basics of conducting such experiments in a lab environment.

Having observed the perfect agreement between the experimental results and the analytical predictions for various cases, it is worth recalling the fundamental assumption of the model, which is the *linear electromechanical system* assumption. If the forcing amplitude (base acceleration) exceeds a certain level, inherent piezoelastic nonlinearities as well as nonlinearities

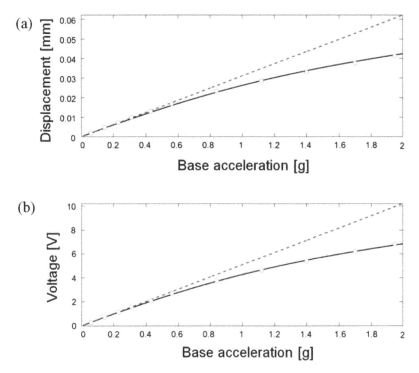

Figure 4.35 Variations of the (a) tip displacement and (b) voltage output of the brass-reinforced PZT-5H bimorph (used in Sections 4.1 and 4.2) with increasing excitation amplitude at its linear resonance frequency for a 100-kΩ load: nonlinear model (solid line); linear model (dashed line); and experimental measurements (circles) [11]

in dissipation can be pronounced and the linear predictions tend to overestimate the experimental results. For relatively flexible configurations (unlike the rather stiff samples tested in this chapter), geometric nonlinearities can also affect the results for high acceleration levels. The experimental acceleration levels used for obtaining the FRFs given in this chapter are less than $0.1g$ (which is low enough to obtain a linear electromechanical response) so that both the experimental system and the model are in the linear regime. Substantial mismatch can be observed between the model and the experimental results if a model is used for predicting the behavior of experiments that violate the modeling assumptions [9, 10].

Recently, the nonlinear piezoelastic behaviors of the PZT-5H and PZT-5A bimorphs used in this chapter have been characterized theoretically and experimentally under high excitation levels [11, 12]. Nonlinear dissipation is also observed at high acceleration levels and modeled in the same work [11,12]. An example graph is shown in Figure 4.35 for excitation at the linear resonance frequency of the PZT-5H bimorph used in Section 4.1 (for a clamping condition with a slightly different overhang length of 24.06 mm and a 100-kΩ electrical load). It is observed that the linear solution is safe to use up to a few hundreds of milli-g acceleration level and this range covers acceleration amplitudes of most ambient vibration energy sources

[8]. If one requires the response of the harvester under relatively high acceleration levels, however, it is necessary to use an appropriate nonlinear electromechanical model [11, 12].

References

1. Newland, D.E. (1993) *Random Vibrations, Spectral and Wavelet Analysis*, John Wiley & Sons, Inc., New York.
2. Bendat, J.S. and Piersol, A.G. (1986) *Random Data Analysis and Measurement Procedures*, John Wiley & Sons, Inc., New York.
3. Lu, F., Lee, H.P., and Lim, S.P. (2004) Modeling and analysis of micro piezoelectric power generators for micro-electromechanical-systems applications. *Smart Materials and Structures*, **13**, 57–63.
4. Chen, S.-N., Wang, G.-J., and Chien, M.-C. (2006) Analytical modeling of piezoelectric vibration-induced micro power generator. *Mechatronics*, **16**, 397–387.
5. Lin, J.H., Wu, X.M., Ren, T.L., and Liu, L.T. (2007) Modeling and simulation of piezoelectric MEMS energy harvesting device. *Integrated Ferroelectrics*, **95**, 128–141.
6. Ajitsaria, J., Choe, S.Y., Shen, D., and Kim, D.J. (2007) Modeling and analysis of a bimorph piezoelectric cantilever beam for voltage generation. *Smart Materials and Structures*, **16**, 447–454.
7. Meriam, J.L. and Kraige, L.G. (2001) *Engineering Mechanics: Dynamics*, John Wiley & Sons, Inc., New York.
8. Roundy, S., Wright, P.K., and Rabaey, J.M. (2003) A study of low level vibrations as a power source for wireless sensor nodes. *Computer Communications*, **26**, 1131–1144.
9. duToit, N.E. and Wardle, B.L. (2007) Experimental verification of models for microfabricated piezoelectric vibration energy harvesters. *AIAA Journal*, **45**, 1126–1137.
10. Kim, M., Hoegen, M., Dugundji, J., and Wardle, B.L. (2010) Modeling and experimental verification of proof mass effects on vibration energy harvester performance. *Smart Materials and Structures*, **19**, 045023.
11. Stanton, S.C., Erturk, A., Mann, B.P., and Inman, D.J. (2010) Nonlinear piezoelectricity in electroelastic energy harvesters: modeling and experimental identification. *Journal of Applied Physics*, **108**, 074903.
12. Stanton, S.C., Erturk, A., Mann, B.P., and Inman, D.J. (2010) Resonant manifestation of intrinsic nonlinearity within electroelastic micropower generators. *Applied Physics Letters*, **97**, 254101.

5

Dimensionless Equations, Asymptotic Analyses, and Closed-Form Relations for Parameter Identification and Optimization

Detailed mathematical analyses of the single-mode electromechanical equations are presented. The focus is placed on the voltage output and the vibration response FRFs per translational base acceleration. The single-mode relations are first expressed in the magnitude–phase form and then they are represented with dimensionless terms for convenience. The asymptotic trends of the voltage output and the tip displacement FRFs are investigated and relations are obtained for the extreme conditions of load resistance. Employing the asymptotic FRFs, the short-circuit and the open-circuit resonance frequencies of the voltage and the tip displacement FRFs are obtained. Equations of the linear voltage and vibration asymptotes are derived at these frequencies. It is mathematically shown that, for excitations at these frequencies, the intersections of the linear voltage asymptotes occur at the respective optimum load resistance. Using this result, a simple technique is introduced to identify the optimum load resistance of a piezoelectric energy harvester using a single resistive load with the open-circuit voltage measurement. Analytical relations are given also for the identification of modal mechanical damping from both the voltage and the vibration FRFs using a single data point. Vibration attenuation and amplification due to resonance frequency shift are also addressed. An experimental case is revisited to validate some of the major equations derived here. Note that the entire discussion of this chapter is given strictly for harmonic excitation at or very close to a resonance frequency.

5.1 Dimensionless Representation of the Single-Mode Electromechanical FRFs

5.1.1 Complex Forms

Recall from the equivalent representations of the single-mode electromechanical equations presented for bimorph energy harvesters given by Equations (3.79) and (3.82) that

$$\hat{v}(t) = \hat{\alpha}(\omega)\left(-\omega^2 W_0 e^{j\omega t}\right) + \hat{\mu}(\omega)\left(-\omega^2 \theta_0 e^{j\omega t}\right) \tag{5.1}$$

$$\hat{w}_{rel}(x,t) = \hat{\beta}(\omega,x)\left(-\omega^2 W_0 e^{j\omega t}\right) + \hat{\psi}(\omega,x)\left(-\omega^2 \theta_0 e^{j\omega t}\right) \tag{5.2}$$

where the respective electromechanical FRFs are as given in Table 5.1. Equations (5.1) and (5.2) are approximately valid for excitations close to the natural frequency of the rth vibration mode, that is, $\omega \approx \omega_r$ (usually the fundamental mode is of interest, hence $r = 1$). In Table 5.1, the electromechanical coupling term $\tilde{\theta}_r$ and the equivalent capacitance term $C_{\bar{p}}^{eq}$ are read from Table 3.1 for the series or parallel connection cases of the piezoceramic layers. According to Equations (5.1) and (5.2), $\hat{\alpha}(\omega)$, $\hat{\mu}(\omega)$, $\hat{\beta}(\omega,x)$, and $\hat{\psi}(\omega,x)$ are the single-mode FRFs for the voltage output to translational base acceleration, voltage output to rotational base acceleration, displacement response to translational base acceleration, and displacement response to rotational base acceleration, respectively. These expressions are the *complex forms* of the single-mode electromechanical FRFs.

For convenience, in the following derivations, the focus is placed on the FRFs for the translational acceleration input, that is, $\hat{\alpha}(\omega)$ and $\hat{\beta}(\omega,x)$, and one can easily obtain the counterparts for $\hat{\mu}(\omega)$ and $\hat{\psi}(\omega,x)$ simply by replacing σ_r with τ_r. It is important to note that the single-mode relations derived in this chapter are approximately valid only around the natural frequency (ω_r) of the respective vibration mode.

5.1.2 Magnitude–Phase Forms

The single-mode voltage FRF can be expressed in magnitude–phase form as

$$\hat{\alpha}(\omega) = |\hat{\alpha}(\omega)| e^{j\Phi(\omega)} \tag{5.3}$$

Table 5.1   Single-mode electromechanical FRFs in complex form for $\omega \approx \omega_r$

FRF	Single-mode expression in complex form
$\hat{\alpha}(\omega)$	$\dfrac{-j\omega R_l \tilde{\theta}_r \sigma_r}{\left(1 + j\omega R_l C_{\bar{p}}^{eq}\right)\left(\omega_r^2 - \omega^2 + j2\zeta_r \omega_r \omega\right) + j\omega R_l \tilde{\theta}_r^2}$
$\hat{\mu}(\omega)$	$\dfrac{-j\omega R_l \tilde{\theta}_r \tau_r}{\left(1 + j\omega R_l C_{\bar{p}}^{eq}\right)\left(\omega_r^2 - \omega^2 + j2\zeta_r \omega_r \omega\right) + j\omega R_l \tilde{\theta}_r^2}$
$\hat{\beta}(\omega,x)$	$\dfrac{\left(1 + j\omega R_l C_{\bar{p}}^{eq}\right)\sigma_r \phi_r(x)}{\left(1 + j\omega R_l C_{\bar{p}}^{eq}\right)\left(\omega_r^2 - \omega^2 + j2\zeta_r \omega_r \omega\right) + j\omega R_l \tilde{\theta}_r^2}$
$\hat{\psi}(\omega,x)$	$\dfrac{\left(1 + j\omega R_l C_{\bar{p}}^{eq}\right)\tau_r \phi_r(x)}{\left(1 + j\omega R_l C_{\bar{p}}^{eq}\right)\left(\omega_r^2 - \omega^2 + j2\zeta_r \omega_r \omega\right) + j\omega R_l \tilde{\theta}_r^2}$

where the magnitude (the modulus) of the FRF is

$$|\hat{\alpha}(\omega)| = \frac{\omega R_l \left| \tilde{\theta}_r \sigma_r \right|}{\left\{ \left[\omega_r^2 - \omega^2 \left(1 + 2R_l C_{\tilde{p}}^{eq} \omega_r \zeta_r \right) \right]^2 + \left[2\zeta_r \omega_r \omega + R_l \omega \left(C_{\tilde{p}}^{eq} \omega_r^2 - C_{\tilde{p}}^{eq} \omega^2 + \tilde{\theta}_r^2 \right) \right]^2 \right\}^{1/2}} \tag{5.4}$$

and the phase (the argument) is

$$\Phi(\omega) = -\frac{\pi}{2} \mathrm{sgn}\left(\tilde{\theta}_r \sigma_r \right) - \tan^{-1}\left(\frac{2\zeta_r \omega_r \omega + R_l \omega \left(C_{\tilde{p}}^{eq} \omega_r^2 - C_{\tilde{p}}^{eq} \omega^2 + \tilde{\theta}_r^2 \right)}{\omega_r^2 - \omega^2 \left(1 + 2R_l C_{\tilde{p}}^{eq} \omega_r \zeta_r \right)} \right) \tag{5.5}$$

where sgn stands for the signum function.

The single-mode tip displacement FRF is then

$$\hat{\beta}(\omega, x) = \left| \hat{\beta}(\omega, x) \right| e^{j\Psi(\omega, x)} \tag{5.6}$$

where its magnitude and the phase are

$$\left| \hat{\beta}(\omega, x) \right| = \frac{\left| \sigma_r \phi_r(x) \right| \left\{ 1 + \left(\omega R_l C_{\tilde{p}}^{eq} \right)^2 \right\}^{1/2}}{\left\{ \left[\omega_r^2 - \omega^2 \left(1 + 2R_l C_{\tilde{p}}^{eq} \omega_r \zeta_r \right) \right]^2 + \left[2\zeta_r \omega_r \omega + R_l \omega \left(C_{\tilde{p}}^{eq} \omega_r^2 - C_{\tilde{p}}^{eq} \omega^2 + \tilde{\theta}_r^2 \right) \right]^2 \right\}^{1/2}} \tag{5.7}$$

$$\Psi(\omega, x) = \tan^{-1}\left(\frac{\omega R_l C_{\tilde{p}}^{eq} \sigma_r \phi_r(x)}{\sigma_r \phi_r(x)} \right) - \tan^{-1}\left(\frac{2\zeta_r \omega_r \omega + R_l \omega \left(C_{\tilde{p}}^{eq} \omega_r^2 - C_{\tilde{p}}^{eq} \omega^2 + \tilde{\theta}_r^2 \right)}{\omega_r^2 - \omega^2 \left(1 + 2R_l C_{\tilde{p}}^{eq} \omega_r \zeta_r \right)} \right) \tag{5.8}$$

5.1.3 Dimensionless Forms

Some of the terms in the magnitude and the phase of the voltage FRF can be put into dimensionless form to give

$$|\hat{\alpha}(\tilde{\omega})| = \frac{\tilde{\omega} \gamma_r \upsilon_r \left| \sigma_r / \tilde{\theta}_r \right|}{\left\{ \left[1 - \tilde{\omega}^2 \left(1 + 2\upsilon_r \zeta_r \right) \right]^2 + \left[\left(2\zeta_r + \upsilon_r \left(1 + \gamma_r \right) \right) \tilde{\omega} - \upsilon_r \tilde{\omega}^3 \right]^2 \right\}^{1/2}} \tag{5.9}$$

$$\Phi(\tilde{\omega}) = -\frac{\pi}{2} \mathrm{sgn}\left(\frac{\sigma_r}{\tilde{\theta}_r} \right) - \tan^{-1}\left(\frac{\left(2\zeta_r + \upsilon_r \left(1 + \gamma_r \right) \right) \tilde{\omega} - \upsilon_r \tilde{\omega}^3}{1 - \tilde{\omega}^2 \left(1 + 2\upsilon_r \zeta_r \right)} \right) \tag{5.10}$$

Similarly, the magnitude and the phase of the tip displacement FRF become

$$|\hat{\beta}(\tilde{\omega}, x)| = \frac{\left|\tilde{f}_r(x)\right| \left\{1 + (\tilde{\omega}\upsilon_r)^2\right\}^{1/2}}{\omega_r^2 \left\{\left[1 - \tilde{\omega}^2 (1 + 2\upsilon_r\zeta_r)\right]^2 + \left[(2\zeta_r + \upsilon_r (1 + \gamma_r))\tilde{\omega} - \upsilon_r\tilde{\omega}^3\right]^2\right\}^{1/2}} \tag{5.11}$$

$$\Psi(\tilde{\omega}, x) = \tan^{-1}\left(\frac{\tilde{\omega}\upsilon_r \tilde{f}_r(x)}{\tilde{f}_r(x)}\right) - \tan^{-1}\left(\frac{(2\zeta_r + \upsilon_r (1 + \gamma_r))\tilde{\omega} - \upsilon_r\tilde{\omega}^3}{1 - \tilde{\omega}^2 (1 + 2\upsilon_r\zeta_r)}\right) \tag{5.12}$$

In Equations (5.9)–(5.12), the dimensionless terms are

$$\upsilon_r = R_l C_{\tilde{p}}^{eq} \omega_r \tag{5.13}$$

$$\gamma_r = \frac{\tilde{\theta}_r^2}{C_{\tilde{p}}^{eq} \omega_r^2} \tag{5.14}$$

$$\tilde{\omega} = \frac{\omega}{\omega_r} \tag{5.15}$$

$$\tilde{f}_r(x) = \sigma_r \phi_r(x) \tag{5.16}$$

where υ_r is the dimensionless resistance, γ_r is the dimensionless electromechanical coupling factor, $\tilde{\omega}$ is the dimensionless excitation frequency, and $\tilde{f}_r(x)$ is the dimensionless modal mechanical forcing function. Note that the voltage modulus given by Equation (5.9) is dimensional due to $\sigma_r/\tilde{\theta}_r$ (with units of $V\,s^2/m$) whereas the tip displacement modulus given by Equation (5.11) is dimensional due to $1/\omega_r^2$ (and has units of s^2).

5.2 Asymptotic Analyses and Resonance Frequencies

5.2.1 Short-Circuit and Open-Circuit Asymptotes of the Voltage FRF

Equation (5.9) can be rewritten as

$$|\hat{\alpha}(\tilde{\omega})| = \frac{\tilde{\omega}\gamma_r \left|\sigma_r/\tilde{\theta}_r\right|}{\left\{\left[(1 - \tilde{\omega}^2)/\upsilon_r - 2\tilde{\omega}^2\zeta_r\right]^2 + \left[2\zeta_r\tilde{\omega}/\upsilon_r + (1 + \gamma_r)\tilde{\omega} - \tilde{\omega}^3\right]^2\right\}^{1/2}} \tag{5.17}$$

Taking the limit as $\upsilon_r \to 0$ yields the following expression for very small values of load resistance:[1]

$$|\hat{\alpha}_{sc}(\tilde{\omega})| = \lim_{\upsilon_r \to 0} |\hat{\alpha}(\tilde{\omega})| = \frac{\tilde{\omega}\gamma_r \upsilon_r \left|\sigma_r/\tilde{\theta}_r\right|}{\left[(1 - \tilde{\omega}^2)^2 + (2\zeta_r\tilde{\omega})^2\right]^{1/2}} \tag{5.18}$$

---

[1] Note that $\upsilon_r \to 0$ implies $R_l \to 0$ (close to short-circuit conditions) and $\upsilon_r \to \infty$ implies $R_l \to \infty$ (close to open-circuit conditions) according to Equation (5.13).

The limit as $v_r \to \infty$ leads to the following relation for very large values of load resistance:

$$|\hat{\alpha}_{oc}(\tilde{\omega})| = \lim_{v_r \to \infty} |\hat{\alpha}(\tilde{\omega})| = \frac{\gamma_r \left|\sigma_r/\tilde{\theta}_r\right|}{\left\{\left[(1+\gamma_r) - \tilde{\omega}^2\right]^2 + (2\zeta_r\tilde{\omega})^2\right\}^{1/2}} \tag{5.19}$$

Here, subscripts *sc* and *oc* stand for the *short-circuit* and *open-circuit* conditions. Equations (5.18) and (5.19) represent the moduli of the voltage FRF for the extreme cases of the load resistance ($R_l \to 0$ and $R_l \to \infty$, respectively). It is useful to note that the short-circuit voltage asymptote depends on the load resistance linearly whereas the open-circuit voltage asymptote does not depend on the load resistance.[2]

5.2.2 Short-Circuit and Open-Circuit Asymptotes of the Tip Displacement FRF

The magnitude of the tip displacement FRF can be re-expressed to give

$$|\hat{\beta}(\tilde{\omega}, x)| = \frac{\left|\tilde{f}_r(x)\right|\left(1/v_r^2 + \tilde{\omega}^2\right)^{1/2}}{\omega_r^2 \left\{\left[(1-\tilde{\omega}^2)/v_r - 2\tilde{\omega}^2\zeta_r\right]^2 + \left[2\zeta_r\tilde{\omega}/v_r + (1+\gamma_r)\tilde{\omega} - \tilde{\omega}^3\right]^2\right\}^{1/2}} \tag{5.20}$$

The asymptotic FRF behaviors in the short-circuit and open-circuit conditions are

$$|\hat{\beta}_{sc}(\tilde{\omega}, x)| = \lim_{v_r \to 0} |\hat{\beta}(\tilde{\omega}, x)| = \frac{\left|\tilde{f}_r(x)\right|}{\omega_r^2 \left[(1-\tilde{\omega}^2)^2 + (2\zeta_r\tilde{\omega})^2\right]^{1/2}} \tag{5.21}$$

$$|\hat{\beta}_{oc}(\tilde{\omega}, x)| = \lim_{v_r \to \infty} |\hat{\beta}(\tilde{\omega}, x)| = \frac{\left|\tilde{f}_r(x)\right|}{\omega_r^2 \left\{\left[(1+\gamma_r) - \tilde{\omega}^2\right]^2 + (2\zeta_r\tilde{\omega})^2\right\}^{1/2}} \tag{5.22}$$

Note that the short-circuit and open-circuit asymptotes obtained for the vibration response of the beam do *not* depend on the load resistance.

---

[2] Asymptotes of the current FRFs are not discussed here but they can easily be derived from the voltage asymptotes.

5.2.3   Short-Circuit and Open-Circuit Resonance Frequencies of the Voltage FRF

Having obtained the magnitude of the voltage FRF for electrical loads close to short-circuit conditions, one can find its dimensionless resonance frequency ($\tilde{\omega}_{vsc}^{res}$) from

$$\frac{\partial \left|\hat{\alpha}_{sc}(\tilde{\omega})\right|}{\partial \tilde{\omega}}\bigg|_{\tilde{\omega}_{vsc}^{res}} = \frac{\partial}{\partial \tilde{\omega}}\left\{\frac{\tilde{\omega}\gamma_r v_r \left|\sigma_r/\tilde{\theta}_r\right|}{\left[\left(1-\tilde{\omega}^2\right)^2 + (2\zeta_r\tilde{\omega})^2\right]^{1/2}}\right\}\bigg|_{\tilde{\omega}_{vsc}^{res}} = 0 \rightarrow \tilde{\omega}_{vsc}^{res} = 1 \qquad (5.23)$$

Therefore the dimensionless resonance frequency ($\tilde{\omega}_{vsc}^{res}$) of the voltage FRF for very low values of load resistance is simply unity.

Similarly, the dimensionless resonance frequency ($\tilde{\omega}_{voc}^{res}$) of the voltage FRF for very large values of load resistance is obtained from

$$\frac{\partial \left|\hat{\alpha}_{oc}(\tilde{\omega})\right|}{\partial \tilde{\omega}}\bigg|_{\tilde{\omega}_{voc}^{res}} = \frac{\partial}{\partial \tilde{\omega}}\left\{\frac{\gamma_r \left|\sigma_r/\tilde{\theta}_r\right|}{\left\{\left[(1+\gamma_r)-\tilde{\omega}^2\right]^2 + (2\zeta_r\tilde{\omega})^2\right\}^{1/2}}\right\}\bigg|_{\tilde{\omega}_{voc}^{res}}$$

$$= 0 \rightarrow \tilde{\omega}_{voc}^{res} = \left(1+\gamma_r - 2\zeta_r^2\right)^{1/2} \qquad (5.24)$$

5.2.4   Short-Circuit and Open-Circuit Resonance Frequencies of the Tip Displacement FRF

Using the short-circuit asymptote of the tip displacement FRF, the dimensionless resonance frequency ($\tilde{\omega}_{wsc}^{res}$) of the tip displacement response for very low values of load resistance is obtained from

$$\frac{\partial \left|\hat{\beta}_{sc}(\tilde{\omega}, x)\right|}{\partial \tilde{\omega}}\bigg|_{\tilde{\omega}_{wsc}^{res}} = \frac{\partial}{\partial \tilde{\omega}}\left\{\frac{\left|\tilde{f}_r(x)\right|}{\omega_r^2\left[\left(1-\tilde{\omega}^2\right)^2 + (2\zeta_r\tilde{\omega})^2\right]^{1/2}}\right\}\bigg|_{\tilde{\omega}_{wsc}^{res}} = 0 \rightarrow \tilde{\omega}_{wsc}^{res} = \left(1-2\zeta_r^2\right)^{1/2}$$

$$(5.25)$$

The dimensionless resonance frequency ($\tilde{\omega}_{woc}^{res}$) of the tip displacement FRF for very large values of load resistance is

$$\frac{\partial \left|\hat{\beta}_{oc}(\tilde{\omega}, x)\right|}{\partial \tilde{\omega}}\bigg|_{\tilde{\omega}_{woc}^{res}} = \frac{\partial}{\partial \tilde{\omega}}\left\{\frac{\left|\tilde{f}_r(x)\right|}{\omega_r^2\left\{\left[(1+\gamma_r)-\tilde{\omega}^2\right]^2 + (2\zeta_r\tilde{\omega})^2\right\}^{1/2}}\right\}\bigg|_{\tilde{\omega}_{woc}^{res}}$$

$$= 0 \rightarrow \tilde{\omega}_{woc}^{res} = \left(1+\gamma_r - 2\zeta_r^2\right)^{1/2} \qquad (5.26)$$

5.2.5 Comparison of the Short-Circuit and Open-Circuit Resonance Frequencies

The dimensionless short-circuit resonance frequencies of the voltage and the tip displacement FRFs are *not* the same. The short-circuit resonance frequency of the voltage FRF does not depend on mechanical damping, whereas the open-circuit resonance frequencies of both FRFs depend on mechanical damping.[3]

The short-circuit and open-circuit resonance frequencies are defined based on the voltage FRF in this text. Therefore, one can express the single-mode approximations of these dimensionless frequencies as[4]

$$\tilde{\omega}_r^{sc} = 1 \tag{5.27}$$

$$\tilde{\omega}_r^{oc} = \left(1 + \gamma_r - 2\zeta_r^2\right)^{1/2} \tag{5.28}$$

The dimensional forms are then

$$\omega_r^{sc} = \omega_r \tag{5.29}$$

$$\omega_r^{oc} = \omega_r \left(1 + \gamma_r - 2\zeta_r^2\right)^{1/2} \tag{5.30}$$

Therefore the resonance frequency shift in the voltage FRF as the load resistance is increased from $R_l \to 0$ to $R_l \to \infty$ is

$$\Delta\omega = \omega_r^{oc} - \omega_r^{sc} = \omega_r \left[\left(1 + \gamma_r - 2\zeta_r^2\right)^{1/2} - 1\right] \tag{5.31}$$

According to Equation (5.31), the resonance frequency shift from the short-circuit to the open-circuit conditions is proportional to γ_r, which means from Equation (5.14) that it is directly proportional to the square of the electromechanical coupling term ($\tilde{\theta}_r^2$) and inversely proportional to the equivalent capacitance (C_p^{eq}) and square of the undamped natural frequency (ω_r^2). The resonance frequency shift is affected by the modal mechanical damping ratio (ζ_r) as well. In Equation (5.31), mechanical losses counteract the effect of electromechanical coupling. Theoretically, the open-circuit resonance frequency of the voltage FRF can be equal to its short-circuit resonance frequency for $2\zeta_r^2 = \gamma_r$. Physically, this condition implies very large mechanical losses and/or small electromechanical coupling. The discussion given here (and throughout the text for the cantilevered configurations) is, however, for *lightly damped* and *strongly coupled* energy harvesters.

---

[3] The mechanically undamped short-circuit and open-circuit natural frequencies of the voltage and tip displacement FRFs are, however, identical, that is, $\tilde{\omega}_r^{sc} = 1$ and $\tilde{\omega}_r^{oc} = (1 + \gamma_r)^{1/2}$ for $\zeta_r = 0$, which is the mechanically lossless case (these undamped forms are obtained through optimizing the power output in duToit *et al.* [1] and Kim *et al.* [2] by ignoring the mechanical damping, which were later improved by Renno *et al.* [3] considering the presence of mechanical damping for calculating the frequencies of the maximum power output).

[4] Note that these frequencies are not necessarily the frequencies of the maximum power output. They are the resonance frequencies of the voltage FRF for very small and very large values of load resistance.

5.3   Identification of Mechanical Damping

In the electromechanical system, one can identify the mechanical damping ratio either using the voltage FRF or using the vibration FRF. The following derivations provide closed-form expressions for the identification of mechanical damping at $\tilde{\omega} = \tilde{\omega}_{sc}$ in the presence of an arbitrary load resistance using both approaches (which can also be given for $\tilde{\omega} = \tilde{\omega}_r^{oc}$ in a similar fashion).

5.3.1   Identification of the Modal Mechanical Damping Ratio from the Voltage FRF

In order to identify the modal mechanical damping ratio for an arbitrary but non-zero value of v_r, one can set $\tilde{\omega} = 1$ in Equation (5.9) to obtain

$$|\hat{\alpha}(1)| = \frac{\gamma_r v_r \left|\sigma_r / \tilde{\theta}_r\right|}{\left[(2v_r \zeta_r)^2 + (2\zeta_r + v_r \gamma_r)^2\right]^{1/2}} \tag{5.32}$$

where $|\hat{\alpha}(1)|$ is known from the experimental measurement (i.e., it is the experimental data point used for damping identification). Equation (5.32) yields the following quadratic relation:

$$A\zeta_r^2 + B\zeta_r + C = 0 \tag{5.33}$$

where

$$A = 4\left(1 + v_r^2\right), \quad B = 4\gamma_r v_r, \quad C = v_r^2 \gamma_r^2 - \left(\frac{\gamma_r v_r \sigma_r}{|\hat{\alpha}(1)|\tilde{\theta}_r}\right)^2 \tag{5.34}$$

The positive root of Equation (5.33) gives the modal mechanical damping ratio as

$$\zeta_r = \frac{\left(B^2 - 4AC\right)^{1/2} - B}{2A} \tag{5.35}$$

Although v_r (the dimensionless measure of load resistance) is arbitrary in identifying the mechanical damping ratio from the mathematical point of view, physically it should be large enough so that $|\hat{\alpha}(1)|$ is a meaningful voltage measurement (i.e., not noise) with acceptable coherence (recall the discussion in Section 4.1.1).

5.3.2 Identification of the Modal Mechanical Damping Ratio from the Tip Displacement FRF

Identification of the modal mechanical damping from the vibration measurement requires the use of Equation (5.11). If the frequency of interest is again $\tilde{\omega} = 1$, Equation (5.11) becomes

$$\left|\hat{\beta}(1, x)\right| = \frac{\left|\tilde{f}_r(x)\right|\left(1 + v_r^2\right)^{1/2}}{\omega_r^2 \left[\left(2v_r\zeta_r\right)^2 + \left(2\zeta_r + v_r\gamma_r\right)^2\right]^{1/2}} \tag{5.36}$$

which yields an alternative quadratic equation of the form given by Equation (5.33), where the coefficients are

$$A = 4\left(1 + v_r^2\right), \quad B = 4\gamma_r v_r, \quad C = v_r^2\gamma_r^2 - \frac{\left|\tilde{f}_r(x)\right|^2\left(1 + v_r^2\right)}{\omega_r^4\left|\hat{\beta}(1, x)\right|^2} \tag{5.37}$$

The modal mechanical damping ratio is the positive root of Equation (5.33) for the coefficients given by Equation (5.37). Note that the value of $\left|\hat{\beta}(1, x)\right|$ at point x on the beam is the experimental data point used in the identification process (for $\tilde{\omega} = 1$). Unlike the case of mechanical damping identification using the voltage FRF, the external load resistance can be chosen very close to zero here (since such a load does not cause noise in the vibration FRF). Thus, for $v_r \to 0$ (shorting the electrodes), Equation (5.36) simplifies to the purely mechanical form of

$$\zeta_r = \frac{\left|\tilde{f}_r(x)\right|}{2\omega_r^2\left|\hat{\beta}(1, x)\right|} \tag{5.38}$$

5.4 Identification of the Optimum Electrical Load for Resonance Excitation

5.4.1 Electrical Power FRF

Using the dimensionless voltage FRF given by Equation (5.9), the electrical power FRF is obtained as

$$\left|\hat{\Pi}(\tilde{\omega})\right| = \frac{\left(\tilde{\omega}\gamma_r v_r \sigma_r / \tilde{\theta}_r\right)^2 / R_l}{\left[1 - \tilde{\omega}^2\left(1 + 2v_r\zeta_r\right)\right]^2 + \left[\left(2\zeta_r + v_r\left(1 + \gamma_r\right)\right)\tilde{\omega} - v_r\tilde{\omega}^3\right]^2} \tag{5.39}$$

For excitation at the short-circuit resonance frequency of the voltage FRF, $\tilde{\omega} = 1$, the electrical power output is

$$\left|\Pi(1)\right| = \frac{\left(\gamma_r\sigma_r / \tilde{\theta}_r\right)^2 / R_l}{\left(2\zeta_r\right)^2 + \left(2\zeta_r / v_r + \gamma_r\right)^2} \tag{5.40}$$

and for excitation at the open-circuit resonance frequency of the voltage FRF, $\tilde{\omega} = \left(1 + \gamma_r - 2\zeta_r^2\right)^{1/2}$, the electrical power output becomes

$$\left|\hat{\Pi}\left(\left(1+\gamma_r-2\zeta_r^2\right)^{1/2}\right)\right|$$

$$= \frac{\left(\gamma_r v_r \sigma_r / \bar{\theta}_r\right)^2 / R_l}{\left[1 - \left(1+\gamma_r-2\zeta_r^2\right)\left(1+2v_r\zeta_r\right)\right]^2 / \left(1+\gamma_r-2\zeta_r^2\right) + \left[2\zeta_r\left(1+\zeta_r v_r\right)\right]^2} \tag{5.41}$$

5.4.2 Optimum Values of Load Resistance at the Short-Circuit and Open-Circuit Resonance Frequencies of the Voltage FRF

Equation (5.39) can be used to obtain the optimum load resistance for the maximum electrical power output at a given excitation frequency $\tilde{\omega}$ (around the respective resonance frequency). The problem of interest is the resonance excitation and one can use Equation (5.40) for excitation at $\tilde{\omega} = 1$ to obtain

$$\left.\frac{\partial \left|\hat{\Pi}\left(1\right)\right|}{\partial R_l}\right|_{R_l^{opt,\tilde{\omega}=1}} = 0 \rightarrow R_l^{opt,\tilde{\omega}=1} = \frac{1}{\omega_r C_{\tilde{p}}^{eq}\left[1 + \left(\gamma_r/2\zeta_r\right)^2\right]^{1/2}} \tag{5.42}$$

which is the optimum load resistance for excitation at the short-circuit resonance frequency of the voltage FRF. A similar approach can be followed for estimating the optimum load resistance for excitation at the open-circuit resonance frequency of the voltage FRF as

$$\left.\frac{\partial \left|\hat{\Pi}\left(\left(1+\gamma_r-2\zeta_r^2\right)^{1/2}\right)\right|}{\partial R_l}\right|_{R_l^{opt,\tilde{\omega}=\left(1+\gamma_r-2\zeta_r^2\right)^{1/2}}} = 0 \tag{5.43}$$

$$\rightarrow R_l^{opt,\tilde{\omega}=\left(1+\gamma_r-2\zeta_r^2\right)^{1/2}} = \frac{1}{\omega_r C_{\tilde{p}}^{eq}}\left[\frac{1 - \zeta_r^2 + \left(\gamma_r/2\zeta_r\right)^2}{\left(1+\gamma_r-\zeta_r^2\right)\left(1+\gamma_r-2\zeta_r^2\right)}\right]^{1/2}$$

For excitations at the short-circuit and open-circuit resonance frequencies, the optimal values of load resistance are inversely proportional to the capacitance and the undamped natural frequency. The electromechanical coupling and the mechanical damping ratio also affect the optimal load resistance. The optimal resistive loads obtained in Equations (5.42) and (5.41) can be back substituted into Equations (5.40) and (5.41) to obtain the maximum power expressions for excitations at these two frequencies.

Recall that the short-circuit and open-circuit resonance frequencies defined here (based on the voltage FRF) are not necessarily the frequencies of maximum power generation. One can first obtain $R_l^{opt}(\tilde{\omega})$ from Equation (5.39) by setting $\partial \left|\hat{\Pi}\left(\tilde{\omega}\right)\right| / \partial R_l = 0$ and then use it in Equation (5.39) to solve $\partial \left|\hat{\Pi}\left(\tilde{\omega}\right)\right| / \partial \tilde{\omega} = 0$ for the frequencies of the maximum power output (see Renno et al. [3] for a detailed analysis of this problem based on lumped-parameter modeling for a piezo-stack).

5.5 Intersection of the Voltage Asymptotes and a Simple Technique for the Experimental Identification of the Optimum Load Resistance

5.5.1 On the Intersection of the Voltage Asymptotes for Resonance Excitation

Recall from the theoretical case study of Chapter 3 and from the experimental case studies of Chapter 4 that the asymptotes of the voltage FRF are linear for the extreme conditions of load resistance in agreement with Equations (5.18) and (5.19). This observation leads to an interesting result if one further examines these linear asymptotes at the short-circuit and open-circuit resonance frequencies.

Substituting $\tilde{\omega} = 1$ into the short-circuit voltage asymptote gives

$$|\hat{\alpha}_{sc}(1)| = \frac{\gamma_r \upsilon_r \left| \sigma_r / \tilde{\theta}_r \right|}{2\zeta_r} \tag{5.44}$$

Likewise, substituting $\tilde{\omega} = 1$ into the open-circuit voltage asymptote yields

$$|\hat{\alpha}_{oc}(1)| = \frac{\gamma_r \left| \sigma_r / \tilde{\theta}_r \right|}{\left[(1 + \gamma_r)^2 + (2\zeta_r)^2 \right]^{1/2}} \tag{5.45}$$

Clearly, these asymptotes intersect at a finite but non-zero value of the dimensionless load resistance

$$|\hat{\alpha}_{sc}(1)| = |\hat{\alpha}_{oc}(1)| \rightarrow \upsilon_r = \frac{2\zeta_r}{\left[\gamma_r^2 + (2\zeta_r)^2 \right]^{1/2}} \tag{5.46}$$

and the dimensional form of this load resistance is

$$R_l = \frac{1}{\omega_r C_p^{eq} \left[1 + (\gamma_r / 2\zeta_r) \right]^{1/2}} \tag{5.47}$$

which is nothing but the optimum load resistance for excitation at the short-circuit resonance frequency ($\tilde{\omega} = 1$) of the voltage FRF as derived in Equation (5.42).

Substituting $\tilde{\omega} = \left(1 + \gamma_r - 2\zeta_r^2 \right)^{1/2}$ into the short-circuit voltage asymptote gives

$$\left| \hat{\alpha}_{sc}\left(\left(1 + \gamma_r - 2\zeta_r^2 \right)^{1/2} \right) \right| = \frac{\left(1 + \gamma_r - 2\zeta_r^2 \right)^{1/2} \gamma_r \upsilon_r \left| \sigma_r / \tilde{\theta}_r \right|}{\left[\left(\gamma_r - 2\zeta_r^2 \right)^2 + (2\zeta_r)^2 \left(1 + \gamma_r - 2\zeta_r^2 \right) \right]^{1/2}}$$

$$= \frac{\left(1 + \gamma_r - 2\zeta_r^2 \right)^{1/2} \gamma_r \upsilon_r \left| \sigma_r / \tilde{\theta}_r \right|}{\left(\gamma_r^2 + 4\zeta_r^2 - 4\zeta_r^4 \right)^{1/2}} \tag{5.48}$$

Similarly, using $\tilde{\omega} = \left(1 + \gamma_r - 2\zeta_r^2\right)^{1/2}$ in the open-circuit voltage asymptote results in

$$\left|\hat{\alpha}_{oc}\left(\left(1 + \gamma_r - 2\zeta_r^2\right)^{1/2}\right)\right| = \frac{\gamma_r \left|\sigma_r/\tilde{\theta}_r\right|}{\left[\left(2\zeta_r^2\right)^2 + (2\zeta_r)^2\left(1 + \gamma_r - 2\zeta_r^2\right)\right]^{1/2}} = \frac{\gamma_r \left|\sigma_r/\tilde{\theta}_r\right|}{\left(4\zeta_r^2 + 4\gamma_r\zeta_r^2 - 4\zeta_r^4\right)^{1/2}}$$

(5.49)

The intersection of the voltage asymptotes for excitation at the open-circuit resonance frequency leads to

$$\left|\hat{\alpha}_{sc}\left(\left(1 + \gamma_r - 2\zeta_r^2\right)^{1/2}\right)\right| = \left|\hat{\alpha}_{oc}\left(\left(1 + \gamma_r - 2\zeta_r^2\right)^{1/2}\right)\right|$$

$$\rightarrow v_r = \left[\frac{1 - \zeta_r^2 + (\gamma_r/2\zeta_r)^2}{\left(1 + \gamma_r - \zeta_r^2\right)\left(1 + \gamma_r - 2\zeta_r^2\right)}\right]^{1/2}$$

(5.50)

which has the dimensional form of

$$R_l = \frac{1}{\omega_r C_p^{eq}}\left[\frac{1 - \zeta_r^2 + (\gamma_r/2\zeta_r)^2}{\left(1 + \gamma_r - \zeta_r^2\right)\left(1 + \gamma_r - 2\zeta_r^2\right)}\right]^{1/2}$$

(5.51)

The load resistance of the intersection point given by Equation (5.51) is the optimum load resistance for excitation at the open-circuit resonance frequency of the voltage FRF as derived in Equation (5.43).

5.5.2   A Simple Technique for the Experimental Identification of the Optimum Load Resistance

It is often required to do a resistor sweep to identify the optimum load resistance in the experiments. Either a variable resistor or a set of several resistors is used for this purpose (as done in Chapter 4). Based on the observation discussed in the previous section, a simple experimental technique for identifying the optimum load resistance using only one resistor along with an open-circuit voltage measurement is proposed here.

It is shown in the previous section that the intersections of the voltage asymptotes correspond to the respective optimum load. Suppose that the experimentalist has fixed the excitation frequency to one of these two frequencies for identifying the optimum load resistance under a certain excitation amplitude. The horizontal asymptote of the open-circuit conditions is simply obtained from the open-circuit voltage measurement (a horizontal line with an amplitude of v_{oc}). Then, a low value of resistance is chosen (R_l^*), and the respective voltage output (v^*) is measured.[5] Since the voltage output for zero load resistance (starting point of the short-circuit asymptote) is zero, one has the short-circuit asymptote, which has a slope of v^*/R_l^*. The value

---

[5] This test resistor of relatively low resistance should be on the linear asymptote. That is, it should be considerably lower than the optimum load (which is unknown). Usually, for configurations similar to the ones covered in the experimental cases of Chapter 4, a resistive load in the range of 10–100 Ω is sufficiently low. One can check with two different (low-valued) resistive loads to guarantee that the slope (v^*/R_l^*) is the same (so that the measurements are noise-free and on the linear asymptote).

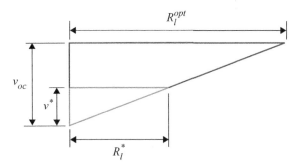

Figure 5.1 Similar triangles describing the relationship between the voltage measurement (v^*) at a low value of load resistance (R_l^*), the open-circuit voltage output (v_{oc}), and the optimum load resistance (R_l^{opt}) for excitation at the short-circuit or open-circuit resonance frequency

of load resistance for which this line intersects the open-circuit voltage amplitude is the optimum load resistance. The problem is reduced to a problem of two similar triangles as depicted in Figure 5.1. The optimum load resistance is obtained from this schematic as

$$R_l^{opt} = \frac{v_{oc}}{v^*} R_l^*$$ (5.52)

For an accurate open-circuit voltage measurement and a resistor chosen in the region of the linear asymptote, the foregoing approach provides predictions with very good accuracy. In order to check the accuracy, one can repeat the experiment for a second resistor and compare the results or do a fine tuning around the identified value.

5.6 Vibration Attenuation/Amplification from the Short-Circuit to Open-Circuit Conditions

Consider the asymptotic vibration response FRFs defined for point x on the beam given by Equations (5.21) and (5.22). The ratio of the response amplitude under open-circuit conditions to the response amplitude under short-circuit conditions at frequency $\tilde{\omega}$ is

$$\frac{\left|\hat{\beta}_{oc}(\tilde{\omega}, x)\right|}{\left|\hat{\beta}_{sc}(\tilde{\omega}, x)\right|} = \frac{\left[\left(1 - \tilde{\omega}^2\right)^2 + (2\zeta_r\tilde{\omega})^2\right]^{1/2}}{\left\{\left[(1 + \gamma_r) - \tilde{\omega}^2\right]^2 + (2\zeta_r\tilde{\omega})^2\right\}^{1/2}}$$ (5.53)

Therefore the percentage variation of the vibration amplitude as the load resistance is increased from $R_l \to 0$ to $R_l \to \infty$ is

$$\delta_w(\tilde{\omega}) = \left(\frac{\left[\left(1 - \tilde{\omega}^2\right)^2 + (2\zeta_r\tilde{\omega})^2\right]^{1/2}}{\left\{\left[(1 + \gamma_r) - \tilde{\omega}^2\right]^2 + (2\zeta_r\tilde{\omega})^2\right\}^{1/2}} - 1\right) \times 100$$ (5.54)

For excitation at the short-circuit resonance frequency of the voltage FRF, $\tilde{\omega} = 1$,

$$\frac{\left|\hat{\beta}_{oc}(1, x)\right|}{\left|\hat{\beta}_{sc}(1, x)\right|} = \frac{1}{\left[1 + (\gamma_r/2\zeta_r)^2\right]^{1/2}} \tag{5.55}$$

yielding a percentage vibration attenuation of

$$\delta_w(1) = \left[\frac{1}{\left(1 + \gamma_r^2/4\zeta_r^2\right)^{1/2}} - 1\right] \times 100 \tag{5.56}$$

Similarly, for excitation at the open-circuit resonance frequency of the voltage FRF, $\tilde{\omega} = \left(1 + \gamma_r - 2\zeta_r^2\right)^{1/2}$,

$$\frac{\left|\hat{\beta}_{oc}\left(\left(1 + \gamma_r - 2\zeta_r^2\right)^{1/2}, x\right)\right|}{\left|\hat{\beta}_{sc}\left(\left(1 + \gamma_r - 2\zeta_r^2\right)^{1/2}, x\right)\right|} = \frac{\left[\left(\gamma_r - 2\zeta_r^2\right)^2 + (2\zeta_r)^2\left(1 + \gamma_r - 2\zeta_r^2\right)\right]^{1/2}}{\left[\left(2\zeta_r^2\right)^2 + (2\zeta_r)^2\left(1 + \gamma_r - 2\zeta_r^2\right)\right]^{1/2}}$$

$$= \left[\frac{(\gamma_r/2\zeta_r)^2 + 1 - \zeta_r^2}{1 + \gamma_r - \zeta_r^2}\right]^{1/2} \tag{5.57}$$

Hence the percentage vibration amplification (assuming a lightly damped and strongly coupled system) at the open-circuit resonance frequency is

$$\delta_w\left(\left(1 + \gamma_r - 2\zeta_r^2\right)^{1/2}\right) = \left\{\left[\frac{(\gamma_r/2\zeta_r)^2 + 1 - \zeta_r^2}{1 + \gamma_r - \zeta_r^2}\right]^{1/2} - 1\right\} \times 100 \tag{5.58}$$

5.7 Experimental Validation for a PZT-5H Bimorph Cantilever

This section revisits the case study for the PZT-5H bimorph cantilever of Section 4.1 (Figure 5.2) for validation of some of the major single-mode relations derived in this chapter. Therefore the details related to this setup and the bimorph can be found in Section 4.1. Note that the following study focuses on the fundamental vibration mode ($r = 1$ in the single-mode equations).

5.7.1 Identification of Mechanical Damping

As an alternative to identifying the mechanical damping ratio graphically by matching the peaks in the vibration FRF (or by using conventional techniques such as the half-power points and the Nyquist plot [4] in frequency domain or the logarithmic decrement [5] in time domain), here the voltage FRF is used along with the closed-form expression given by Equation (5.35). For a resistive load of 470 Ω, the experimental voltage amplitude at the short-circuit resonance frequency is 0.148 V/g (where g is the gravitational acceleration: $g = 9.81$ m/s^2). Therefore the experimental data point is $|\hat{\alpha}(1)| = 0.148/9.81 = 0.0151$ V s^2/m. For this resistive load

PZT-5H
bimorph
cantilever

Small
accelerometer

Resistive
loads

Electromagnetic
shaker

Figure 5.2   PZT-5H bimorph cantilever without a tip mass under base excitation (revisited)

and the remaining system parameters, the coefficients in Equation (5.33) are $A = 4.0005$, $B = 0.004226$, and $C = -0.0003469$. For these numbers, the modal mechanical damping ratio is obtained as $\zeta_1 = 0.00880$, which is very close to the one obtained in Section 4.1 by matching the peak amplitude of the vibration FRF using the multi-mode solution ($\zeta_1 = 0.00874$).

5.7.2   Fundamental Short-Circuit and Open-Circuit Resonance Frequencies

The fundamental short-circuit resonance frequency (of the voltage FRF) is simply the undamped natural frequency of the cantilever as obtained in Equation (5.29): $\omega_1^{sc} = \omega_1$. The model predicts this frequency as $f_1^{sc} = 502.6\,\text{Hz}$ (where $f_r^{sc} = \omega_r^{sc}/2\pi$). According to Equation (5.30), the fundamental open-circuit resonance frequency depends on γ_1 in addition to ζ_1. For the given system parameters, one obtains $\gamma_1 = 0.0940$ from Equation (5.14). The fundamental open-circuit resonance frequency is then $f_1^{oc} = 525.7\,\text{Hz}$ (from $\tilde{\omega}_1^{oc} = 1.0459$). This frequency overestimates the experimental value (524.7 Hz) and the multi-mode analytical value (524.5 Hz). The reason that the single-mode solution deviates from the multi-mode solution is the effect of the residuals of the neighboring modes which are excluded in this approximation (and this is what makes the single-mode solution an approximation). The relative errors in the single-mode short-circuit and the open-circuit resonance frequencies are $+0.02\%$ and $+0.2\%$, respectively.

5.7.3   Magnitude and Phase of the Voltage FRF

The magnitude and the phase diagrams of the voltage FRFs for three different resistive loads ($1.2\,\text{k}\Omega$, $44.9\,\text{k}\Omega$, and $995\,\text{k}\Omega$) are obtained and compared against the experimental and the multi-mode model results. Figures 5.3a and 5.3b, respectively, are plotted using Equations (5.4) and (5.5) with the damping ratio identified using a single data point of the voltage FRF for $470\,\Omega$ (Section 5.7.1). The modulus expression given by Equation (5.4) successfully predicts the experimental voltage amplitude in Figure 5.3a, and the slight inaccuracy is for the largest resistive load (due to the 0.2% overestimation of the open-circuit resonance frequency). The phase diagrams of these curves are also predicted very well for all of these three resistive loads in Figure 5.3b. It should be noted that these phase curves intersect each other at the fundamental short-circuit and open-circuit resonance frequencies.

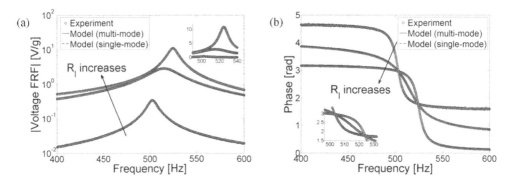

Figure 5.3 (a) Amplitude and (b) phase diagrams of the voltage FRF for three different resistive loads: 1.2 kΩ, 44.9 kΩ, and 995 kΩ

5.7.4 Voltage Asymptotes for Resonance Excitation

For excitation at the fundamental short-circuit resonance frequency, the linear short-circuit and open-circuit voltage asymptotes are obtained using Equations (5.44) and (5.45), respectively. Variation of the single-mode voltage output with load resistance predicted using $\tilde{\omega} = 1$ in Equation (5.17) is plotted in Figure 5.4a along with the linear asymptotes. The prediction of the single-mode expressions agrees very well with the multi-mode solution and the experimental data. The short-circuit and open-circuit asymptotes successfully represent the limiting trends as $R_l \to 0$ and $R_l \to \infty$. Equation (5.17) is plotted versus load resistance for $\tilde{\omega} = 1.0459$ in Figure 5.4b. The resulting single-mode curve exhibits good agreement with the multi-mode solution and the experimental data points. Both the single-mode and multi-mode solutions as well as the experimental data points follow the linear single-mode asymptotes (obtained from Equations (5.48) and (5.49)) closely for $R_l \to 0$ and $R_l \to \infty$.

The accuracy of the single-mode prediction is better in Figure 5.4a compared to its accuracy in Figure 5.4b (because the single-mode prediction of the open-circuit resonance is slightly

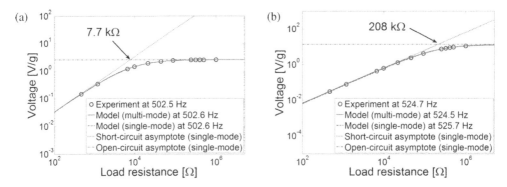

Figure 5.4 Voltage versus load resistance diagrams for excitations (a) at the short-circuit resonance frequency and (b) at the open-circuit resonance frequency along with the linear asymptotes for $R_l \to 0$ and $R_l \to \infty$

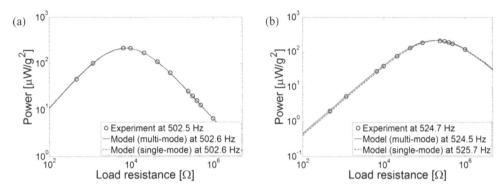

Figure 5.5 Power versus load resistance diagrams for excitations (a) at the short-circuit resonance frequency and (b) at the open-circuit resonance frequency of the voltage FRF

less accurate). The values of the load resistance at the intersections of the short-circuit and open-circuit asymptotes in Figures 5.4a and 5.4b are 7.7 kΩ and 208 kΩ, respectively. These are expected to be the single-mode estimates of the optimum load resistance for excitations at the fundamental short-circuit and open-circuit resonance frequencies of the voltage FRF as mathematically shown in Section 5.5.1.

5.7.5　Power vs. Load Resistance Diagrams and the Optimum Loads

Variation of the single-mode power output with load resistance for excitation at the fundamental short-circuit resonance frequency is obtained by using $\tilde{\omega} = 1$ in Equation (5.39). Similarly, substituting $\tilde{\omega} = 1.0459$ into the same equation gives the single-mode power prediction at the fundamental open-circuit resonance frequency. These predictions are plotted in Figures 5.5a and 5.5b, respectively, and both of them exhibit good agreement with the multi-mode solutions and the experimental data points. The slight inaccuracy in Figure 5.5b is barely noticeable since the single-mode predictions around the open-circuit resonance are less accurate.

The optimum values of load resistance for excitations at the short-circuit and open-circuit resonance frequencies are calculated by using Equations (5.42) and (5.43) as 7.7 kΩ and 208 kΩ, respectively. Therefore, for excitation at the fundamental short-circuit resonance frequency, the single-mode estimate of the optimum load resistance is 7.7 kΩ, whereas for excitation at the fundamental open-circuit resonance frequency, the single-mode estimate of the optimum load resistance is 208 kΩ (which simply correspond to the intersection points of the voltage asymptotes in Figure 5.4). These values overestimate the multi-mode estimates of 7.6 kΩ and 189 kΩ, respectively (obtained in Section 4.1.3).

5.7.6　Comment on the Optimum Load Resistance Obtained from the Simplified Circuit Representations of a Piezoceramic Layer

A frequently quoted expression [6,7] for the optimum load resistance ($R_l^{opt} = 1/\omega_r C_{\tilde{p}}^{eq}$) gives the optimum values of load resistance for this cantilever at the fundamental short-circuit and open-circuit resonance frequencies as $41.8\,\text{k}\Omega$ and $40.1\,\text{k}\Omega$ (which are highly inaccurate compared to multi-mode predictions of $7.6\,\text{k}\Omega$ and $189\,\text{k}\Omega$ or the single-mode predictions of $7.7\,\text{k}\Omega$ and $208\,\text{k}\Omega$, respectively). The reason for this inaccuracy is the thermodynamic inconsistency behind the equation $R_l^{opt} = 1/\omega_r C_{\bar{p}}^{eq}$.[6] This expression is obtained for the simple electrical engineering representation of a piezoceramic layer as an electrical source: either a *current source in parallel with its internal capacitance* or a *voltage source in series with its internal capacitance*. One should note that the former representation was obtained in the derivation steps of the analytical model (Figure 3.2b) for a *dependent* current source. In Figure 3.2b, if one assumes a constant amplitude oscillating at frequency ω_r (i.e., $i_p(t) = I_p e^{j\omega_r t}$ where I_p is independent) and proceeds to obtain the optimum load for maximum power generation using the circuit equation, that load turns out to be $R_l^{opt} = 1/\omega_r C_{\bar{p}}^{eq}$. However, the electric current in Equation (3.28) is a function of the velocity response of the structure due to the second expression in Equations (3.29). Therefore, the current source in this representation does *not* have a constant amplitude and it strongly depends on the load resistance because the vibration response strongly depends on the load resistance as shown both theoretically and experimentally in Chapters 3 and 4. Therefore, using $R_l^{opt} = 1/\omega_r C_{\bar{p}}^{eq}$ implies that there is no shunt damping effect in the structure; that is, there is no converse piezoelectric effect, although one generates electricity with the direct piezoelectric effect. A figure demonstrating the meaning of ignoring piezoelectric coupling in the mechanical equation is shown in Figure 5.6 for excitation at the fundamental short-circuit resonance frequency of the cantilever discussed here. As the load resistance is changed from the smallest load to the largest load, the correct peak of the power in the FRF increases, moves to the right (on the frequency axis), and then decreases (Figure 5.6a). This trend is completely due to the feedback received in the mechanical domain. If the electrical term in the mechanical domain is artificially set equal to zero to simulate the aforementioned inconsistent scenario, the dashed curves in Figures 5.6a and 5.6b are obtained. Therefore, if the piezoelectric coupling in the mechanical equation is ignored, the power amplitude in Figure 5.6a increases much more than the real case (and there is no frequency shift). Indeed, when the variation of the peak power with load resistance is plotted as shown in Figure 5.6b, the optimum load resistance turns out to be the incorrect estimate ($R_l^{opt} = 1/\omega_r C_{\bar{p}}^{eq}$) and the peak power is orders of magnitude larger than the experimental measurement as well as the model prediction with backward coupling in the mechanical equation.

5.8 Summary

In this chapter, mathematical analyses of the distributed-parameter electromechanical equations are presented for parameter identification and optimization. The focus is placed on the voltage output and the vibration response FRFs per translational base acceleration input. The single-mode relations derived in the complex form in Chapter 3 (based on the multi-mode solutions) are first expressed in the magnitude–phase form and then they are represented by dimensionless terms. After expressing the asymptotic trends of the single voltage and tip

---

[6] This expression provides a good approximation in limited cases where the effect of piezoelectric coupling on the mechanical response is negligible (e.g., consider the problem of generating electricity by attaching a small piezoelectric patch to a large civil engineering structure to utilize surface strain fluctuations – a problem that is addressed in Section 7.5).

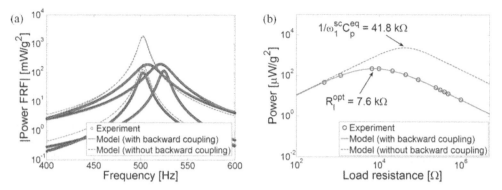

Figure 5.6 Comparison of the coupled and uncoupled distributed-parameter model predictions: (a) electrical power FRFs for four different resistive loads and (b) variation of the electrical power amplitude with load resistance for resonance excitation

displacement FRFs, closed-form expressions are obtained for their short-circuit and open-circuit resonance frequencies accounting for the mechanical losses. It is shown that the short-circuit resonance frequencies of the voltage FRF and the tip displacement FRF are slightly different due to the presence of mechanical damping. The linear asymptotes of the voltage and tip displacement FRFs for the extreme conditions of the load resistance (observed in the previous chapters) are mathematically verified here. Closed-form expressions for the optimum electrical loads of the maximum power output at the short-circuit and the open-circuit resonance frequencies of the voltage FRF are extracted. It is shown that, for excitations at these two frequencies, intersections of the linear voltage asymptotes correspond to the respective optimum load resistance values. Based on this observation, a simple technique is introduced to identify the optimum load resistance of a piezoelectric energy harvester using a single resistive load along with an open-circuit voltage measurement. As an alternative to conventional techniques of damping identification, closed-form expressions are given for the identification of modal mechanical damping either using the voltage FRF or using the tip displacement FRF. A single data point of the voltage FRF is successfully used for identification of the modal mechanical damping ratio. This approach can be preferred for identifying mechanical damping especially for micro-scale cantilevers in the absence of vibration testing equipment (such as a laser vibrometer). Analytical relations are given also to estimate the variation of the vibration response as the load resistance is changed between its two extrema. The experimental case study of a PZT-5H bimorph is revisited and the major closed-form relations derived here are validated. An issue related to estimation of the optimum load resistance using the simplified circuit representations of the piezoceramic is addressed.

5.9 Chapter Notes

The single-mode analytical derivations given here can be used for parameter identification and optimization for modal vibrations in piezoelectric energy harvesting. Obviously not all the possible derivations are given in this chapter, but the reader can use the resulting expressions to optimize or identify the system parameter(s) of interest.

Although the equations are derived for excitations close to the natural frequency of the rth mode ($\omega \approx \omega_r$), in practice the mode of interest is usually the fundamental vibration mode, that is, $r = 1$ (hence $\omega \approx \omega_1$ is the main concern). One can recall from Figures 4.33 and 4.34 that while the single-mode expressions are derived for $\omega \approx \omega_r$, the single-mode solution for the fundamental mode is valid over a wider frequency range compared to the valid frequency range of the single-mode solutions for higher modes. Indeed the single-mode approximation $\omega \approx \omega_1$ can successfully represent the electromechanical response over the range $0 < \omega < \omega_1$ as well (see Section 4.3.3). Therefore, as far as the fundamental vibration mode is concerned ($r = 1$), the expressions derived in this chapter are valid not only around the fundamental natural frequency, but also at several other frequencies less than the fundamental natural frequency.

Early derivations for the optimum conditions can be found in duToit *et al.* [1] based on the lumped-parameter modeling of a piezo-stack configuration in longitudinal vibrations (see Figure 2.15), where the *resonance* and *anti-resonance* frequencies are obtained by optimizing the power FRF. However, the presence of mechanical damping is ignored in duToit *et al.* [1] and later in Kim *et al.* [2]. As a result, expressions similar to the undamped form of the short-circuit and open-circuit resonance frequencies of the voltage FRF (given in Section 5.2.5) are obtained. For the same lumped-parameter configuration, Renno *et al.* [3] present a rigorous analysis by keeping the mechanical damping ratio in optimizing the power output. They [3] introduce bifurcations diagrams versus mechanical damping and show the coalescence of two frequencies of the maximum power output for large mechanical damping. Therefore the short-circuit and open-circuit resonance frequencies of the voltage FRF are the frequencies of maximum power output only for sufficiently light mechanical damping and strong electromechanical coupling (which is often the case for a well-designed energy harvester, as in the experimental results of Chapter 4). An analysis similar to the one given by Renno *et al.* [3] must be conducted for the case with heavy mechanical damping. Other than these efforts on the AC problem, the reader is referred to Shu and Lien [8] for parameter optimization in the presence of an AC–DC converter.

References

1. duToit, N.E., Wardle, B.L., and Kim, S. (2005) Design considerations for MEMS-scale piezoelectric mechanical vibration energy harvesters. *Journal of Integrated Ferroelectrics*, **71**, 121–160.
2. Kim, M., Hoegen, M., Dugundji, J., and Wardle, B.L. (2010) Modeling and experimental verification of proof mass effects on vibration energy harvester performance. *Smart Materials and Structures*, **19**, 045023.
3. Renno, J.M., Daqaq, F.M., and Inman, D.J. (2009) On the optimal energy harvesting from a vibration source. *Journal of Sound and Vibration*, **320**, 386–405.
4. Ewins, D.J. (2000) *Modal Testing: Theory, Practice and Application*, Research Studies Press, Baldock, Hertford-shire.
5. Clough, R.W. and Penzien, J. (1975) *Dynamics of Structures*, John Wiley & Sons, Inc., New York.
6. Lu, F., Lee, H.P., and Lim, S.P. (2004) Modeling and analysis of micro piezoelectric power generators for micro-electromechanical-systems applications. *Smart Materials and Structures*, **13**, 57–63.
7. Beeby, S.P., Tudor, M.J., and White, N.M. (2006) Energy harvesting vibration sources for microsystems applications. *Measurement Science and Technology*, **17**, R175–R195.
8. Shu, Y.C. and Lien, I.C. (2006) Analysis of power outputs for piezoelectric energy harvesting systems. *Smart Materials and Structures*, **15**, 1499–1502.

6

Approximate Analytical Distributed-Parameter Electromechanical Modeling of Cantilevered Piezoelectric Energy Harvesters

This chapter covers approximate analytical distributed-parameter modeling of cantilevered piezoelectric energy harvesters for which the analytical solution is either more involved or does not exist (such as configurations with varying cross-section and/or material properties in the longitudinal direction). The technique used here is an electromechanical version of the assumed-modes method, which is based on the extended Hamilton's principle for electromechanical continua. After obtaining the distributed-parameter energy expressions, the extended Hamilton's principle is used in order to derive the spatially discretized Lagrange's equations with electromechanical coupling. An axial displacement variable is introduced to account for the coupling between the axial and transverse displacement variables due to asymmetric laminates and the focus is placed on the unimorph configuration. The derivations are given in this chapter under three sections: (1) the Euler–Bernoulli formulation, (2) the Rayleigh formulation, and (3) the Timoshenko formulation. The first type of formulation neglects the rotary inertia and the transverse shear deformation effects, whereas the second type includes the rotary inertia effect but neglects the transverse shear deformation. The third solution type accounts for the influences of both the transverse shear deformation and rotary inertia on the resulting electromechanical behavior. Therefore the Euler–Bernoulli solution is valid for thin beams, whereas the Timoshenko solution is valid for moderately thick beams (and the Rayleigh solution lies in between). Experimental validations are given for the thin-beam case by comparing the assumed-modes predictions to the experimental and analytical results for different numbers of modes. An experimental case study for a two-segment cantilever is also summarized.

Piezoelectric Energy Harvesting, First Edition. Alper Erturk and Daniel J. Inman.
© 2011 John Wiley & Sons, Ltd. Published 2011 by John Wiley & Sons, Ltd.

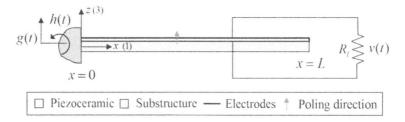

Figure 6.1 Unimorph piezoelectric energy harvester with varying cross-section

6.1   Unimorph Piezoelectric Energy Harvester Configuration

Consider the piezoelectric energy harvester configuration shown in Figure 6.1. The configuration has a single piezoceramic layer bonded onto a substructure layer, therefore it is a unimorph cantilever. The perfectly conductive electrodes (of negligible thickness) fully cover the respective (upper and lower) faces of the piezoceramic and they are connected to a resistive electrical load. The layers are perfectly bonded to each other so there is no relative sliding at the interface. The purpose is to find the voltage output of the piezoceramic ($v(t)$) across the load resistance in response to imposed base excitation. As in the analytical derivations of Chapter 3 (where the analytical solutions for symmetric bimorph configurations are given), the base motion is represented as translation in the transverse direction with superimposed small rotation. Linear material behavior is assumed for the piezoceramic and the substructure layers. Note that the x-, y-, and z-directions, respectively, are coincident with the 1-, 2-, and 3-directions of piezoelectricity (the y-direction is into the page). These directional subscripts are used interchangeably as the former is preferred for mechanical derivations, whereas the latter is used in the piezoelectric constitutive relations (Appendix A). Note that the following Euler–Bernoulli, Rayleigh, and Timoshenko types of derivations based on the assumed-modes method[1] can handle the modeling of thin and moderately thick beams with varying cross-section and changing material properties in the x-direction (as long as the cross-section is symmetric with respect to the xz-plane since it is assumed that there is no torsion). The coupling between the axial and the transverse displacement components due to structural asymmetry is taken into consideration in all cases for completeness. The harvester beam shown in Figure 6.1 has no tip mass. The effect of a tip mass (with non-zero mass moment of inertia) on the following formulation is discussed after the basic derivations (in Section 6.6). The derivation given in the following is for a unimorph (Figure 6.1) and the general expressions presented here can be used for the modeling of bimorphs as well as multimorphs with symmetric and asymmetric laminates.

---

[1] As discussed by Meirovitch [1], the assumed-modes method for distributed-parameter systems is closely related to the Rayleigh–Ritz method [2–5]. Indeed, the resulting discretized equations obtained in both methods are identical for the same *admissible functions*. The difference is that the Rayleigh–Ritz method is concerned with spatial discretization of the differential eigenvalue problem whereas the assumed-modes method directly begins with spatial discretization of the boundary-value problem in an implicit manner.

6.2 Electromechanical Euler–Bernoulli Model with Axial Deformations

6.2.1 Distributed-Parameter Electromechanical Energy Formulation

The effective base displacement acting on the beam is given by Equation (3.1) as

$$w_b(x, t) = g(t) + xh(t) \tag{6.1}$$

where $g(t)$ is the translation in the transverse direction and $h(t)$ is the superimposed small rotation.

The non-zero displacement field in the beam relative to the moving base is given by

$$u(x, z, t) = u^0(x, t) - z\frac{\partial w^0(x, t)}{\partial x} \tag{6.2}$$

$$w(x, t) = w^0(x, t) \tag{6.3}$$

where $u^0(x, t)$ and $w^0(x, t)$ are the axial displacement and the transverse displacement of the neutral axis at point x and time t relative to the moving base.[2]

The vector form of the displacement field is

$$\mathbf{u} = \left[u^0(x, t) - z\frac{\partial w^0(x, t)}{\partial x} \quad 0 \quad w^0(x, t) \right]^t \tag{6.4}$$

where the superscript t stands for the transpose (otherwise it is time). From this displacement field, the only non-zero strain component can be expressed as

$$S_{xx}(x, z, t) = \frac{\partial u(x, z, t)}{\partial x} = \frac{\partial u^0(x, t)}{\partial x} - z\frac{\partial^2 w^0(x, t)}{\partial x^2} \tag{6.5}$$

The total potential energy in the structure is

$$U = \frac{1}{2}\left(\int_{V_s} \mathbf{S}^t\mathbf{T}\,dV_s + \int_{V_p} \mathbf{S}^t\mathbf{T}\,dV_p \right) \tag{6.6}$$

where $\mathbf{S}$ is the vector of engineering strain components, $\mathbf{T}$ is the vector of engineering stress components, the subscripts s and p stand for substructure and piezoceramic, respectively, and the integrations are performed over the volume (V) of the respective material.

The isotropic substructure obeys Hooke's law (Appendix F.2)

$$T_{xx}(x, z, t) = Y_s S_{xx}(x, z, t) \tag{6.7}$$

where Y_s is the elastic modulus of the substructure layer.

---

[2] The subscript *rel* used in Chapters 2 and 3 to denote *relative motion* is removed here to avoid notational complexity. These terms are defined relative to the moving base throughout this chapter.

The constitutive equation for the stress component in the piezoceramic is (Appendix A.2)

$$T_{xx}(x, z, t) = T_1 = \bar{c}_{11}^E S_1 - \bar{e}_{31} E_3 = \bar{c}_{11}^E S_{xx}(x, z, t) + \bar{e}_{31} \frac{v(t)}{h_p} \tag{6.8}$$

Here, the electric field is given in terms of the voltage output (i.e., $E_3(t) = -v(t)/h_p$ where $v(t)$ is the voltage across the electrodes and h_p is the thickness of the piezoceramic) and the reduced elastic modulus, piezoelectric stress constant, and permittivity constant expressions are $\bar{c}_{11}^E = 1/s_{11}^E$, $\bar{e}_{31} = d_{31}/s_{11}^E$, and $\bar{\varepsilon}_{33}^S = \varepsilon_{33}^T - d_{31}^2/s_{11}^E$, respectively (Appendix A.2).

Equation (6.6) becomes

$$U = \frac{1}{2} \int_{V_s} Y_s \left(\frac{\partial u^0(x, t)}{\partial x} - z \frac{\partial^2 w^0(x, t)}{\partial x^2} \right)^2 dV_s$$

$$+ \frac{1}{2} \int_{V_p} \left[\bar{c}_{11}^E \left(\frac{\partial u^0(x, t)}{\partial x} - z \frac{\partial^2 w^0(x, t)}{\partial x^2} \right)^2 + \bar{e}_{31} \frac{v(t)}{h_p} \left(\frac{\partial u^0(x, t)}{\partial x} - z \frac{\partial^2 w^0(x, t)}{\partial x^2} \right) \right] dV_p$$

$$\tag{6.9}$$

The total kinetic energy of the system can be given by

$$T = \frac{1}{2} \left(\int_{V_s} \rho_s \frac{\partial \mathbf{u}_m^t}{\partial t} \frac{\partial \mathbf{u}_m}{\partial t} dV_s + \int_{V_p} \rho_p \frac{\partial \mathbf{u}_m^t}{\partial t} \frac{\partial \mathbf{u}_m}{\partial t} dV_p \right) \tag{6.10}$$

where ρ_s and ρ_p are the mass densities of the substructure and piezoceramic layers, respectively, and $\mathbf{u}_m$ is the modified displacement vector that is the superposition of the base displacement input given by Equation (6.1) and the displacement vector $\mathbf{u}$:[3]

$$\mathbf{u}_m = \left[u^0(x, t) - z \frac{\partial w^0(x, t)}{\partial x} \quad 0 \quad w^0(x, t) + w_b(x, t) \right]^t \tag{6.11}$$

Then the total kinetic energy becomes

$$T = \frac{1}{2} \int_{V_s} \rho_s \left[\left(\frac{\partial u^0(x, t)}{\partial t} - z \frac{\partial^2 w^0(x, t)}{\partial t \partial x} \right)^2 + \left(\frac{\partial w^0(x, t)}{\partial t} + \frac{\partial w_b(x, t)}{\partial t} \right)^2 \right] dV_s$$

$$+ \frac{1}{2} \int_{V_p} \rho_p \left[\left(\frac{\partial u^0(x, t)}{\partial t} - z \frac{\partial^2 w^0(x, t)}{\partial t \partial x} \right)^2 + \left(\frac{\partial w^0(x, t)}{\partial t} + \frac{\partial w_b(x, t)}{\partial t} \right)^2 \right] dV_p \tag{6.12}$$

---

[3] Note that the external forcing on the structure is due to the base motion. Since the total kinetic energy is written for the velocity field of the beam relative to the absolute reference frame, the forcing function on the beam (in terms of $w_b(x, t)$) will emerge from this kinetic energy expression. Alternatively, one could express the total kinetic energy relative to the moving base and include the (virtual) work done by the excitation term (due to base displacement $w_b(x, t)$) as a non-conservative effect.

The total potential energy can be re-expressed as follows:

$$
\begin{aligned}
U = \frac{1}{2} \int_0^L \Bigg\{ & Y_s \Bigg[A_s \left(\frac{\partial u^0(x,t)}{\partial x} \right)^2 + I_s \left(\frac{\partial^2 w^0(x,t)}{\partial x^2} \right)^2 - 2H_s \frac{\partial u^0(x,t)}{\partial x} \frac{\partial^2 w^0(x,t)}{\partial x^2} \Bigg] \\
& + \bar{c}_{11}^E \Bigg[A_p \left(\frac{\partial u^0(x,t)}{\partial x} \right)^2 + I_p \left(\frac{\partial^2 w^0(x,t)}{\partial x^2} \right)^2 - 2H_p \frac{\partial u^0(x,t)}{\partial x} \frac{\partial^2 w^0(x,t)}{\partial x^2} \Bigg] \\
& + B_p v(t) \frac{\partial u^0(x,t)}{\partial x} - J_p v(t) \frac{\partial^2 w^0(x,t)}{\partial x^2} \Bigg\} dx
\end{aligned}
\tag{6.13}
$$

The total kinetic energy in the expanded form is

$$
\begin{aligned}
T = \frac{1}{2} \int_0^L \Bigg\{ & \rho_s \Bigg[A_s \left(\frac{\partial u^0(x,t)}{\partial t} \right)^2 + A_s \left(\frac{\partial w^0(x,t)}{\partial t} \right)^2 + 2A_s \frac{\partial w^0(x,t)}{\partial t} \frac{\partial w_b(x,t)}{\partial t} \\
& + A_s \left(\frac{\partial w_b(x,t)}{\partial t} \right)^2 - 2H_s \frac{\partial u^0(x,t)}{\partial t} \frac{\partial^2 w^0(x,t)}{\partial t \partial x} \Bigg] \\
& + \rho_p \Bigg[A_p \left(\frac{\partial u^0(x,t)}{\partial t} \right)^2 + A_p \left(\frac{\partial w^0(x,t)}{\partial t} \right)^2 + 2A_p \frac{\partial w^0(x,t)}{\partial t} \frac{\partial w_b(x,t)}{\partial t} \\
& + A_p \left(\frac{\partial w_b(x,t)}{\partial t} \right)^2 - 2H_p \frac{\partial u^0(x,t)}{\partial t} \frac{\partial^2 w^0(x,t)}{\partial t \partial x} \Bigg] \Bigg\} dx
\end{aligned}
\tag{6.14}
$$

where the kinetic energy contributions from the rotary inertia are neglected but the terms which might cause coupling between the axial and the transverse displacement components due to asymmetric laminates are kept for completeness. Here, the zeroth, first, and second moments of area for the substructure and piezoceramic cross-sections at an arbitrary point x are

$$
(A_s, H_s, I_s) = \iint_s (1, z, z^2) \, dy \, dz
\tag{6.15}
$$

$$
(A_p, H_p, I_p) = \iint_p (1, z, z^2) \, dy \, dz
\tag{6.16}
$$

where H_s and H_p vanish for a structure that is symmetric with respect to the neutral axis of the beam (just like a symmetric bimorph or any symmetric multimorph). Therefore, for a symmetric structure, there is no coupling between $u^0(x,t)$ and $w^0(x,t)$. As a result, in a symmetric structure, the axial displacement $u^0(x,t)$ cannot be excited by the imposed base displacement $w_b(x,t)$, which simplifies the problem (and allows analytical solutions as in

Chapter 3).[4] It is worth adding that, in the foregoing expressions as well as in the following, $A_s = A_s(x)$, $A_p = A_p(x)$, $H_s = H_s(x)$, $H_p = H_p(x)$, $I_s = I_s(x)$, and $I_p = I_p(x)$ are allowed.

The terms related to piezoelectric coupling are

$$B_p = \iint_p \frac{\bar{e}_{31}}{h_p} \, dy \, dz \tag{6.17}$$

$$J_p = \iint_p \frac{\bar{e}_{31}}{h_p} z \, dy \, dz \tag{6.18}$$

where B_p couples the voltage and the extension component and J_p couples the voltage and the curvature component according to Equation (6.13).

The internal electrical energy in the piezoceramic layer is

$$W_{ie} = \frac{1}{2} \int_{V_p} \mathbf{E}' \mathbf{D} \, dV_p \tag{6.19}$$

where $\mathbf{E}$ is the vector of electric field components and $\mathbf{D}$ is the vector of electric displacement components. Substituting the respective terms (the non-zero electric field and the electric displacement components from Appendix A.2) gives

$$W_{ie} = -\frac{1}{2} \int_{V_p} \frac{v(t)}{h_p} \left[\bar{e}_{31} \left(\frac{\partial u^0(x,t)}{\partial x} - z \frac{\partial^2 w^0(x,t)}{\partial x^2} \right) - \bar{\varepsilon}_{33}^S \frac{v(t)}{h_p} \right] dV_p$$

$$= -\frac{1}{2} \int_0^L \left(B_p v(t) \frac{\partial u^0(x,t)}{\partial x} - J_p v(t) \frac{\partial^2 w^0(x,t)}{\partial x^2} \right) dx + \frac{1}{2} C_p v^2(t) \tag{6.20}$$

Here, C_p is the internal capacitance of the piezoceramic given by

$$C_p = \bar{\varepsilon}_{33}^S \frac{A_p}{h_p} \tag{6.21}$$

where A_p is the electrode area.

In order to account for mechanical damping, Rayleigh's dissipation function [1] can be used. Alternatively, a damping matrix proportional to the resulting mass and stiffness matrices can be introduced later (i.e., after the system is spatially discretized in the assumed-modes solution procedure). The latter approach is preferred here for simplicity.

---

[4] For the unimorph configuration, therefore, an analytical solution of the form given in Chapter 3 requires neglecting the contribution of the axial displacement $u^0(x,t)$ to the mechanical strain field in Equation (6.5) as a first approximation [6].

In the absence of mechanical dissipative effects, the extended Hamilton's principle with internal electrical energy is

$$\int_{t_1}^{t_2} (\delta T - \delta U + \delta W_{ie} + \delta W_{nc}) \, dt = 0 \tag{6.22}$$

where δT, δU, and δW_{ie} are the first variations [7] of the total kinetic energy, the total potential energy, and the internal electrical energy while δW_{nc} is the virtual work of the non-conservative mechanical force and electric charge components. Since the effect of base excitation is considered in the total kinetic energy term and the mechanical damping effect is to be introduced later, the only non-conservative virtual work is due to the electric charge output ($Q(t)$):

$$\delta W_{nc} = \delta W_{nce} = Q(t)\delta v(t) \tag{6.23}$$

Following the regular procedure of the assumed-modes method [1], the next step is to discretize the components of the extended Hamilton's principle, which are the total kinetic energy, the total potential energy, the internal electrical energy, and the virtual work of the non-conservative electric charge.

6.2.2 Spatial Discretization of the Energy Equations

The distributed-parameter variables in the mechanical domain are $w^0(x, t)$ and $u^0(x, t)$ while the electrical variable is $v(t)$. Let the following two finite series expansions represent the two components of vibration response (assuming the same number of modes for simplicity):

$$w^0(x, t) = \sum_{r=1}^{N} a_r(t)\phi_r(x) \tag{6.24}$$

$$u^0(x, t) = \sum_{r=1}^{N} b_r(t)\alpha_r(x) \tag{6.25}$$

where $\phi_r(x)$ and $\alpha_r(x)$ are the kinematically admissible trial functions which satisfy the respective essential boundary conditions (Appendix G.1), $a_r(t)$ and $b_r(t)$ are the unknown generalized coordinates, and N is the number of modes considered in the solution. Using Equations (6.24) and (6.25) in Equation (6.13), the total potential energy equation becomes

$$U = \frac{1}{2} \sum_{r=1}^{N} \sum_{l=1}^{N} \left\{ a_r a_l \int_0^L \left(Y_s I_s + \bar{c}_{11}^E I_p\right) \phi_r''(x)\phi_l''(x) \, dx + b_r b_l \int_0^L \left(Y_s A_s + \bar{c}_{11}^E A_p\right) \alpha_r'(x)\alpha_l'(x) \, dx \right.$$
$$- 2a_r b_l \int_0^L \left(Y_s H_s + \bar{c}_{11}^E H_p\right) \phi_r''(x)\alpha_l'(x) \, dx + b_r v(t) \int_0^L B_p \alpha_r'(x) \, dx - a_r v(t) \int_0^L J_p \phi_r''(x) \, dx \right\}$$

$$\tag{6.26}$$

where the primes represent ordinary differentiation with respect to space (variable x).

Similarly, the total kinetic energy expression given by Equation (6.14) can be discretized to give

$$
\begin{aligned}
T = \frac{1}{2} \sum_{r=1}^{N} \sum_{l=1}^{N} \Bigg\{ & \dot{a}_r \dot{a}_l \int_0^L \left(\rho_s A_s + \rho_p A_p \right) \phi_r(x) \phi_l(x) \, dx \\
& + 2\dot{a}_r \int_0^L \left(\rho_s A_s + \rho_p A_p \right) \phi_r(x) \frac{\partial w_b(x,t)}{\partial t} \, dx + \int_0^L \left(\rho_s A_s + \rho_p A_p \right) \left(\frac{\partial w_b(x,t)}{\partial t} \right)^2 dx \\
& + \dot{b}_r \dot{b}_l \int_0^L \left(\rho_s A_s + \rho_p A_p \right) \alpha_r(x)\alpha_l(x) \, dx - 2\dot{a}_r \dot{b}_l \int_0^L \left(\rho_s H_s + \rho_p H_p \right) \phi_r'(x)\alpha_l(x) \, dx \Bigg\}
\end{aligned}
$$

(6.27)

where the overdot represents ordinary differentiation with respect to time (variable t).

Substituting Equations (6.24) and (6.25) into Equation (6.20), the internal electrical energy expression becomes

$$
W_{ie} = -\frac{1}{2} \sum_{r=1}^{N} \left\{ b_r v(t) \int_0^L B_p \alpha_r'(x) \, dx - a_r v(t) \int_0^L J_p \phi_r''(x) \, dx - C_p v^2(t) \right\}
$$

(6.28)

Equations (6.26)–(6.28) can be written as

$$
U = \frac{1}{2} \sum_{r=1}^{N} \sum_{l=1}^{N} \left(a_r a_l k_{rl}^{aa} + b_r b_l k_{rl}^{bb} - 2a_r b_l k_{rl}^{ab} - a_r v \tilde{\theta}_r^a + b_r v \tilde{\theta}_r^b \right)
$$

(6.29)

$$
\begin{aligned}
T = \frac{1}{2} \sum_{r=1}^{N} \sum_{l=1}^{N} & \left(\dot{a}_r \dot{a}_l m_{rl}^{aa} + \dot{b}_r \dot{b}_l m_{rl}^{bb} - 2\dot{a}_r \dot{b}_l m_{rl}^{ab} + 2\dot{a}_r p_r \right) \\
& + \frac{1}{2} \int_0^L \left(\rho_s A_s + \rho_p A_p \right) \left(\frac{\partial w_b(x,t)}{\partial t} \right)^2 dx
\end{aligned}
$$

(6.30)

$$
W_{ie} = -\frac{1}{2} \sum_{r=1}^{N} \left(-a_r v \tilde{\theta}_r^a + b_r v \tilde{\theta}_r^b - C_p v^2 \right)
$$

(6.31)

Here,

$$
m_{rl}^{aa} = \int_0^L \left(\rho_s A_s + \rho_p A_p \right) \phi_r(x) \phi_l(x) \, dx
$$

(6.32)

$$m_{rl}^{bb} = \int_0^L \left(\rho_s A_s + \rho_p A_p \right) \alpha_r(x) \alpha_l(x) \, dx \tag{6.33}$$

$$m_{rl}^{ab} = \int_0^L \left(\rho_s H_s + \rho_p H_p \right) \phi_r'(x) \alpha_l(x) \, dx \tag{6.34}$$

$$p_r = \int_0^L \left(\rho_s A_s + \rho_p A_p \right) \phi_r(x) \frac{\partial w_b(x, t)}{\partial t} dx \tag{6.35}$$

$$k_{rl}^{aa} = \int_0^L \left(Y_s I_s + \bar{c}_{11}^E I_p \right) \phi_r''(x) \phi_l''(x) \, dx \tag{6.36}$$

$$k_{rl}^{bb} = \int_0^L \left(Y_s I_s + \bar{c}_{11}^E I_p \right) \alpha_r'(x) \alpha_l'(x) \, dx \tag{6.37}$$

$$k_{rl}^{ab} = \int_0^L \left(Y_s H_s + \bar{c}_{11}^E H_p \right) \phi_r''(x) \alpha_l'(x) \, dx \tag{6.38}$$

$$\tilde{\theta}_r^a = \int_0^L J_p \phi_r''(x) \, dx \tag{6.39}$$

$$\tilde{\theta}_r^b = \int_0^L B_p \alpha_r'(x) \, dx \tag{6.40}$$

where $r = 1, \ldots, N$ and $l = 1, \ldots, N$.

6.2.3 Electromechanical Lagrange Equations

The electromechanical Lagrange equations (Appendix H) based on the extended Hamilton's principle given by Equation (6.22) are

$$\frac{d}{dt} \left(\frac{\partial T}{\partial \dot{a}_i} \right) - \frac{\partial T}{\partial a_i} + \frac{\partial U}{\partial a_i} - \frac{\partial W_{ie}}{\partial a_i} = 0 \tag{6.41}$$

$$\frac{d}{dt}\left(\frac{\partial T}{\partial \dot{b}_i}\right) - \frac{\partial T}{\partial b_i} + \frac{\partial U}{\partial b_i} - \frac{\partial W_{ie}}{\partial b_i} = 0 \tag{6.42}$$

$$\frac{d}{dt}\left(\frac{\partial T}{\partial \dot{v}}\right) - \frac{\partial T}{\partial v} + \frac{\partial U}{\partial v} - \frac{\partial W_{ie}}{\partial v} = Q \tag{6.43}$$

where Q is the electric charge output of the piezoceramic layer. Recall that the mechanical forcing function due to base excitation is expected to emerge from the total kinetic energy and the mechanical dissipation effects will be introduced in the form of proportional damping later in this section.

The non-zero components of Equation (6.41) are

$$\frac{\partial T}{\partial \dot{a}_i} = \frac{1}{2}\sum_{r=1}^{N}\sum_{l=1}^{N}\left[\left(\frac{\partial \dot{a}_r}{\partial \dot{a}_i}\dot{a}_l + \frac{\partial \dot{a}_l}{\partial \dot{a}_i}\dot{a}_r\right)m_{rl}^{aa} - 2\frac{\partial \dot{a}_r}{\partial \dot{a}_i}\dot{b}_l m_{rl}^{ab} + 2\frac{\partial \dot{a}_r}{\partial \dot{a}_i}p_r\right]$$

$$= \frac{1}{2}\sum_{r=1}^{N}\sum_{l=1}^{N}\left[(\delta_{ri}\dot{a}_l + \delta_{li}\dot{a}_r)m_{rl}^{aa} - 2\delta_{ri}\dot{b}_l m_{rl}^{ab} + 2\delta_{ri}p_r\right] = \sum_{l=1}^{N}\left(m_{il}^{aa}\dot{a}_l - m_{il}^{ab}\dot{b}_l + p_i\right) \tag{6.44}$$

$$\frac{\partial U}{\partial a_i} = \frac{1}{2}\sum_{r=1}^{N}\sum_{l=1}^{N}\left[\left(\frac{\partial a_r}{\partial a_i}a_l + \frac{\partial a_l}{\partial a_i}a_r\right)k_{rl}^{aa} - 2\frac{\partial a_r}{\partial a_i}b_l k_{rl}^{ab} - \frac{\partial a_r}{\partial a_i}v\tilde{\theta}_r^a\right]$$

$$= \frac{1}{2}\sum_{r=1}^{N}\sum_{l=1}^{N}\left[(\delta_{ri}a_l + \delta_{li}a_r)k_{rl}^{aa} - 2\delta_{ri}b_l k_{rl}^{ab} - \delta_{ri}v\tilde{\theta}_r^a\right] = \sum_{l=1}^{N}\left(k_{il}^{aa}a_l - k_{il}^{ab}b_l\right) - \frac{1}{2}\tilde{\theta}_i^a v \tag{6.45}$$

$$\frac{\partial W_{ie}}{\partial a_i} = \frac{1}{2}\sum_{r=1}^{N}\frac{\partial a_r}{\partial a_i}\tilde{\theta}_r^a v = \frac{1}{2}\sum_{r=1}^{N}\delta_{ri}\tilde{\theta}_r^a v = \frac{1}{2}\tilde{\theta}_i^a v \tag{6.46}$$

Then the first set of Lagrange equations (for the generalized coordinate a_l) becomes

$$\sum_{l=1}^{N}\left(m_{il}^{aa}\ddot{a}_l - m_{il}^{ab}\ddot{b}_l + k_{il}^{aa}a_l - k_{il}^{ab}b_l - \tilde{\theta}_i^a v - f_i\right) = 0 \tag{6.47}$$

where f_i is the forcing component due to base excitation

$$f_i = -\frac{\partial p_i}{\partial t} = -\int_0^L (\rho_s A_s + \rho_p A_p)\phi_i(x)\frac{\partial^2 w_b(x,t)}{\partial t^2}dx$$

$$= -\frac{d^2 g(t)}{dt^2}\int_0^L (\rho_s A_s + \rho_p A_p)\phi_i(x)dx - \frac{d^2 h(t)}{dt^2}\int_0^L (\rho_s A_s + \rho_p A_p)x\phi_i(x)dx \tag{6.48}$$

Similarly, the non-zero components in Equation (6.42) are

$$
\begin{aligned}
\frac{\partial T}{\partial \dot{b}_i} &= \frac{1}{2} \sum_{r=1}^{N} \sum_{l=1}^{N} \left[\left(\frac{\partial \dot{b}_r}{\partial \dot{b}_i} \dot{b}_l + \frac{\partial \dot{b}_l}{\partial \dot{b}_i} \dot{b}_r \right) m_{rl}^{bb} - 2\frac{\partial \dot{b}_r}{\partial \dot{b}_i} \dot{a}_l m_{rl}^{ab} \right] \\
&= \frac{1}{2} \sum_{r=1}^{N} \sum_{l=1}^{N} \left[(\delta_{ri}\dot{b}_l + \delta_{li}\dot{b}_r) m_{rl}^{bb} - 2\delta_{ri}\dot{a}_l m_{rl}^{ab} \right] = \sum_{l=1}^{N} \left(m_{il}^{bb} \dot{b}_l - m_{il}^{ab} \dot{a}_l \right) \quad (6.49)
\end{aligned}
$$

$$
\begin{aligned}
\frac{\partial U}{\partial b_i} &= \frac{1}{2} \sum_{r=1}^{N} \sum_{l=1}^{N} \left[\left(\frac{\partial b_r}{\partial b_i} b_l + \frac{\partial b_l}{\partial b_i} b_r \right) k_{rl}^{bb} - 2\frac{\partial b_r}{\partial b_i} a_l k_{rl}^{ab} + \frac{\partial b_r}{\partial b_i} v\tilde{\theta}_r^b \right] \\
&= \frac{1}{2} \sum_{r=1}^{N} \sum_{l=1}^{N} \left[(\delta_{ri}b_l + \delta_{li}b_r) k_{rl}^{bb} - 2\delta_{ri}a_l k_{rl}^{ab} + \delta_{ri} v\tilde{\theta}_r^b \right] = \sum_{l=1}^{N} \left(k_{il}^{bb} b_l - k_{il}^{ab} a_l \right) + \frac{1}{2}\tilde{\theta}_i^b v
\end{aligned}
$$

$$(6.50)$$

$$
\frac{\partial W_{ie}}{\partial b_i} = -\frac{1}{2} \sum_{r=1}^{N} \frac{\partial b_r}{\partial b_i} \tilde{\theta}_r^b v = -\frac{1}{2} \sum_{r=1}^{N} \delta_{ri}\tilde{\theta}_r^b v = -\frac{1}{2}\tilde{\theta}_i^b v \quad (6.51)
$$

The Lagrange equation for the generalized coordinate b_l is

$$
\sum_{l=1}^{N} \left(m_{il}^{bb} \ddot{b}_l - m_{il}^{ab} \ddot{a}_l + k_{il}^{bb} b_l - k_{il}^{ab} a_l + \tilde{\theta}_i^b v \right) = 0 \quad (6.52)
$$

The non-zero components on the left hand side of Equation (6.43) are

$$
\frac{\partial U}{\partial v} = \frac{1}{2} \sum_{r=1}^{N} \left(-a_r \tilde{\theta}_r^a + b_r \tilde{\theta}_r^b \right) \quad (6.53)
$$

$$
\frac{\partial W_{ie}}{\partial v} = C_p v - \frac{1}{2} \sum_{r=1}^{N} \left(-a_r \tilde{\theta}_r^a + b_r \tilde{\theta}_r^b \right) \quad (6.54)
$$

yielding

$$
C_p v + Q + \sum_{r=1}^{N} \left(a_r \tilde{\theta}_r^a - b_r \tilde{\theta}_r^b \right) = 0 \quad (6.55)
$$

Taking the time derivative of Equation (6.55) gives

$$
C_p \dot{v} + \dot{Q} + \sum_{r=1}^{N} \left(\dot{a}_r \tilde{\theta}_r^a - \dot{b}_r \tilde{\theta}_r^b \right) = 0 \quad (6.56)
$$

where the time rate of change of charge is the electric current passing through the resistor:

$$\dot{Q} = \frac{v}{R_l} \tag{6.57}$$

The Lagrange equation for v becomes

$$C_p \dot{v} + \frac{v}{R_l} + \sum_{r=1}^{N} \left(\dot{a}_r \tilde{\theta}_r^a - \dot{b}_r \tilde{\theta}_r^b \right) = 0 \tag{6.58}$$

The first two Lagrange equations (Equations (6.47) and (6.52)) can be given in matrix form as

$$\begin{bmatrix} \mathbf{m}^{aa} & -\mathbf{m}^{ab} \\ -\mathbf{m}^{ab} & \mathbf{m}^{bb} \end{bmatrix} \begin{bmatrix} \ddot{\mathbf{a}} \\ \ddot{\mathbf{b}} \end{bmatrix} + \begin{bmatrix} \mathbf{k}^{aa} & -\mathbf{k}^{ab} \\ -\mathbf{k}^{ab} & \mathbf{k}^{bb} \end{bmatrix} \begin{bmatrix} \mathbf{a} \\ \mathbf{b} \end{bmatrix} + \begin{bmatrix} -\tilde{\theta}^a \\ \tilde{\theta}^b \end{bmatrix} v = \begin{bmatrix} \mathbf{f} \\ \mathbf{0} \end{bmatrix} \tag{6.59}$$

Introducing Rayleigh damping to represent the dissipative electromechanical system as a normal-mode system, Equation (6.59) becomes

$$\begin{bmatrix} \mathbf{m}^{aa} & -\mathbf{m}^{ab} \\ -\mathbf{m}^{ab} & \mathbf{m}^{bb} \end{bmatrix} \begin{bmatrix} \ddot{\mathbf{a}} \\ \ddot{\mathbf{b}} \end{bmatrix} + \begin{bmatrix} \mathbf{d}^{aa} & -\mathbf{d}^{ab} \\ -\mathbf{d}^{ab} & \mathbf{d}^{bb} \end{bmatrix} \begin{bmatrix} \dot{\mathbf{a}} \\ \dot{\mathbf{b}} \end{bmatrix} + \begin{bmatrix} \mathbf{k}^{aa} & -\mathbf{k}^{ab} \\ -\mathbf{k}^{ab} & \mathbf{k}^{bb} \end{bmatrix} \begin{bmatrix} \mathbf{a} \\ \mathbf{b} \end{bmatrix} + \begin{bmatrix} -\tilde{\theta}^a \\ \tilde{\theta}^b \end{bmatrix} v = \begin{bmatrix} \mathbf{f} \\ \mathbf{0} \end{bmatrix}$$

$$\tag{6.60}$$

where the damping matrix is

$$\begin{bmatrix} \mathbf{d}^{aa} & -\mathbf{d}^{ab} \\ -\mathbf{d}^{ab} & \mathbf{d}^{bb} \end{bmatrix} = \mu \begin{bmatrix} \mathbf{m}^{aa} & -\mathbf{m}^{ab} \\ -\mathbf{m}^{ab} & \mathbf{m}^{bb} \end{bmatrix} + \gamma \begin{bmatrix} \mathbf{k}^{aa} & -\mathbf{k}^{ab} \\ -\mathbf{k}^{ab} & \mathbf{k}^{bb} \end{bmatrix} \tag{6.61}$$

Here, μ and γ are the constants of mass and stiffness proportionality, respectively.

The electrical circuit equation given by Equation (6.58) becomes

$$C_p \dot{v} + \frac{v}{R_l} + \left(\tilde{\theta}^a \right)^t \dot{\mathbf{a}} - \left(\tilde{\theta}^b \right)^t \dot{\mathbf{b}} = 0 \tag{6.62}$$

Equations (6.60) and (6.62) are the discretized equations of the distributed-parameter electromechanical system. Here, the $N \times 1$ vectors of generalized coordinates are

$$\mathbf{a} = \begin{bmatrix} a_1 & a_2 & \dots & a_N \end{bmatrix}^t, \quad \mathbf{b} = \begin{bmatrix} b_1 & b_2 & \dots & b_N \end{bmatrix}^t \tag{6.63}$$

and the $N \times 1$ vectors of electromechanical coupling are

$$\tilde{\theta}^a = \begin{bmatrix} \tilde{\theta}_1^a & \tilde{\theta}_2^a & \dots & \tilde{\theta}_N^a \end{bmatrix}^t, \quad \tilde{\theta}^b = \begin{bmatrix} \tilde{\theta}_1^b & \tilde{\theta}_2^b & \dots & \tilde{\theta}_N^b \end{bmatrix}^t \tag{6.64}$$

The mass, stiffness, and damping submatrices ($\mathbf{m}^{aa}$, $\mathbf{m}^{bb}$, $\mathbf{m}^{ab}$, $\mathbf{k}^{aa}$, $\mathbf{k}^{bb}$, $\mathbf{k}^{ab}$, $\mathbf{d}^{aa}$, $\mathbf{d}^{bb}$, and $\mathbf{d}^{ab}$) are $N \times N$ matrices and the forcing vector $\mathbf{f}$ is an $N \times 1$ vector whose elements are given by Equation (6.48).

6.2.4   Solution of the Electromechanical Lagrange Equations

The electromechanical Lagrange equations given by Equations (6.60) and (6.62) can be re-expressed as

$$\mathbf{m}^{aa}\ddot{\mathbf{a}} - \mathbf{m}^{ab}\ddot{\mathbf{b}} + \mathbf{d}^{aa}\dot{\mathbf{a}} - \mathbf{d}^{ab}\dot{\mathbf{b}} + \mathbf{k}^{aa}\mathbf{a} - \mathbf{k}^{ab}\mathbf{b} - \tilde{\theta}^{a}v = \mathbf{f} \tag{6.65}$$

$$-\mathbf{m}^{ab}\ddot{\mathbf{a}} + \mathbf{m}^{bb}\ddot{\mathbf{b}} - \mathbf{d}^{ab}\dot{\mathbf{a}} + \mathbf{d}^{bb}\dot{\mathbf{b}} - \mathbf{k}^{ab}\mathbf{a} + \mathbf{k}^{bb}\mathbf{b} + \tilde{\theta}^{b}v = 0 \tag{6.66}$$

$$C_p\dot{v} + \frac{v}{R_l} + \left(\tilde{\theta}^{a}\right)^{t}\dot{\mathbf{a}} - \left(\tilde{\theta}^{b}\right)^{t}\dot{\mathbf{b}} = 0 \tag{6.67}$$

If the base displacement is harmonic of the form $g(t) = W_0 e^{j\omega t}$ and $h(t) = \theta_0 e^{j\omega t}$, then the components of the forcing vector become

$$\mathbf{f} = \mathbf{F} e^{j\omega t} \tag{6.68}$$

Here,

$$F_r = -\sigma_r \omega^2 W_0 - \tau_r \omega^2 \theta_0 \tag{6.69}$$

where

$$\sigma_r = -\int_0^L \left(\rho_s A_s + \rho_p A_p\right)\phi_r(x)\,dx \tag{6.70}$$

$$\tau_r = -\int_0^L \left(\rho_s A_s + \rho_p A_p\right)x\phi_r(x)\,dx \tag{6.71}$$

Based on the linear electromechanical system assumption, the generalized coordinates and the voltage output at steady state are $\mathbf{a} = \mathbf{A}e^{j\omega t}$, $\mathbf{b} = \mathbf{B}e^{j\omega t}$, and $v = Ve^{j\omega t}$ (where $\mathbf{A} = [A_1\ A_2\ \dots\ A_N]^t$, $\mathbf{B} = [B_1\ B_2\ \dots\ B_N]^t$, and V are complex valued). Equations (6.65)–(6.67) become

$$\left(-\omega^2\mathbf{m}^{aa} + j\omega\mathbf{d}^{aa} + \mathbf{k}^{aa}\right)\mathbf{A} - \left(-\omega^2\mathbf{m}^{ab} + j\omega\mathbf{d}^{ab} + \mathbf{k}^{ab}\right)\mathbf{B} - \tilde{\theta}^{a}V = \mathbf{F} \tag{6.72}$$

$$-\left(-\omega^2\mathbf{m}^{ab} + j\omega\mathbf{d}^{ab} + \mathbf{k}^{ab}\right)\mathbf{A} + \left(-\omega^2\mathbf{m}^{bb} + j\omega\mathbf{d}^{bb} + \mathbf{k}^{bb}\right)\mathbf{B} + \tilde{\theta}^{b}V = 0 \tag{6.73}$$

$$\left(j\omega C_p + \frac{1}{R_l}\right)V = j\omega\left[-\left(\tilde{\theta}^{a}\right)^{t}\mathbf{A} + \left(\tilde{\theta}^{b}\right)^{t}\mathbf{B}\right] \tag{6.74}$$

From Equation (6.74),

$$V = j\omega \left(j\omega C_p + \frac{1}{R_l} \right)^{-1} \left[-\left(\tilde{\theta}^a \right)^t A + \left(\tilde{\theta}^b \right)^t B \right] \tag{6.75}$$

Substituting Equation (6.75) into Equations (6.72) and (6.73) gives

$$\Gamma^{aa} A - \Gamma^{ab} B = F \tag{6.76}$$

$$-\Gamma^{ba} A + \Gamma^{bb} B = 0 \tag{6.77}$$

where

$$\Gamma^{aa} = -\omega^2 m^{aa} + j\omega d^{aa} + k^{aa} + j\omega \left(j\omega C_p + \frac{1}{R_l} \right)^{-1} \tilde{\theta}^a \left(\tilde{\theta}^a \right)^t \tag{6.78}$$

$$\Gamma^{bb} = -\omega^2 m^{bb} + j\omega d^{bb} + k^{bb} + j\omega \left(j\omega C_p + \frac{1}{R_l} \right)^{-1} \tilde{\theta}^b \left(\tilde{\theta}^b \right)^t \tag{6.79}$$

$$\Gamma^{ab} = -\omega^2 m^{ab} + j\omega d^{ab} + k^{ab} + j\omega \left(j\omega C_p + \frac{1}{R_l} \right)^{-1} \tilde{\theta}^a \left(\tilde{\theta}^b \right)^t \tag{6.80}$$

$$\Gamma^{ba} = -\omega^2 m^{ab} + j\omega d^{ab} + k^{ab} + j\omega \left(j\omega C_p + \frac{1}{R_l} \right)^{-1} \tilde{\theta}^b \left(\tilde{\theta}^a \right)^t \tag{6.81}$$

From Equation (6.77),

$$B = \left(\Gamma^{bb} \right)^{-1} \Gamma^{ba} A \tag{6.82}$$

Substituting Equation (6.82) into Equation (6.76) gives the complex generalized coordinates A_r:

$$A = \left[\Gamma^{aa} - \Gamma^{ab} \left(\Gamma^{bb} \right)^{-1} \Gamma^{ba} \right]^{-1} F \tag{6.83}$$

Back substitution of Equation (6.83) into Equation (6.82) gives the complex generalized coordinates B_r:

$$B = \left(\Gamma^{bb} \right)^{-1} \Gamma^{ba} \left[\Gamma^{aa} - \Gamma^{ab} \left(\Gamma^{bb} \right)^{-1} \Gamma^{ba} \right]^{-1} F \tag{6.84}$$

Eventually, the complex voltage V is obtained from Equation (6.75) as

$$V = j\omega \left(j\omega C_p + \frac{1}{R_l}\right)^{-1} \left[-\left(\tilde{\theta}^a\right)^t + \left(\tilde{\theta}^b\right)^t \left(\Gamma^{bb}\right)^{-1} \Gamma^{ba}\right] \left[\Gamma^{aa} - \Gamma^{ab} \left(\Gamma^{bb}\right)^{-1} \Gamma^{ba}\right]^{-1} F$$

(6.85)

Substituting the elements of $\mathbf{A}$ and $\mathbf{B}$ into Equations (6.24) and (6.25) gives the transverse displacement and axial displacement response expressions at steady state as

$$w^0(x,t) = \sum_{r=1}^{N} A_r e^{j\omega t} \phi_r(x)$$

(6.86)

$$u^0(x,t) = \sum_{r=1}^{N} B_r e^{j\omega t} \alpha_r(x)$$

(6.87)

The steady-state voltage response is

$$v(t) = V e^{j\omega t} = j\omega \left(j\omega C_p + \frac{1}{R_l}\right)^{-1} \left[-\left(\tilde{\theta}^a\right)^t + \left(\tilde{\theta}^b\right)^t \left(\Gamma^{bb}\right)^{-1} \Gamma^{ba}\right]$$
$$\left[\Gamma^{aa} - \Gamma^{ab} \left(\Gamma^{bb}\right)^{-1} \Gamma^{ba}\right]^{-1} F e^{j\omega t}$$

(6.88)

Following the procedure of Chapter 3, one can use the split form of the base excitation given by Equation (6.69) ($F = -\sigma\omega^2 W_0 - \tau\omega^2\theta_0$) and define six electromechanical FRFs for the three steady-state response expressions given by Equations (6.86)–(6.88) per base acceleration components. These FRFs are

$$\frac{w^0(x,t)}{-\omega^2 W_0 e^{j\omega t}}, \quad \frac{u^0(x,t)}{-\omega^2 W_0 e^{j\omega t}}, \quad \frac{v(t)}{-\omega^2 W_0 e^{j\omega t}}, \quad \frac{w^0(x,t)}{-\omega^2 \theta_0 e^{j\omega t}}, \quad \frac{u^0(x,t)}{-\omega^2 \theta_0 e^{j\omega t}}, \quad \frac{v(t)}{-\omega^2 \theta_0 e^{j\omega t}}$$

(6.89)

It is worth mentioning that the short-circuit and open-circuit natural frequencies of the system can be obtained from the homogeneous form of Equations (6.76) and (6.77) by setting the mechanical damping terms equal to zero ($\mathbf{d} \to \mathbf{0}$) and considering the $R_l \to 0$ and $R_l \to \infty$ cases, respectively:

$$\begin{bmatrix} \Gamma^{aa} & -\Gamma^{ab} \\ -\Gamma^{ba} & \Gamma^{bb} \end{bmatrix} \begin{bmatrix} \mathbf{A} \\ \mathbf{B} \end{bmatrix} = \begin{bmatrix} \mathbf{0} \\ \mathbf{0} \end{bmatrix}$$

(6.90)

The characteristic equation is obtained from the determinant

$$\begin{vmatrix} \Gamma^{aa} & -\Gamma^{ab} \\ -\Gamma^{ba} & \Gamma^{bb} \end{vmatrix} = 0$$

(6.91)

After removing the mechanical damping terms in the submatrices Γ^{aa}, Γ^{ab}, Γ^{ba}, Γ^{bb}, if one sets $R_l \to 0$, the electrical coupling terms simply vanish and the resulting natural frequencies obtained from Equation (6.91) are the short-circuit natural frequencies (natural frequencies at the constant electric field condition). If one sets $R_l \rightarrow \infty$, the resulting natural frequencies are the open-circuit natural frequencies (natural frequencies at the constant electric displacement condition) as the effective stiffness increases due to electromechanical coupling.

6.3 Electromechanical Rayleigh Model with Axial Deformations

6.3.1 Distributed-Parameter Electromechanical Energy Formulation

The form of the total potential energy is the same as that of the Euler–Bernoulli formulation:

$$
\begin{aligned}
U = \frac{1}{2} \int_0^L \Bigg\{ & Y_s \left[A_s \left(\frac{\partial u^0(x,t)}{\partial x} \right)^2 + I_s \left(\frac{\partial^2 w^0(x,t)}{\partial x^2} \right)^2 - 2H_s \frac{\partial u^0(x,t)}{\partial x} \frac{\partial^2 w^0(x,t)}{\partial x^2} \right] \\
& + \bar{c}_{11}^E \left[A_p \left(\frac{\partial u^0(x,t)}{\partial x} \right)^2 + I_p \left(\frac{\partial^2 w^0(x,t)}{\partial x^2} \right)^2 - 2H_p \frac{\partial u^0(x,t)}{\partial x} \frac{\partial^2 w^0(x,t)}{\partial x^2} \right] \\
& + B_p v(t) \frac{\partial u^0(x,t)}{\partial x} - J_p v(t) \frac{\partial^2 w^0(x,t)}{\partial x^2} \Bigg\} \, dx
\end{aligned}
\tag{6.92}
$$

The main difference in the Rayleigh formulation is that the rotary inertia terms are kept in the total kinetic energy expression:

$$
\begin{aligned}
T = \frac{1}{2} \int_0^L \Bigg\{ & \rho_s \Bigg[A_s \left(\frac{\partial u^0(x,t)}{\partial t} \right)^2 + A_s \left(\frac{\partial w^0(x,t)}{\partial t} \right)^2 + 2A_s \frac{\partial w^0(x,t)}{\partial t} \frac{\partial w_b(x,t)}{\partial t} \\
& + A_s \left(\frac{\partial w_b(x,t)}{\partial t} \right)^2 - 2H_s \frac{\partial u^0(x,t)}{\partial t} \frac{\partial^2 w^0(x,t)}{\partial t \partial x} + I_s \left(\frac{\partial^2 w^0(x,t)}{\partial t \partial x} \right)^2 \Bigg] \\
& + \rho_p \Bigg[A_p \left(\frac{\partial u^0(x,t)}{\partial t} \right)^2 + A_p \left(\frac{\partial w^0(x,t)}{\partial t} \right)^2 + 2A_p \frac{\partial w^0(x,t)}{\partial t} \frac{\partial w_b(x,t)}{\partial t} \\
& + A_p \left(\frac{\partial w_b(x,t)}{\partial t} \right)^2 - 2H_p \frac{\partial u^0(x,t)}{\partial t} \frac{\partial^2 w^0(x,t)}{\partial t \partial x} + I_p \left(\frac{\partial^2 w^0(x,t)}{\partial t \partial x} \right)^2 \Bigg] \Bigg\} \, dx
\end{aligned}
\tag{6.93}
$$

The internal electrical energy and the non-conservative virtual work of the charge output are also the same as in the previous discussion:

$$
\begin{aligned}
W_{ie} & = -\frac{1}{2} \int_{V_p} \frac{v(t)}{h_p} \left[\bar{e}_{31} \left(\frac{\partial u^0(x,t)}{\partial x} - z \frac{\partial^2 w^0(x,t)}{\partial x^2} \right) - \bar{\varepsilon}_{33}^S \frac{v(t)}{h_p} \right] dV_p \\
& = -\frac{1}{2} \int_0^L \left(B_p v(t) \frac{\partial u^0(x,t)}{\partial x} - J_p v(t) \frac{\partial^2 w^0(x,t)}{\partial x^2} \right) dx + \frac{1}{2} C_p v^2(t)
\end{aligned}
\tag{6.94}
$$

$$
\delta W_{nc} = \delta W_{nce} = Q(t) \delta v(t)
\tag{6.95}
$$

6.3.2 Spatial Discretization of the Energy Equations

The assumed vibration response expressions are given by Equations (6.24) and (6.25), which discretize Equations (6.92)–(6.94) into

$$U = \frac{1}{2} \sum_{r=1}^{N} \sum_{l=1}^{N} \left(a_r a_l k_{rl}^{aa} + b_r b_l k_{rl}^{bb} - 2a_r b_l k_{rl}^{ab} - a_r v \tilde{\theta}_r^a + b_r v \tilde{\theta}_r^b \right) \tag{6.96}$$

$$T = \frac{1}{2} \sum_{r=1}^{N} \sum_{l=1}^{N} \left(\dot{a}_r \dot{a}_l m_{rl}^{aa} + \dot{b}_r \dot{b}_l m_{rl}^{bb} - 2\dot{a}_r \dot{b}_l m_{rl}^{ab} + 2\dot{a}_r \, p_r \right)$$

$$+ \frac{1}{2} \int_0^L (\rho_s A_s + \rho_p A_p) \left(\frac{\partial w_b(x, t)}{\partial t} \right)^2 dx \tag{6.97}$$

$$W_{ie} = -\frac{1}{2} \sum_{r=1}^{N} \left(-a_r v \tilde{\theta}_r^a + b_r v \tilde{\theta}_r^b - C_p v^2 \right) \tag{6.98}$$

where

$$m_{rl}^{aa} = \int_0^L \left[(\rho_s A_s + \rho_p A_p) \, \phi_r(x) \phi_l(x) + (\rho_s I_s + \rho_p I_p) \, \phi_r'(x) \phi_l'(x) \right] dx \tag{6.99}$$

and the rest of the mass, stiffness, and damping terms are as given by Equations (6.33)–(6.40). Therefore, the rotary inertia effect modifies m_{rl}^{aa} only.

6.3.3 Electromechanical Lagrange Equations

Since the discretized form of the electromechanical equations does not change (other than the modification in the m_{rl}^{aa} term), the resulting Lagrange equations are

$$\mathbf{m^{aa} \ddot{a}} - \mathbf{m^{ab} \ddot{b}} + \mathbf{d^{aa} \dot{a}} - \mathbf{d^{ab} \dot{b}} + \mathbf{k^{aa} a} - \mathbf{k^{ab} b} - \tilde{\theta}^a v = \mathbf{f} \tag{6.100}$$

$$-\mathbf{m^{ab} \ddot{a}} + \mathbf{m^{bb} \ddot{b}} - \mathbf{d^{ab} \dot{a}} + \mathbf{d^{bb} \dot{b}} - \mathbf{k^{ab} a} + \mathbf{k^{bb} b} + \tilde{\theta}^b v = \mathbf{0} \tag{6.101}$$

$$C_p \dot{v} + \frac{v}{R_l} + \left(\tilde{\theta}^a \right)^t \dot{\mathbf{a}} - \left(\tilde{\theta}^b \right)^t \dot{\mathbf{b}} = 0 \tag{6.102}$$

6.3.4   Solution of the Electromechanical Lagrange Equations

For harmonic base excitation ($g(t) = W_0 e^{j\omega t}$ and $h(t) = \theta_0 e^{j\omega t}$ so that $\mathbf{f} = \mathbf{F} e^{j\omega t}$), the vectors of complex generalized coordinates are

$$\mathbf{A} = \left[\mathbf{\Gamma}^{aa} - \mathbf{\Gamma}^{ab} \left(\mathbf{\Gamma}^{bb} \right)^{-1} \mathbf{\Gamma}^{ba} \right]^{-1} \mathbf{F} \tag{6.103}$$

$$\mathbf{B} = \left(\mathbf{\Gamma}^{bb} \right)^{-1} \mathbf{\Gamma}^{ba} \left[\mathbf{\Gamma}^{aa} - \mathbf{\Gamma}^{ab} \left(\mathbf{\Gamma}^{bb} \right)^{-1} \mathbf{\Gamma}^{ba} \right]^{-1} \mathbf{F} \tag{6.104}$$

The complex voltage is

$$V = j\omega \left(j\omega C_p + \frac{1}{R_l} \right)^{-1} \left[-\left(\tilde{\theta}^a \right)^t + \left(\tilde{\theta}^b \right)^t \left(\mathbf{\Gamma}^{bb} \right)^{-1} \mathbf{\Gamma}^{ba} \right] \left[\mathbf{\Gamma}^{aa} - \mathbf{\Gamma}^{ab} \left(\mathbf{\Gamma}^{bb} \right)^{-1} \mathbf{\Gamma}^{ba} \right]^{-1} \mathbf{F} \tag{6.105}$$

Equations (6.103)–(6.105) can be used in Equations (6.86)–(6.88) to express the steady-state response of the system. The discussion related to obtaining the short-circuit and open-circuit natural frequencies is the same as in the Euler–Bernoulli formulation. Hence, the only difference in the foregoing equations is due to m_{rl}^{aa} which includes the rotary inertia effect based on the Rayleigh formulation.

6.4   Electromechanical Timoshenko Model with Axial Deformations

6.4.1   Distributed-Parameter Electromechanical Energy Formulation

The non-zero displacement field in the Timoshenko formulation is

$$u(x, z, t) = u^0(x, t) - z\psi^0(x, t) \tag{6.106}$$

$$w(x, t) = w^0(x, t) \tag{6.107}$$

where $u^0(x, t)$ and $w^0(x, t)$ are the axial displacement and transverse displacement at point x on the neutral axis relative to the moving base and $\psi^0(x, t)$ is the cross-section rotation.

The vector form of the displacement field is then

$$\mathbf{u} = \left[u^0(x, t) - z\psi^0(x, t) \quad 0 \quad w^0(x, t) \right]^t \tag{6.108}$$

The two non-zero strain components based on the given displacement field are

$$S_{xx}(x, z, t) = \frac{\partial u(x, z, t)}{\partial x} = \frac{\partial u^0(x, t)}{\partial x} - z\frac{\partial \psi^0(x, t)}{\partial x} \tag{6.109}$$

$$S_{xz}(x, t) = \frac{\partial u(x, z, t)}{\partial z} + \frac{\partial w(x, t)}{\partial x} = \frac{\partial w^0(x, t)}{\partial x} - \psi^0(x, t) \tag{6.110}$$

where $S_{xx}(x, z, t)$ is the axial strain component and $S_{xz}(x, t)$ is the transverse engineering shear strain component.

The isotropic substructure has the following constitutive equations (Appendix F.3):

$$T_{xx}(x, z, t) = Y_s S_{xx}(x, z, t) \tag{6.111}$$

$$T_{xz}(x, t) = \kappa G_s S_{xz}(x, t) \tag{6.112}$$

where κ is the cross-section-dependent shear correction factor introduced by Timoshenko [8,9] and it accounts for the non-uniform distribution of the shear stress over the cross-section [8–18].[5] Moreover, G_s is the shear modulus of the substructure layer and is related to the elastic modulus of the substructure layer through

$$G_s = \frac{Y_s}{2(1 + v_s)} \tag{6.113}$$

where v_s is Poisson's ratio for the substructure layer.

The constitutive equations for the piezoceramic layer are (Appendix A.3)

$$T_{xx}(x, z, t) = T_1 = \bar{c}_{11}^E S_1 - \bar{e}_{31} E_3 = \bar{c}_{11}^E S_{xx}(x, z, t) + \bar{e}_{31} \frac{v(t)}{h_p} \tag{6.114}$$

$$T_{xz}(x, t) = T_5 = \kappa \bar{c}_{55}^E S_5 = \kappa \bar{c}_{55}^E S_{xz}(x, t) \tag{6.115}$$

where $\bar{c}_{55}^E$ is the shear modulus of the piezoceramic at constant electric field ($\bar{c}_{55}^E = 1/s_{55}^E$).

The total potential energy in the structure is

$$U = \frac{1}{2} \left(\int_{V_s} \mathbf{S}^t \mathbf{T} \, dV_s + \int_{V_p} \mathbf{S}^t \mathbf{T} \, dV_p \right) \tag{6.116}$$

yielding

$$U = \frac{1}{2} \int_{V_s} \left[Y_s \left(\frac{\partial u^0(x, t)}{\partial x} - z \frac{\partial \psi^0(x, t)}{\partial x} \right)^2 + \kappa G_s \left(\frac{\partial w^0(x, t)}{\partial x} - \psi^0(x, t) \right)^2 \right] dV_s$$

---

[5] Various expressions [8–18] for the shear correction factor κ have been derived in the literature since Timoshenko's beam theory [8] was established in 1921. A review of the shear correction factors proposed in 1921–1975 was presented by Kaneko [13], concluding that the expressions derived by Timoshenko [9] should be preferred. For rectangular cross-sections, Timoshenko [9] derived $\kappa = (5 + 5v_s)/(6 + 5v_s)$ theoretically, whereas Mindlin [10,11] obtained $\kappa = \pi^2/12$ experimentally (for crystal plates). Cowper's [12] solution is also widely used and it differs slightly from Timoshenko's solution: $\kappa = (10 + 10v_s)/(12 + 11v_s)$. The effect of width-to-depth ratio of the cross-section has been taken into account by Stephen [15,16] and more recently by Hutchinson [17]. Recently, an experimental study on the effect of width-to-depth ratio of the cross-section has been presented by Puchegger *et al.* [18].

$$
+\frac{1}{2}\int_{V_p}\left[\bar{c}_{11}^E\left(\frac{\partial u^0(x,t)}{\partial x}-z\frac{\partial \psi^0(x,t)}{\partial x}\right)^2+\kappa\bar{c}_{55}^E\left(\frac{\partial w^0(x,t)}{\partial x}-\psi^0(x,t)\right)^2\right.
$$

$$
\left.+\bar{e}_{31}\frac{v(t)}{h_p}\left(\frac{\partial u^0(x,t)}{\partial x}-z\frac{\partial \psi^0(x,t)}{\partial x}\right)\right]dV_p \tag{6.117}
$$

The total kinetic energy of the system is given by

$$
T=\frac{1}{2}\left(\int_{V_s}\rho_s\frac{\partial \mathbf{u}_m^t}{\partial t}\frac{\partial \mathbf{u}_m}{\partial t}dV_s+\int_{V_p}\rho_p\frac{\partial \mathbf{u}_m^t}{\partial t}\frac{\partial \mathbf{u}_m}{\partial t}dV_p\right) \tag{6.118}
$$

where the modified displacement vector (including the base displacement) is

$$
\mathbf{u}_m=\left[u^0(x,t)-z\psi^0(x,t)\quad 0\quad w^0(x,t)+w_b(x,t)\right]^t \tag{6.119}
$$

Then the total kinetic energy becomes

$$
T=\frac{1}{2}\int_{V_s}\rho_s\left[\left(\frac{\partial u^0(x,t)}{\partial t}-z\frac{\partial \psi^0(x,t)}{\partial t}\right)^2+\left(\frac{\partial w^0(x,t)}{\partial t}+\frac{\partial w_b(x,t)}{\partial t}\right)^2\right]dV_s
$$

$$
+\frac{1}{2}\int_{V_p}\rho_p\left[\left(\frac{\partial u^0(x,t)}{\partial t}-z\frac{\partial \psi^0(x,t)}{\partial t}\right)^2+\left(\frac{\partial w^0(x,t)}{\partial t}+\frac{\partial w_b(x,t)}{\partial t}\right)^2\right]dV_p \tag{6.120}
$$

Equation (6.117) can be written as

$$
U=\frac{1}{2}\int_0^L\left\{Y_s\left[A_s\left(\frac{\partial u^0(x,t)}{\partial x}\right)^2+I_s\left(\frac{\partial \psi^0(x,t)}{\partial x}\right)^2-2H_s\frac{\partial u^0(x,t)}{\partial x}\frac{\partial \psi^0(x,t)}{\partial x}\right]\right.
$$

$$
+\kappa G_s A_s\left[\left(\frac{\partial w^0(x,t)}{\partial x}\right)^2+(\psi^0(x,t))^2-2\psi^0(x,t)\frac{\partial w^0(x,t)}{\partial x}\right]
$$

$$
+\bar{c}_{11}^E\left[A_p\left(\frac{\partial u^0(x,t)}{\partial x}\right)^2+I_p\left(\frac{\partial \psi^0(x,t)}{\partial x}\right)^2-2H_p\frac{\partial u^0(x,t)}{\partial x}\frac{\partial \psi^0(x,t)}{\partial x}\right] \tag{6.121}
$$

$$
+\kappa\bar{c}_{55}^E A_p\left[\left(\frac{\partial w^0(x,t)}{\partial x}\right)^2+(\psi^0(x,t))^2-2\psi^0(x,t)\frac{\partial w^0(x,t)}{\partial x}\right]
$$

$$
\left.+B_p v(t)\frac{\partial u^0(x,t)}{\partial x}-J_p v(t)\frac{\partial \psi^0(x,t)}{\partial x}\right\}dx
$$

where B_p and J_p are as given by Equations (6.17) and (6.18), respectively.

Equation (6.120) becomes

$$
T = \frac{1}{2} \int_0^L \left\{ \rho_s \left[A_s \left(\frac{\partial u^0(x,t)}{\partial t} \right)^2 + A_s \left(\frac{\partial w^0(x,t)}{\partial t} \right)^2 + 2A_s \frac{\partial w^0(x,t)}{\partial t} \frac{\partial w_b(x,t)}{\partial t} \right. \right.
$$

$$
+ A_s \left(\frac{\partial w_b(x,t)}{\partial t} \right)^2 - 2H_s \frac{\partial u^0(x,t)}{\partial t} \frac{\partial \psi^0(x,t)}{\partial t} + I_s \left(\frac{\partial \psi^0(x,t)}{\partial t} \right)^2 \right]
$$

$$
+ \rho_p \left[A_p \left(\frac{\partial u^0(x,t)}{\partial t} \right)^2 + A_p \left(\frac{\partial w^0(x,t)}{\partial t} \right)^2 + 2A_p \frac{\partial w^0(x,t)}{\partial t} \frac{\partial w_b(x,t)}{\partial t} \right.
$$

$$
\left. \left. + A_p \left(\frac{\partial w_b(x,t)}{\partial t} \right)^2 - 2H_p \frac{\partial u^0(x,t)}{\partial t} \frac{\partial \psi^0(x,t)}{\partial t} + I_p \left(\frac{\partial \psi^0(x,t)}{\partial t} \right)^2 \right] \right\} dx \quad (6.122)
$$

where the kinetic energy contributions from the rotary inertia are kept. The area moments in the foregoing equations are as given by Equations (6.15) and (6.16) and the argument related to the decoupling of $u^0(x,t)$ from the equations mentioned in the Euler–Bernoulli formulation is still valid. That is, for a symmetric structure (with respect to the neutral axis), $H_s = H_p = 0$ and $u^0(x,t)$ is not excited by the transverse excitation of the beam.

The internal electrical energy in the piezoceramic layer is

$$
W_{ie} = \frac{1}{2} \int_{V_p} \mathbf{E}^t \mathbf{D} \, dV_p = -\frac{1}{2} \int_{V_p} \frac{v(t)}{h_p} \left[\bar{e}_{31} \left(\frac{\partial u^0(x,t)}{\partial x} - z \frac{\partial \psi^0(x,t)}{\partial x} \right) - \bar{\varepsilon}_{33}^S \frac{v(t)}{h_p} \right] dV_p
$$

$$
= -\frac{1}{2} \int_0^L \left(B_p v(t) \frac{\partial u^0(x,t)}{\partial x} - J_p v(t) \frac{\partial \psi^0(x,t)}{\partial x} \right) dx + \frac{1}{2} C_p v^2(t) \quad (6.123)
$$

where the internal capacitance of the piezoceramic is given by Equation (6.21).

As in Section 6.2, the mechanical dissipation effects will be included in the form of proportional damping after the equations are discretized. Therefore the only virtual work of non-conservative forces in the extended Hamilton's principle given by Equation (6.22) is that of the electric charge given by Equation (6.23).

6.4.2 Spatial Discretization of the Energy Equations

The distributed-parameter variables in the mechanical domain are $w^0(x,t)$, $u^0(x,t)$, and $\psi^0(x,t)$, and the electrical variable is $v(t)$. Let the following finite series expansions represent the components of vibration response:

$$
w^0(x,t) = \sum_{r=1}^N a_r(t)\phi_r(x) \quad (6.124)
$$

$$u^0(x,t) = \sum_{r=1}^{N} b_r(t)\alpha_r(x) \qquad (6.125)$$

$$\psi^0(x,t) = \sum_{r=1}^{N} c_r(t)\beta_r(x) \qquad (6.126)$$

where $\phi_r(x)$, $\alpha_r(x)$, and $\beta_r(x)$ are the kinematically admissible trial functions which satisfy the essential boundary conditions (Appendix G.2), and $a_r(t)$, $b_r(t)$, and $c_r(t)$ are the unknown generalized coordinates.

Using Equations (6.124)–(6.126) in Equation (6.121), the total potential energy equation becomes

$$
\begin{aligned}
U = \frac{1}{2}\sum_{r=1}^{N}\sum_{l=1}^{N} & \left\{ a_r a_l \int_0^L \kappa \left(G_s A_s + \bar{c}_{55}^E A_p\right)\phi_r'(x)\phi_l'(x)\,dx \right. \\
& + b_r b_l \int_0^L \left(Y_s A_s + \bar{c}_{11}^E A_p\right)\alpha_r'(x)\alpha_l'(x)\,dx \\
& + c_r c_l \int_0^L \left[\left(Y_s I_s + \bar{c}_{11}^E I_p\right)\beta_r'(x)\beta_l'(x) + \kappa\left(G_s A_s + \bar{c}_{55}^E A_p\right)\beta_r(x)\beta_l(x)\right]dx \\
& - 2a_r c_l \int_0^L \kappa\left(G_s A_s + \bar{c}_{55}^E A_p\right)\phi_r'(x)\beta_l(x)\,dx - 2b_r c_l \int_0^L \left(Y_s H_s + \bar{c}_{11}^E H_p\right)\alpha_r'(x)\beta_l'(x)\,dx \\
& \left. + b_r v(t)\int_0^L B_p\alpha_r'(x)\,dx - c_r v(t)\int_0^L J_p\beta_r'(x)\,dx \right\}
\end{aligned}
\qquad (6.127)
$$

Similarly, the total kinetic energy expression is discretized into

$$
\begin{aligned}
T = \frac{1}{2}\sum_{r=1}^{N}\sum_{l=1}^{N} & \left\{ \dot{a}_r\dot{a}_l \int_0^L \left(\rho_s A_s + \rho_p A_p\right)\phi_r(x)\phi_l(x)\,dx \right. \\
& + 2\dot{a}_r \int_0^L \left(\rho_s A_s + \rho_p A_p\right)\phi_r(x)\frac{\partial w_b(x,t)}{\partial t}\,dx + \int_0^L \left(\rho_s A_s + \rho_p A_p\right)\left(\frac{\partial w_b(x,t)}{\partial t}\right)^2 dx \\
& + \dot{b}_r\dot{b}_l \int_0^L \left(\rho_s A_s + \rho_p A_p\right)\alpha_r(x)\alpha_l(x)\,dx + \dot{c}_r\dot{c}_l \int_0^L \left(\rho_s I_s + \rho_p I_p\right)\beta_r(x)\beta_l(x)\,dx \\
& \left. - 2\dot{b}_r\dot{c}_l \int_0^L \left(\rho_s H_s + \rho_p H_p\right)\alpha_r(x)\beta_l(x)\,dx \right\}
\end{aligned}
\qquad (6.128)
$$

The discretized internal electrical energy is then

$$
W_{ie} = -\frac{1}{2} \sum_{r=1}^{N} \left\{ b_r v(t) \int_0^L B_p \alpha_r'(x) \, dx - c_r v(t) \int_0^L J_p \beta_r'(x) \, dx \right\} + \frac{1}{2} C_p v^2(t)
\tag{6.129}
$$

Equations (6.127)–(6.129) can be written as

$$
U = \frac{1}{2} \sum_{r=1}^{N} \sum_{l=1}^{N} \left(a_r a_l k_{rl}^{aa} + b_r b_l k_{rl}^{bb} + c_r c_l k_{rl}^{cc} - 2 a_r c_l k_{rl}^{ac} - 2 b_r c_l k_{rl}^{bc} + b_r v \tilde{\theta}_r^b - c_r v \tilde{\theta}_r^c \right)
$$
$$
\tag{6.130}
$$

$$
T = \frac{1}{2} \sum_{r=1}^{N} \sum_{l=1}^{N} \left(\dot{a}_r \dot{a}_l m_{rl}^{aa} + \dot{b}_r \dot{b}_l m_{rl}^{bb} + \dot{c}_r \dot{c}_l m_{rl}^{cc} - 2 \dot{b}_r \dot{c}_l m_{rl}^{bc} + 2 \dot{a}_r p_r \right)
$$
$$
+ \frac{1}{2} \int_0^L (\rho_s A_s + \rho_p A_p) \left(\frac{\partial w_b(x, t)}{\partial t} \right)^2 dx
\tag{6.131}
$$

$$
W_{ie} = -\frac{1}{2} \sum_{r=1}^{N} \left(b_r v \tilde{\theta}_r^b - c_r v \tilde{\theta}_r^c - C_p v^2 \right)
\tag{6.132}
$$

Here,

$$
m_{rl}^{aa} = \int_0^L (\rho_s A_s + \rho_p A_p) \, \phi_r(x) \phi_l(x) \, dx
\tag{6.133}
$$

$$
m_{rl}^{bb} = \int_0^L (\rho_s A_s + \rho_p A_p) \, \alpha_r(x) \alpha_l(x) \, dx
\tag{6.134}
$$

$$
m_{rl}^{cc} = \int_0^L (\rho_s I_s + \rho_p I_p) \, \beta_r(x) \beta_l(x) \, dx
\tag{6.135}
$$

$$
m_{rl}^{bc} = \int_0^L (\rho_s H_s + \rho_p H_p) \, \alpha_r(x) \beta_l(x) \, dx
\tag{6.136}
$$

$$
p_r = \int_0^L (\rho_s A_s + \rho_p A_p) \, \phi_r(x) \frac{\partial w_b(x, t)}{\partial t} dx
\tag{6.137}
$$

$$k_{rl}^{aa} = \int_0^L \kappa \left(G_s A_s + \bar{c}_{55}^E A_p \right) \phi_r'(x)\phi_l'(x)\, dx \tag{6.138}$$

$$k_{rl}^{bb} = \int_0^L \left(Y_s A_s + \bar{c}_{11}^E A_p \right) \alpha_r'(x)\alpha_l'(x)\, dx \tag{6.139}$$

$$k_{rl}^{cc} = \int_0^L \left[\left(Y_s I_s + \bar{c}_{11}^E I_p \right) \beta_r'(x)\beta_l'(x) + \kappa \left(G_s A_s + \bar{c}_{55}^E A_p \right) \beta_r(x)\beta_l(x) \right] dx \tag{6.140}$$

$$k_{rl}^{ac} = \int_0^L \kappa \left(G_s A_s + \bar{c}_{55}^E A_p \right) \phi_r'(x)\beta_l(x)\, dx \tag{6.141}$$

$$k_{rl}^{bc} = \int_0^L \left(Y_s H_s + \bar{c}_{11}^E H_p \right) \alpha_r'(x)\beta_l'(x)\, dx \tag{6.142}$$

$$\tilde{\theta}_r^b = \int_0^L B_p \alpha_r'(x)\, dx \tag{6.143}$$

$$\tilde{\theta}_r^c = \int_0^L J_p \beta_r'(x)\, dx \tag{6.144}$$

where $r = 1, \ldots, N$ and $l = 1, \ldots, N$.

6.4.3 Electromechanical Lagrange Equations

The electromechanical Lagrange equations (Appendix H), based on the extended Hamilton's principle, are expressed as

$$\frac{d}{dt}\left(\frac{\partial T}{\partial \dot{a}_i}\right) - \frac{\partial T}{\partial a_i} + \frac{\partial U}{\partial a_i} - \frac{\partial W_{ie}}{\partial a_i} = 0 \tag{6.145}$$

$$\frac{d}{dt}\left(\frac{\partial T}{\partial \dot{b}_i}\right) - \frac{\partial T}{\partial b_i} + \frac{\partial U}{\partial b_i} - \frac{\partial W_{ie}}{\partial b_i} = 0 \tag{6.146}$$

$$\frac{d}{dt}\left(\frac{\partial T}{\partial \dot{c}_i}\right) - \frac{\partial T}{\partial c_i} + \frac{\partial U}{\partial c_i} - \frac{\partial W_{ie}}{\partial c_i} = 0 \tag{6.147}$$

$$\frac{d}{dt}\left(\frac{\partial T}{\partial \dot{v}}\right) - \frac{\partial T}{\partial v} + \frac{\partial U}{\partial v} - \frac{\partial W_{ie}}{\partial v} = Q \tag{6.148}$$

where Q is the electric charge output and the forcing component due to base excitation will be obtained from the total kinetic energy (more precisely from $\partial T/\partial \dot{a}_i$).

The non-zero components of Equation (6.145) are

$$\frac{\partial T}{\partial \dot{a}_i} = \frac{1}{2}\sum_{r=1}^{N}\sum_{l=1}^{N}\left[\left(\frac{\partial \dot{a}_r}{\partial \dot{a}_i}\dot{a}_l + \frac{\partial \dot{a}_l}{\partial \dot{a}_i}\dot{a}_r\right)m_{rl}^{aa} + 2\frac{\partial \dot{a}_r}{\partial \dot{a}_i}p_r\right]$$

$$= \frac{1}{2}\sum_{r=1}^{N}\sum_{l=1}^{N}\left[(\delta_{ri}\dot{a}_l + \delta_{li}\dot{a}_r)m_{rl}^{aa} + 2\delta_{ri}p_r\right] = \sum_{l=1}^{N}\left(m_{il}^{aa}\dot{a}_l + p_i\right) \tag{6.149}$$

$$\frac{\partial U}{\partial a_i} = \frac{1}{2}\sum_{r=1}^{N}\sum_{l=1}^{N}\left[\left(\frac{\partial a_r}{\partial a_i}a_l + \frac{\partial a_l}{\partial a_i}a_r\right)k_{rl}^{aa} - 2\frac{\partial a_r}{\partial a_i}c_l k_{rl}^{ac}\right]$$

$$= \frac{1}{2}\sum_{r=1}^{N}\sum_{l=1}^{N}\left[(\delta_{ri}a_l + \delta_{li}a_r)k_{rl}^{aa} - 2\delta_{ri}c_l k_{rl}^{ac}\right] = \sum_{l=1}^{N}\left(k_{il}^{aa}a_l - k_{il}^{ac}c_l\right) \tag{6.150}$$

Then the first set of Lagrange equations (for the generalized coordinate a_l) becomes

$$\sum_{l=1}^{N}\left(m_{il}^{aa}\ddot{a}_l + k_{il}^{aa}a_l - k_{il}^{ac}c_l - f_i\right) = 0 \tag{6.151}$$

where the forcing component due to base excitation is

$$f_i = -\frac{\partial p_i}{\partial t} = -\int_0^L (\rho_s A_s + \rho_p A_p)\,\phi_i(x)\frac{\partial^2 w_b(x,t)}{\partial t^2}dx$$

$$= -\frac{d^2g(t)}{dt^2}\int_0^L (\rho_s A_s + \rho_p A_p)\,\phi_i(x)\,dx - \frac{d^2h(t)}{dt^2}\int_0^L (\rho_s A_s + \rho_p A_p)\,x\phi_i(x)\,dx \tag{6.152}$$

The non-zero components in Equation (6.146) are

$$\frac{\partial T}{\partial \dot{b}_i} = \frac{1}{2}\sum_{r=1}^{N}\sum_{l=1}^{N}\left[\left(\frac{\partial \dot{b}_r}{\partial \dot{b}_i}\dot{b}_l + \frac{\partial \dot{b}_l}{\partial \dot{b}_i}\dot{b}_r\right)m_{rl}^{bb} - 2\frac{\partial \dot{b}_r}{\partial \dot{b}_i}\dot{c}_l m_{rl}^{bc}\right]$$

$$= \frac{1}{2}\sum_{r=1}^{N}\sum_{l=1}^{N}\left[(\delta_{ri}\dot{b}_l + \delta_{li}\dot{b}_r)m_{rl}^{bb} - 2\delta_{ri}\dot{c}_l m_{rl}^{bc}\right] = \sum_{l=1}^{N}\left(m_{il}^{bb}\dot{b}_l - m_{il}^{bc}\dot{c}_l\right) \tag{6.153}$$

$$\frac{\partial U}{\partial b_i} = \frac{1}{2} \sum_{r=1}^{N} \sum_{l=1}^{N} \left[\left(\frac{\partial b_r}{\partial b_i} b_l + \frac{\partial b_l}{\partial b_i} b_r \right) k_{rl}^{bb} - 2 \frac{\partial b_r}{\partial b_i} c_l k_{rl}^{bc} + \frac{\partial b_r}{\partial b_i} v \tilde{\theta}_r^b \right]$$

$$= \frac{1}{2} \sum_{r=1}^{N} \sum_{l=1}^{N} \left[(\delta_{ri} b_l + \delta_{li} b_r) k_{rl}^{bb} - 2 \delta_{ri} c_l k_{rl}^{bc} + \delta_{ri} v \tilde{\theta}_r^b \right] = \sum_{l=1}^{N} \left(k_{il}^{bb} b_l - k_{il}^{bc} c_l \right) + \frac{1}{2} \tilde{\theta}_i^b v$$

$$(6.154)$$

$$\frac{\partial W_{ie}}{\partial b_i} = -\frac{1}{2} \sum_{r=1}^{N} \frac{\partial b_r}{\partial b_i} v \tilde{\theta}_r^b = -\frac{1}{2} \sum_{r=1}^{N} \delta_{ri} v \tilde{\theta}_r^b = -\frac{1}{2} \tilde{\theta}_i^b v \qquad (6.155)$$

Then the Lagrange equation for the generalized coordinate b_l is

$$\sum_{l=1}^{N} \left(m_{il}^{bb} \ddot{b}_l - m_{il}^{bc} \ddot{c}_l + k_{il}^{bb} b_l - k_{il}^{bc} c_l + \tilde{\theta}_i^b v \right) = 0 \qquad (6.156)$$

The non-zero components in Equation (6.147) are

$$\frac{\partial T}{\partial \dot{b}_i} = \frac{1}{2} \sum_{r=1}^{N} \sum_{l=1}^{N} \left[\left(\frac{\partial \dot{c}_r}{\partial \dot{c}_i} \dot{c}_l + \frac{\partial \dot{c}_l}{\partial \dot{c}_i} \dot{c}_r \right) m_{rl}^{cc} - 2 \frac{\partial \dot{c}_r}{\partial \dot{c}_i} \dot{b}_l m_{rl}^{bc} \right]$$

$$= \frac{1}{2} \sum_{r=1}^{N} \sum_{l=1}^{N} \left[(\delta_{ri} \dot{c}_l + \delta_{li} \dot{c}_r) m_{rl}^{cc} - 2 \delta_{ri} \dot{b}_l m_{rl}^{bc} \right] = \sum_{l=1}^{N} \left(m_{il}^{cc} \dot{c}_l - m_{il}^{bc} \dot{b}_l \right) \quad (6.157)$$

$$\frac{\partial U}{\partial b_i} = \frac{1}{2} \sum_{r=1}^{N} \sum_{l=1}^{N} \left[\left(\frac{\partial c_r}{\partial c_i} c_l + \frac{\partial c_l}{\partial c_i} c_r \right) k_{rl}^{cc} - 2 \frac{\partial c_r}{\partial c_i} a_l k_{rl}^{ac} - 2 \frac{\partial c_r}{\partial c_i} b_l k_{rl}^{bc} - \frac{\partial c_r}{\partial c_i} v \tilde{\theta}_r^c \right]$$

$$= \frac{1}{2} \sum_{r=1}^{N} \sum_{l=1}^{N} \left[(\delta_{ri} c_l + \delta_{li} c_r) k_{rl}^{cc} - 2 \delta_{ri} a_l k_{rl}^{ac} - 2 \delta_{ri} b_l k_{rl}^{bc} - \delta_{ri} v \tilde{\theta}_r^c \right]$$

$$= \sum_{l=1}^{N} \left(k_{il}^{cc} c_l - k_{il}^{ac} a_l - k_{il}^{bc} b_l \right) - \frac{1}{2} \tilde{\theta}_i^c v \qquad (6.158)$$

$$\frac{\partial W_{ie}}{\partial c_i} = \frac{1}{2} \sum_{r=1}^{N} \frac{\partial c_r}{\partial c_i} \tilde{\theta}_r^c v = \frac{1}{2} \sum_{r=1}^{N} \delta_{ri} \tilde{\theta}_r^c v = \frac{1}{2} \tilde{\theta}_i^c v \qquad (6.159)$$

Hence the Lagrange equation for the generalized coordinate c_l is

$$\sum_{l=1}^{N} \left(m_{il}^{cc} \ddot{c}_l - m_{il}^{bc} \ddot{b}_l + k_{il}^{cc} c_l - k_{il}^{ac} a_l - k_{il}^{bc} b_l - \tilde{\theta}_i^c v \right) = 0 \qquad (6.160)$$

The non-zero components on the left hand side of Equation (6.148) are

$$\frac{\partial U}{\partial v} = \frac{1}{2}\sum_{r=1}^{N}\left(b_r\tilde{\theta}_r^b - c_r\tilde{\theta}_r^c\right) \tag{6.161}$$

$$\frac{\partial W_{ie}}{\partial v} = C_p v - \frac{1}{2}\sum_{r=1}^{N}\left(b_r\tilde{\theta}_r^b - c_r\tilde{\theta}_r^c\right) \tag{6.162}$$

yielding

$$C_p v + Q - \sum_{r=1}^{N}\left(b_r\tilde{\theta}_r^b - c_r\tilde{\theta}_r^c\right) = 0 \tag{6.163}$$

Taking the time derivative of Equation (6.163) and using $\dot{Q} = v/R_l$ gives the Lagrange equation for v as

$$C_p\dot{v} + \frac{v}{R_l} - \sum_{r=1}^{N}\left(\dot{b}_r\tilde{\theta}_r^b - \dot{c}_r\tilde{\theta}_r^c\right) = 0 \tag{6.164}$$

The first three Lagrange equations can be given in matrix form as

$$
\begin{bmatrix} m^{aa} & 0 & 0 \\ 0 & m^{bb} & -m^{bc} \\ 0 & -m^{bc} & m^{cc} \end{bmatrix}
\begin{bmatrix} \ddot{a} \\ \ddot{b} \\ \ddot{c} \end{bmatrix}
+
\begin{bmatrix} k^{aa} & 0 & -k^{ac} \\ 0 & k^{bb} & -k^{bc} \\ -k^{ac} & -k^{bc} & k^{cc} \end{bmatrix}
\begin{bmatrix} a \\ b \\ c \end{bmatrix}
+
\begin{bmatrix} 0 \\ \tilde{\theta}^b \\ -\tilde{\theta}^c \end{bmatrix} v
=
\begin{bmatrix} f \\ 0 \\ 0 \end{bmatrix} \tag{6.165}
$$

Rayleigh damping is then introduced to account for the mechanical dissipative effects by preserving the normal-mode system:

$$
\begin{bmatrix} m^{aa} & 0 & 0 \\ 0 & m^{bb} & -m^{bc} \\ 0 & -m^{bc} & m^{cc} \end{bmatrix}
\begin{bmatrix} \ddot{a} \\ \ddot{b} \\ \ddot{c} \end{bmatrix}
+
\begin{bmatrix} d^{aa} & 0 & -d^{ac} \\ 0 & d^{bb} & -d^{bc} \\ -d^{ac} & -d^{bc} & d^{cc} \end{bmatrix}
\begin{bmatrix} \dot{a} \\ \dot{b} \\ \dot{c} \end{bmatrix}
$$
$$
+
\begin{bmatrix} k^{aa} & 0 & -k^{ac} \\ 0 & k^{bb} & -k^{bc} \\ -k^{ac} & -k^{bc} & k^{cc} \end{bmatrix}
\begin{bmatrix} a \\ b \\ c \end{bmatrix}
+
\begin{bmatrix} 0 \\ \tilde{\theta}^b \\ -\tilde{\theta}^c \end{bmatrix} v
=
\begin{bmatrix} f \\ 0 \\ 0 \end{bmatrix} \tag{6.166}
$$

Here the damping matrix is

$$
\begin{bmatrix} d^{aa} & 0 & -d^{ac} \\ 0 & d^{bb} & -d^{bc} \\ -d^{ac} & -d^{bc} & d^{cc} \end{bmatrix}
= \mu
\begin{bmatrix} m^{aa} & 0 & 0 \\ 0 & m^{bb} & -m^{bc} \\ 0 & -m^{bc} & m^{cc} \end{bmatrix}
+ \gamma
\begin{bmatrix} k^{aa} & 0 & -k^{ac} \\ 0 & k^{bb} & -k^{bc} \\ -k^{ac} & -k^{bc} & k^{cc} \end{bmatrix} \tag{6.167}
$$

where μ and γ are the constants of mass and stiffness proportionality.

Equation (6.164) can be expressed as

$$C_p \dot{v} + \frac{v}{R_l} - \left(\tilde{\theta}^b\right)^t \dot{b} + \left(\tilde{\theta}^c\right)^t \dot{c} = 0 \tag{6.168}$$

Equations (6.166) and (6.168) are the discretized equations of the distributed-parameter electromechanical system where the $N \times 1$ vectors of generalized coordinates are

$$\mathbf{a} = \begin{bmatrix} a_1 \ a_2 \ \dots \ a_N \end{bmatrix}^t, \quad \mathbf{b} = \begin{bmatrix} b_1 \ b_2 \ \dots \ b_N \end{bmatrix}^t, \quad \mathbf{c} = \begin{bmatrix} c_1 \ c_2 \ \dots \ c_N \end{bmatrix}^t \tag{6.169}$$

and the $N \times 1$ vectors of electromechanical coupling are

$$\tilde{\theta}^b = \begin{bmatrix} \tilde{\theta}_1^b \ \tilde{\theta}_2^b \ \dots \ \tilde{\theta}_N^b \end{bmatrix}^t, \quad \tilde{\theta}^c = \begin{bmatrix} \tilde{\theta}_1^c \ \tilde{\theta}_2^c \ \dots \ \tilde{\theta}_N^c \end{bmatrix}^t \tag{6.170}$$

6.4.4 Solution of the Electromechanical Lagrange Equations

The discretized equations of the electromechanical system are

$$\mathbf{m}^{aa}\ddot{\mathbf{a}} + \mathbf{d}^{aa}\dot{\mathbf{a}} - \mathbf{d}^{ac}\dot{\mathbf{c}} + \mathbf{k}^{aa}\mathbf{a} - \mathbf{k}^{ac}\mathbf{c} = \mathbf{f} \tag{6.171}$$

$$\mathbf{m}^{bb}\ddot{\mathbf{b}} - \mathbf{m}^{bc}\ddot{\mathbf{c}} + \mathbf{d}^{bb}\dot{\mathbf{b}} - \mathbf{d}^{bc}\dot{\mathbf{c}} + \mathbf{k}^{bb}\mathbf{b} - \mathbf{k}^{bc}\mathbf{c} + \tilde{\theta}^b v = 0 \tag{6.172}$$

$$-\mathbf{m}^{bc}\ddot{\mathbf{b}} + \mathbf{m}^{cc}\ddot{\mathbf{c}} - \mathbf{d}^{ac}\dot{\mathbf{a}} - \mathbf{d}^{bc}\dot{\mathbf{b}} + \mathbf{d}^{cc}\dot{\mathbf{c}} - \mathbf{k}^{ac}\mathbf{a} - \mathbf{k}^{bc}\mathbf{b} + \mathbf{k}^{cc}\mathbf{c} - \tilde{\theta}^c v = 0 \tag{6.173}$$

$$C_p \dot{v} + \frac{v}{R_l} - \left(\tilde{\theta}^b\right)^t \dot{b} + \left(\tilde{\theta}^c\right)^t \dot{c} = 0 \tag{6.174}$$

For harmonic base displacement of the form $g(t) = W_0 e^{j\omega t}$ and $h(t) = \theta_0 e^{j\omega t}$, the components of the forcing vector become

$$\mathbf{f} = \mathbf{F} e^{j\omega t} \tag{6.175}$$

Here,

$$F_r = -\sigma_r \omega^2 W_0 - \tau_r \omega^2 \theta_0 \tag{6.176}$$

where

$$\sigma_r = -\int_0^L \left(\rho_s A_s + \rho_p A_p\right) \phi_r(x) \, dx \tag{6.177}$$

$$\tau_r = -\int_0^L \left(\rho_s A_s + \rho_p A_p\right) x \phi_r(x) \, dx \tag{6.178}$$

The steady-state generalized coordinates and the voltage output are $\mathbf{a} = \mathbf{A}e^{j\omega t}$, $\mathbf{b} = \mathbf{B}e^{j\omega t}$, $\mathbf{c} = \mathbf{C}e^{j\omega t}$, and $v = Ve^{j\omega t}$ (where the elements of $\mathbf{A}$, $\mathbf{B}$, and $\mathbf{C}$ as well as V are complex valued).

The steady-state forms of Equations (6.171)–(6.174) are then

$$\left(-\omega^2 \mathbf{m}^{\mathrm{aa}} + j\omega \mathbf{d}^{\mathrm{aa}} + \mathbf{k}^{\mathrm{aa}}\right)\mathbf{A} - \left(j\omega \mathbf{d}^{\mathrm{ac}} + \mathbf{k}^{\mathrm{ac}}\right)\mathbf{C} = \mathbf{F} \tag{6.179}$$

$$\left(-\omega^2 \mathbf{m}^{\mathrm{bb}} + j\omega \mathbf{d}^{\mathrm{bb}} + \mathbf{k}^{\mathrm{bb}}\right)\mathbf{B} - \left(-\omega^2 \mathbf{m}^{\mathrm{bc}} + j\omega \mathbf{d}^{\mathrm{bc}} + \mathbf{k}^{\mathrm{bc}}\right)\mathbf{C} + \tilde{\theta}^{\mathrm{b}} V = 0 \tag{6.180}$$

$$- \left(j\omega \mathbf{d}^{\mathrm{ac}} + \mathbf{k}^{\mathrm{ac}}\right)\mathbf{A} - \left(-\omega^2 \mathbf{m}^{\mathrm{bc}} + j\omega \mathbf{d}^{\mathrm{bc}} + \mathbf{k}^{\mathrm{bc}}\right)\mathbf{B} + \left(-\omega^2 \mathbf{m}^{\mathrm{cc}} + j\omega \mathbf{d}^{\mathrm{cc}} + \mathbf{k}^{\mathrm{cc}}\right)\mathbf{C} - \tilde{\theta}^{\mathrm{c}} V = 0 \tag{6.181}$$

$$\left(j\omega C_p + \frac{1}{R_l}\right) V = j\omega \left[\left(\tilde{\theta}^{\mathrm{b}}\right)^t \mathbf{B} - \left(\tilde{\theta}^{\mathrm{c}}\right)^t \mathbf{C}\right] \tag{6.182}$$

From Equation (6.182),

$$V = j\omega \left(j\omega C_p + \frac{1}{R_l}\right)^{-1} \left[\left(\tilde{\theta}^{\mathrm{b}}\right)^t \mathbf{B} - \left(\tilde{\theta}^{\mathrm{c}}\right)^t \mathbf{C}\right] \tag{6.183}$$

Substituting Equation (6.183) into Equations (6.180) and (6.181) gives

$$\mathbf{\Gamma}^{\mathrm{aa}}\mathbf{A} - \mathbf{\Gamma}^{\mathrm{ac}}\mathbf{C} = \mathbf{F} \tag{6.184}$$

$$-\mathbf{\Gamma}^{\mathrm{bc}}\mathbf{C} + \mathbf{\Gamma}^{\mathrm{bb}}\mathbf{B} = 0 \tag{6.185}$$

$$-\mathbf{\Gamma}^{\mathrm{ca}}\mathbf{A} - \mathbf{\Gamma}^{\mathrm{cb}}\mathbf{B} + \mathbf{\Gamma}^{\mathrm{cc}}\mathbf{C} = 0 \tag{6.186}$$

where

$$\mathbf{\Gamma}^{\mathrm{aa}} = -\omega^2 \mathbf{m}^{\mathrm{aa}} + j\omega \mathbf{d}^{\mathrm{aa}} + \mathbf{k}^{\mathrm{aa}} \tag{6.187}$$

$$\mathbf{\Gamma}^{\mathrm{bb}} = -\omega^2 \mathbf{m}^{\mathrm{bb}} + j\omega \mathbf{d}^{\mathrm{bb}} + \mathbf{k}^{\mathrm{bb}} + j\omega \left(j\omega C_p + \frac{1}{R_l}\right)^{-1} \tilde{\theta}^{\mathrm{b}} \left(\tilde{\theta}^{\mathrm{b}}\right)^t \tag{6.188}$$

$$\mathbf{\Gamma}^{\mathrm{cc}} = -\omega^2 \mathbf{m}^{\mathrm{cc}} + j\omega \mathbf{d}^{\mathrm{cc}} + \mathbf{k}^{\mathrm{cc}} + j\omega \left(j\omega C_p + \frac{1}{R_l}\right)^{-1} \tilde{\theta}^{\mathrm{c}} \left(\tilde{\theta}^{\mathrm{c}}\right)^t \tag{6.189}$$

$$\boldsymbol{\Gamma}^{ac} = \boldsymbol{\Gamma}^{ca} = j\omega \mathbf{d}^{ac} + \mathbf{k}^{ac} \tag{6.190}$$

$$\boldsymbol{\Gamma}^{bc} = -\omega^2 \mathbf{m}^{bc} + j\omega \mathbf{d}^{bc} + \mathbf{k}^{bc} + j\omega \left(j\omega C_p + \frac{1}{R_l} \right)^{-1} \tilde{\theta}^b \left(\tilde{\theta}^c \right)^t \tag{6.191}$$

$$\boldsymbol{\Gamma}^{cb} = -\omega^2 \mathbf{m}^{bc} + j\omega \mathbf{d}^{bc} + \mathbf{k}^{bc} + j\omega \left(j\omega C_p + \frac{1}{R_l} \right)^{-1} \tilde{\theta}^c \left(\tilde{\theta}^b \right)^t \tag{6.192}$$

From Equation (6.185),

$$\mathbf{B} = \left(\boldsymbol{\Gamma}^{bb} \right)^{-1} \boldsymbol{\Gamma}^{bc} \mathbf{C} \tag{6.193}$$

Substituting Equation (6.193) into Equation (6.186) gives

$$\mathbf{A} = \left(\boldsymbol{\Gamma}^{ca} \right)^{-1} \left[\boldsymbol{\Gamma}^{cc} - \boldsymbol{\Gamma}^{cb} \left(\boldsymbol{\Gamma}^{bb} \right)^{-1} \boldsymbol{\Gamma}^{bc} \right] \mathbf{C} \tag{6.194}$$

Using Equation (6.194) in Equation (6.184) gives the vector of complex generalized coordinates $\mathbf{C}$ as

$$\mathbf{C} = \left\{ \boldsymbol{\Gamma}^{aa} \left(\boldsymbol{\Gamma}^{ca} \right)^{-1} \left[\boldsymbol{\Gamma}^{cc} - \boldsymbol{\Gamma}^{cb} \left(\boldsymbol{\Gamma}^{bb} \right)^{-1} \boldsymbol{\Gamma}^{bc} \right] - \boldsymbol{\Gamma}^{ac} \right\}^{-1} \mathbf{F} \tag{6.195}$$

which can be substituted into Equation (6.193) to give vector $\mathbf{B}$:

$$\mathbf{B} = \left(\boldsymbol{\Gamma}^{bb} \right)^{-1} \boldsymbol{\Gamma}^{bc} \left\{ \boldsymbol{\Gamma}^{aa} \left(\boldsymbol{\Gamma}^{ca} \right)^{-1} \left[\boldsymbol{\Gamma}^{cc} - \boldsymbol{\Gamma}^{cb} \left(\boldsymbol{\Gamma}^{bb} \right)^{-1} \boldsymbol{\Gamma}^{bc} \right] - \boldsymbol{\Gamma}^{ac} \right\}^{-1} \mathbf{F} \tag{6.196}$$

Similarly, the vector of complex generalized coordinates $\mathbf{A}$ is obtained as

$$\mathbf{A} = \left(\boldsymbol{\Gamma}^{ca} \right)^{-1} \left[\boldsymbol{\Gamma}^{cc} - \boldsymbol{\Gamma}^{cb} \left(\boldsymbol{\Gamma}^{bb} \right)^{-1} \boldsymbol{\Gamma}^{bc} \right] \left\{ \boldsymbol{\Gamma}^{aa} \left(\boldsymbol{\Gamma}^{ca} \right)^{-1} \left[\boldsymbol{\Gamma}^{cc} - \boldsymbol{\Gamma}^{cb} \left(\boldsymbol{\Gamma}^{bb} \right)^{-1} \boldsymbol{\Gamma}^{bc} \right] - \boldsymbol{\Gamma}^{ac} \right\}^{-1} \mathbf{F} \tag{6.197}$$

and the complex voltage is

$$V = j\omega \left(j\omega C_p + \frac{1}{R_l} \right)^{-1} \left[\left(\tilde{\theta}^b \right)^t \left(\boldsymbol{\Gamma}^{bb} \right)^{-1} \boldsymbol{\Gamma}^{bc} - \left(\tilde{\theta}^c \right)^t \right]$$
$$\left\{ \boldsymbol{\Gamma}^{aa} \left(\boldsymbol{\Gamma}^{ca} \right)^{-1} \left[\boldsymbol{\Gamma}^{cc} - \boldsymbol{\Gamma}^{cb} \left(\boldsymbol{\Gamma}^{bb} \right)^{-1} \boldsymbol{\Gamma}^{bc} \right] - \boldsymbol{\Gamma}^{ac} \right\}^{-1} \mathbf{F} \tag{6.198}$$

Substituting the elements of **A**, **B**, and **C** into Equations (6.124)–(6.126) gives the transverse displacement, the axial displacement, and the cross-section rotation response at steady state as

$$w^0(x, t) = \sum_{r=1}^{N} A_r e^{j\omega t} \phi_r(x) \tag{6.199}$$

$$u^0(x, t) = \sum_{r=1}^{N} B_r e^{j\omega t} \alpha_r(x) \tag{6.200}$$

$$\psi^0(x, t) = \sum_{r=1}^{N} C_r e^{j\omega t} \beta_r(x) \tag{6.201}$$

The steady-state voltage response is

$$v(t) = j\omega \left(j\omega C_p + \frac{1}{R_l} \right)^{-1} \left[\left(\tilde{\theta}^b \right)^t \left(\Gamma^{bb} \right)^{-1} \Gamma^{bc} - \left(\tilde{\theta}^c \right)^t \right]$$
$$\left\{ \Gamma^{aa} \left(\Gamma^{ca} \right)^{-1} \left[\Gamma^{cc} - \Gamma^{cb} \left(\Gamma^{bb} \right)^{-1} \Gamma^{bc} \right] - \Gamma^{ac} \right\}^{-1} F e^{j\omega t} \tag{6.202}$$

One can use the form of the base excitation given by Equation (6.176) ($F = -\sigma \omega^2 W_0 - \tau \omega^2 \theta_0$) and define the following eight electromechanical FRFs for the four steady-state response expressions given by Equations (6.199)–(6.202):

$$\frac{w^0(x, t)}{-\omega^2 W_0 e^{j\omega t}}, \quad \frac{u^0(x, t)}{-\omega^2 W_0 e^{j\omega t}}, \quad \frac{\psi^0(x, t)}{-\omega^2 W_0 e^{j\omega t}}, \quad \frac{v(t)}{-\omega^2 W_0 e^{j\omega t}}, \quad \frac{w^0(x, t)}{-\omega^2 \theta_0 e^{j\omega t}},$$
$$\frac{u^0(x, t)}{-\omega^2 \theta_0 e^{j\omega t}}, \quad \frac{\psi^0(x, t)}{-\omega^2 \theta_0 e^{j\omega t}}, \quad \frac{v(t)}{-\omega^2 \theta_0 e^{j\omega t}} \tag{6.203}$$

The short-circuit and open-circuit natural frequencies of the system can be obtained by setting the mechanical damping terms equal to zero ($d \rightarrow 0$) and considering the $R_l \rightarrow 0$ and $R_l \rightarrow \infty$ cases, respectively, in the following characteristic equation obtained from the free vibration ($F = 0$) problem:

$$\begin{vmatrix} \Gamma^{aa} & 0 & -\Gamma^{ac} \\ 0 & \Gamma^{bb} & -\Gamma^{bc} \\ -\Gamma^{ca} & -\Gamma^{cb} & \Gamma^{cc} \end{vmatrix} = 0 \tag{6.204}$$

6.5   Modeling of Symmetric Configurations

6.5.1   Euler–Bernoulli and Rayleigh Models

For a geometrically symmetric configuration (i.e., symmetric laminates) such as the symmetric bimorph configurations shown in Figure 3.1, the H_s and H_p terms causing the coupling between the transverse displacement and the axial displacement vanish, reducing the governing discretized equations of the system (given by Equations (6.65)–(6.67)) to

$$\mathbf{m^{aa}\ddot{a}} + \mathbf{d^{aa}\dot{a}} + \mathbf{k^{aa}a} - \tilde{\theta}^{\mathbf{a}}v = \mathbf{f} \tag{6.205}$$

$$C_p\dot{v} + \frac{v}{R_l} + \left(\tilde{\theta}^{\mathbf{a}}\right)^t \dot{\mathbf{a}} = 0 \tag{6.206}$$

This simplest form is similar to the symmetric thin-bimorph equations derived using the Rayleigh–Ritz technique [2–5]. Following the well-known procedure with the assumption of harmonic excitation ($\mathbf{f} = \mathbf{F}e^{j\omega t}$), the steady-state response expressions are harmonic of the forms $\mathbf{a} = \mathbf{A}e^{j\omega t}$ and $v = Ve^{j\omega t}$, yielding

$$\left(-\omega^2\mathbf{m^{aa}} + j\omega\mathbf{d^{aa}} + \mathbf{k^{aa}}\right)\mathbf{A} - \tilde{\theta}^{\mathbf{a}}V = \mathbf{F} \tag{6.207}$$

$$V = -j\omega\left(j\omega C_p + \frac{1}{R_l}\right)^{-1}\left(\tilde{\theta}^{\mathbf{a}}\right)^t\mathbf{A} \tag{6.208}$$

These last expressions give the complex generalized coordinate vector $\mathbf{A}$ and the complex voltage as

$$\mathbf{A} = \left(\mathbf{\Gamma^{aa}}\right)^{-1}\mathbf{F} \tag{6.209}$$

$$V = -j\omega\left(j\omega C_p + \frac{1}{R_l}\right)^{-1}\left(\tilde{\theta}^{\mathbf{a}}\right)^t\left(\mathbf{\Gamma^{aa}}\right)^{-1}\mathbf{F} \tag{6.210}$$

where $\tilde{\theta}^{\mathbf{a}}$ and $\mathbf{\Gamma^{aa}}$ are as given by Equations (6.39) and (6.78), respectively.

6.5.2 Timoshenko Model

For a symmetric cantilever, Equation (6.172) decouples from the electromechanical equations of the system, reducing the governing equations to

$$\mathbf{m^{aa}\ddot{a}} + \mathbf{d^{aa}\dot{a}} - \mathbf{d^{ac}\dot{c}} + \mathbf{k^{aa}a} - \mathbf{k^{ac}c} = \mathbf{f} \tag{6.211}$$

$$\mathbf{m^{cc}\ddot{c}} - \mathbf{d^{ac}\dot{a}} + \mathbf{d^{cc}\dot{c}} - \mathbf{k^{ac}a} + \mathbf{k^{cc}c} - \tilde{\theta}^{\mathbf{c}}v = \mathbf{0} \tag{6.212}$$

$$C_p\dot{v} + \frac{v}{R_l} + \left(\tilde{\theta}^{\mathbf{c}}\right)^t\dot{\mathbf{c}} = 0 \tag{6.213}$$

With the assumption of harmonic excitation ($\mathbf{f} = \mathbf{F}e^{j\omega t}$) the steady-state response expressions are $\mathbf{a} = \mathbf{A}e^{j\omega t}$, $\mathbf{c} = \mathbf{C}e^{j\omega t}$, and $v = Ve^{j\omega t}$. It follows from Equations (6.211)–(6.213) that

$$\mathbf{A} = \left(\boldsymbol{\Gamma}^{\mathbf{ca}}\right)^{-1}\boldsymbol{\Gamma}^{\mathbf{cc}}\left[\boldsymbol{\Gamma}^{\mathbf{aa}}\left(\boldsymbol{\Gamma}^{\mathbf{ca}}\right)^{-1}\boldsymbol{\Gamma}^{\mathbf{cc}} - \boldsymbol{\Gamma}^{\mathbf{ac}}\right]^{-1}\mathbf{F} \tag{6.214}$$

$$\mathbf{C} = \left[\boldsymbol{\Gamma}^{\mathbf{aa}}\left(\boldsymbol{\Gamma}^{\mathbf{ca}}\right)^{-1}\boldsymbol{\Gamma}^{\mathbf{cc}} - \boldsymbol{\Gamma}^{\mathbf{ac}}\right]^{-1}\mathbf{F} \tag{6.215}$$

$$V = -j\omega\left(j\omega C_p + \frac{1}{R_l}\right)^{-1}\left(\tilde{\boldsymbol{\theta}}^{\mathbf{c}}\right)^t\left[\boldsymbol{\Gamma}^{\mathbf{aa}}\left(\boldsymbol{\Gamma}^{\mathbf{ca}}\right)^{-1}\boldsymbol{\Gamma}^{\mathbf{cc}} - \boldsymbol{\Gamma}^{\mathbf{ac}}\right]^{-1}\mathbf{F} \tag{6.216}$$

where $\tilde{\boldsymbol{\theta}}^{\mathbf{c}}$, $\boldsymbol{\Gamma}^{\mathbf{aa}}$, $\boldsymbol{\Gamma}^{\mathbf{cc}}$, and $\boldsymbol{\Gamma}^{\mathbf{ac}} = \boldsymbol{\Gamma}^{\mathbf{ca}}$ are as given by Equations (6.144), (6.187), (6.189), and (6.190), respectively.

6.6   Presence of a Tip Mass in the Euler–Bernoulli, Rayleigh, and Timoshenko Models

If the energy harvester configuration shown in Figure 6.1 has a tip mass of M_t with a mass moment of inertia of I_t (about $x = L$), the total kinetic energy expressions should be modified in the electromechanical models derived in this chapter.

In the Euler–Bernoulli and Rayleigh models, the total kinetic energy expression with the kinetic energy contribution of a tip mass located at $x = L$ becomes

$$T = \frac{1}{2}\int_{V_s}\rho_s\left[\left(\frac{\partial u^0(x,t)}{\partial t} - z\frac{\partial^2 w^0(x,t)}{\partial t\partial x}\right)^2 + \left[\frac{\partial w^0(x,t)}{\partial t} + \frac{\partial w_b(x,t)}{\partial t}\right]^2\right]dV_s$$

$$+ \frac{1}{2}\int_{V_p}\rho_p\left[\left(\frac{\partial u^0(x,t)}{\partial t} - z\frac{\partial^2 w^0(x,t)}{\partial t\partial x}\right)^2 + \left[\frac{\partial w^0(x,t)}{\partial t} + \frac{\partial w_b(x,t)}{\partial t}\right]^2\right]dV_p$$

$$+ \frac{1}{2}M_t\left[\left(\frac{\partial w^0(x,t)}{\partial t} + \frac{\partial w_b(x,t)}{\partial t}\right)\Bigg|_{x=L}\right]^2 + \frac{1}{2}I_t\left[\frac{\partial^2 w^0(x,t)}{\partial t\partial x}\Bigg|_{x=L}\right]^2 \tag{6.217}$$

which modifies the submatrix m_{rl}^{aa} in the Euler–Bernoulli model (where the distributed rotary inertia is neglected) to

$$m_{rl}^{aa} = \int_0^L\left(\rho_s A_s + \rho_p A_p\right)\phi_r(x)\phi_l(x)\,dx + M_t\phi_r(L)\phi_l(L) + I_t\phi_r'(L)\phi_l'(L) \tag{6.218}$$

and the submatrix m_{rl}^{aa} in the Rayleigh model becomes

$$m_{rl}^{aa} = \int_0^L \left[\left(\rho_s A_s + \rho_p A_p \right) \phi_r(x) \phi_l(x) + \left(\rho_s I_s + \rho_p I_p \right) \phi_r'(x) \phi_l'(x) \right] dx$$

$$+ M_t \phi_r(L) \phi_l(L) + I_t \phi_r'(L) \phi_l'(L) \qquad (6.219)$$

In both the Euler–Bernoulli and Rayleigh models, the term p_r derived from the total kinetic energy expression becomes

$$p_r = \int_0^L \left(\rho_s A_s + \rho_p A_p \right) \phi_r(x) \frac{\partial w_b(x, t)}{\partial t} dx + M_t \phi_r(L) \left. \frac{\partial w_b(x, t)}{\partial t} \right|_{x=L} \qquad (6.220)$$

which alters the effective force due to base excitation as follows:

$$f_r = -\frac{\partial p_r}{\partial t} = -\int_0^L \left(\rho_s A_s + \rho_p A_p \right) \phi_r(x) \frac{\partial^2 w_b(x, t)}{\partial t^2} dx - M_t \phi_r(L) \left. \frac{\partial^2 w_b(x, t)}{\partial t^2} \right|_{x=L}$$

$$= -\frac{d^2 g(t)}{dt^2} \left[\int_0^L \left(\rho_s A_s + \rho_p A_p \right) \phi_r(x) dx + M_t \phi_r(L) \right]$$

$$- \frac{d^2 h(t)}{dt^2} \left[\int_0^L \left(\rho_s A_s + \rho_p A_p \right) x \phi_r(x) dx + M_t L \phi_r(L) \right] \qquad (6.221)$$

It also modifies the σ_r and τ_r terms accordingly in the split representation of the forcing term $(F_r = -\sigma_r \omega^2 W_0 - \tau_r \omega^2 \theta_0)$.

After the inclusion of a tip mass, the total kinetic energy expression in the Timoshenko model becomes

$$T = \frac{1}{2} \int_{V_s} \rho_s \left[\left(\frac{\partial u^0(x, t)}{\partial t} - z \frac{\partial \psi^0(x, t)}{\partial t} \right)^2 + \left[\frac{\partial w^0(x, t)}{\partial t} + \frac{\partial w_b(x, t)}{\partial t} \right]^2 \right] dV_s$$

$$+ \frac{1}{2} \int_{V_p} \rho_p \left[\left(\frac{\partial u^0(x, t)}{\partial t} - z \frac{\partial \psi^0(x, t)}{\partial t} \right)^2 + \left[\frac{\partial w^0(x, t)}{\partial t} + \frac{\partial w_b(x, t)}{\partial t} \right]^2 \right] dV_p$$

$$+ \frac{1}{2} M_t \left[\left(\frac{\partial w^0(x, t)}{\partial t} + \frac{\partial w_b(x, t)}{\partial t} \right) \Big|_{x=L} \right]^2 + \frac{1}{2} I_t \left[\frac{\partial \psi^0(x, t)}{\partial t} \Big|_{x=L} \right]^2 \qquad (6.222)$$

The submatrices altered in the Timoshenko model due to this modification are m_{rl}^{aa} and m_{rl}^{cc}:

$$m_{rl}^{aa} = \int_0^L \left(\rho_s A_s + \rho_p A_p \right) \phi_r(x) \phi_l(x) \, dx + M_t \phi_r(L) \phi_l(L) \tag{6.223}$$

$$m_{rl}^{cc} = \int_0^L \left(\rho_s I_s + \rho_p I_p \right) \beta_r(x) \beta_l(x) \, dx + I_t \beta_r(L) \beta_l(L) \tag{6.224}$$

The base excitation-related terms derived from the total kinetic energy take the forms given by Equations (6.220) and (6.221) (identical in the Euler–Bernoulli, Rayleigh, and Timoshenko models).

6.7   Comments on the Kinematically Admissible Trial Functions

6.7.1   Euler–Bernoulli and Rayleigh Models

The essential boundary conditions of a clamped–free beam in the Euler–Bernoulli and Rayleigh models are given in Appendix G.1. According to the kinematic boundary conditions at the clamped end, the admissible functions in Equations (6.24) and (6.25) should satisfy

$$\phi_r(0) = 0 \tag{6.225}$$

$$\phi_r'(0) = 0 \tag{6.226}$$

$$\alpha_r(0) = 0 \tag{6.227}$$

For the admissible functions $\phi_r(x)$ of the transverse displacement, one can use the eigenfunctions of the respective symmetric and uniform structure given in Chapter 3 (or in Appendix C.1). Therefore,

$$\phi_r(x) = \cos \frac{\lambda_r}{L} x - \cosh \frac{\lambda_r}{L} x + \varsigma_r \left(\sin \frac{\lambda_r}{L} x - \sinh \frac{\lambda_r}{L} x \right) \tag{6.228}$$

where ς_r is obtained from

$$\varsigma_r = \frac{\sin \lambda_r - \sinh \lambda_r + \lambda_r \frac{M_t}{mL} (\cos \lambda_r - \cosh \lambda_r)}{\cos \lambda_r + \cosh \lambda_r - \lambda_r \frac{M_t}{mL} (\sin \lambda_r - \sinh \lambda_r)} \tag{6.229}$$

Here, λ_r is the rth root of the transcendental equation for the rth vibration mode:

$$1 + \cos \lambda \cosh \lambda + \lambda \frac{M_t}{mL} (\cos \lambda \sinh \lambda - \sin \lambda \cosh \lambda) - \frac{\lambda^3 I_t}{mL^3} (\cosh \lambda \sin \lambda + \sinh \lambda \cos \lambda)$$
$$+ \frac{\lambda^4 M_t I_t}{m^2 L^4} (1 - \cos \lambda \cosh \lambda) = 0 \tag{6.230}$$

where m is the mass per length of the beam, which can be given for a uniform cantilever as

$$m = \rho_s A_s + \rho_p A_p \tag{6.231}$$

The foregoing expressions simplify considerably in the absence of a tip mass ($M_t = I_t = 0$). Indeed, even in the presence of a tip mass, one can use the form of the $\phi_r(x)$ for $M_t = I_t = 0$,

$$\phi_r(x) = \cos \frac{\lambda_r}{L} x - \cosh \frac{\lambda_r}{L} x + \frac{\sin \lambda_r - \sinh \lambda_r}{\cos \lambda_r + \cosh \lambda_r} \left(\sin \frac{\lambda_r}{L} x - \sinh \frac{\lambda_r}{L} x \right) \tag{6.232}$$

which is still kinematically admissible with λ_r obtained from

$$1 + \cos \lambda \cosh \lambda = 0 \tag{6.233}$$

However, in the presence of a tip mass, using Equation (6.228) with the eigenvalues obtained from Equation (6.230) can lead to faster convergence (with fewer modes) in the discretized system. Note that the foregoing admissible functions become the eigenfunctions for a symmetric and uniform structure.

If one prefers to avoid the hyperbolic functions appearing in the eigenfunctions of the symmetric and uniform structure, the following is a typical admissible function used for clamped–free boundary conditions [19]:

$$\phi_r(x) = 1 - \cos \left[\frac{(2r - 1) \pi x}{2L} \right] \tag{6.234}$$

which satisfies Equations (6.225) and (6.226) ($r = 1, 2, \ldots, N$). Polynomial forms and static solutions can also be used to satisfy Equations (6.225) and (6.226).

Similarly, the eigenfunctions of the uniform structure under longitudinal vibrations can be used as admissible functions of the asymmetric structure here. From Chapter 2 (or Appendix C.2),

$$\alpha_r(x) = \sin \frac{\eta_r}{L} x \tag{6.235}$$

where η_r is the rth root of the transcendental equation for the rth vibration mode:

$$\frac{M_t}{mL} \eta_r \sin \eta_r - \cos \eta_r = 0 \tag{6.236}$$

Alternatively, the roots of $\cos \eta_r = 0$ (i.e., $\eta_r = (2r - 1)\pi/2, r = 1, 2, \ldots, N$) can be used in Equation (6.235) for simplicity.

6.7.2   Timoshenko Model

The essential boundary conditions of a clamped–free Timoshenko beam are given in Appendix G.2. Based on the kinematic boundary conditions, the admissible functions in Equations

(6.124)–(6.126) should satisfy

$$\phi_r(0) = 0 \tag{6.237}$$

$$\alpha_r(0) = 0 \tag{6.238}$$

$$\beta_r(0) = 0 \tag{6.239}$$

According to Equation (6.237), one can use the form of $\phi_r(x)$ given by Equation (6.228) since it satisfies Equation (6.237). However, it is useful to note that $\phi'_r(0) = 0$ implies zero shear strain at the root (due to $\beta_r(0) = 0$), which is not realistic for a clamped boundary. A simple trigonometric function similar to Equation (6.235) could be a better alternative compared to Equation (6.228). Equations (6.238) and (6.239) also accept trigonometric forms. However, using similar trigonometric functions might result in cancellations of the cross-integrals in Equations (6.133)–(6.144) due to the orthogonality of trigonometric functions. Polynomial forms can be employed as an alternative. Several other alternatives exist in the literature, such as implementations of the Chebyshev polynomials [20] and static solutions [21].

6.8   Experimental Validation of the Assumed-Modes Solution for a Bimorph Cantilever

6.8.1   PZT-5H Bimorph Cantilever without a Tip Mass

The experimental case study given for the PZT-5H bimorph cantilever without a tip mass in Section 4.1 is revisited here for validation of the electromechanical assumed-modes solution. The assumed-modes counterpart of the analytical thin-beam solution given in Chapter 3 is the Euler–Bernoulli formulation given in Section 6.2. Note that the structure is thin enough to neglect the effects of shear deformation and rotary inertia in modeling as far as the fundamental vibration mode is concerned. For a device with moderate thickness, the Timoshenko formulation should be used. According to the geometric and materials properties of the cantilever given in Table 4.1, the coupling between the transverse and the longitudinal displacement components vanishes (i.e., due to the structural symmetry, $H_s = H_p = 0$). The admissible function used in all simulations is the trigonometric admissible function given by Equation (6.234) (therefore the exact eigenfunction, although available, is not used). All the comparisons here are given against the experimental measurements and the analytical solutions for the entire set of resistors used in Section 4.1.

Figure 6.2 shows the assumed-modes predictions with only one mode ($N = 1$). Both the voltage and the tip velocity predictions are highly inaccurate (especially in terms of the resonance frequency). If the number of modes in the assume-modes procedure is increased to $N = 3$, the predictions are improved substantially, as observed in Figure 6.3. Further increase in the number of modes up to $N = 5$ (Figure 6.4) and then to $N = 10$ (Figure 6.5) provides convergence to the analytical frequencies (but not as dramatic as the improvement from one mode to three modes). Although further increasing the number of modes does not seem to improve the model predictions significantly, including more modes improves the predictions

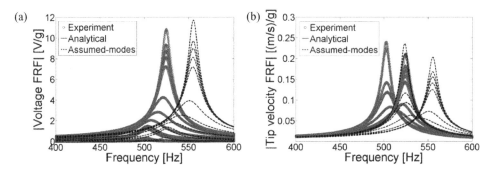

Figure 6.2 (a) Voltage FRFs and (b) tip velocity FRFs of the PZT-5H bimorph cantilever without a tip mass ($N = 1$ in the assumed-modes solution)

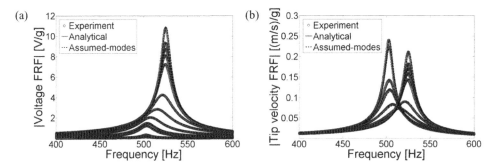

Figure 6.3 (a) Voltage FRFs and (b) tip velocity FRFs of the PZT-5H bimorph cantilever without a tip mass ($N = 3$ in the assumed-modes solution)

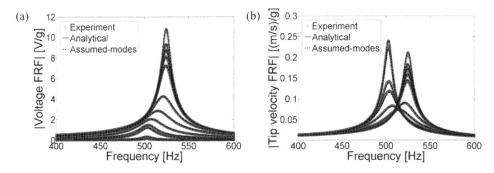

Figure 6.4 (a) Voltage FRFs and (b) tip velocity FRFs of the PZT-5H bimorph cantilever without a tip mass ($N = 5$ in the assumed-modes solution)

of higher vibration modes which are not discussed here. Table 6.1 lists the assumed-modes prediction of the short-circuit and open-circuit resonance frequencies with increasing number of modes along with the analytical and the experimental results. Note that the fundamental natural frequency estimated using this technique gives an upper bound of the lowest natural frequency [1] (as in the Rayleigh–Ritz method) and the approximate fundamental

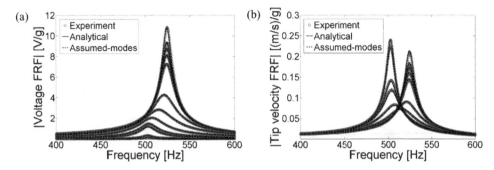

Figure 6.5 (a) Voltage FRFs and (b) tip velocity FRFs of the PZT-5H bimorph cantilever without a tip mass ($N = 10$ in the assumed-modes solution)

Table 6.1 Predictions of the short-circuit and the open-circuit resonance frequencies of the voltage FRF for the PZT-5H bimorph cantilever without a tip mass using the assumed-modes method

	f_1^{sc} (Hz)	f_1^{oc} (Hz)
Experimental	502.5	524.7
Analytical	502.6	524.5
Assumed-modes ($N = 1$)	523.8	555.3
Assumed-modes ($N = 3$)	503.2	525.5
Assumed-modes ($N = 5$)	502.7	524.7
Assumed-modes ($N = 10$)	502.6	524.5

natural frequency cannot underestimate the analytical value regardless of the number of modes used.

6.8.2 PZT-5H Bimorph Cantilever with a Tip Mass

The configuration tested in Section 4.2 (the same cantilever of Section 6.8.1 with a tip mass attachment) is revisited next. The only difference in the formulation (compared to that of Section 6.8.1) is due to the contribution of the tip mass information and its mass moment of inertia to the mass matrix and the forcing vector as discussed in Section 6.6. Therefore, Equations (6.118) and (6.221) should be used in order to calculate the submatrix m_{rl}^{aa} and the forcing vector f_r in the assume-modes solution procedure of Section 6.2. All resistors are considered in the comparisons against the analytical solutions and the experimental results. The same admissible function given by Equation (6.234) is used in the assumed-modes simulations.

The assumed-modes prediction for only one mode ($N = 1$) gives highly inaccurate predictions as shown in Figure 6.6 (as in the previous case). Just like the case without the

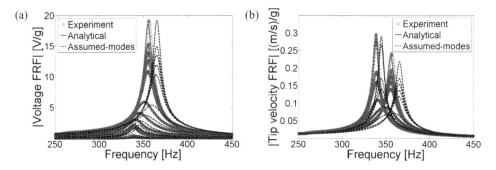

Figure 6.6 (a) Voltage FRFs and (b) tip velocity FRFs of the PZT-5H bimorph cantilever with a tip mass ($N = 1$ in the assumed-modes solution)

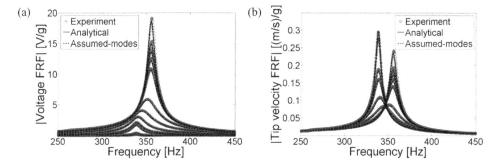

Figure 6.7 Voltage FRFs and (b) tip velocity FRFs of the PZT-5H bimorph cantilever with a tip mass ($N = 3$ in the assumed-modes solution)

Table 6.2 Assumed-mode predictions of the short-circuit and the open-circuit resonance frequencies of the voltage FRF for the PZT-5H bimorph cantilever with a tip mass

	f_1^{sc} (Hz)	f_1^{oc} (Hz)
Experimental	338.4	356.3
Analytical	338.5	355.4
Assumed-modes ($N = 1$)	344.6	365.3
Assumed-modes ($N = 3$)	338.7	355.8
Assumed-modes ($N = 5$)	338.5	355.5
Assumed-modes ($N = 10$)	338.5	355.4

tip mass, if two more modes are added to the solution (so that $N = 3$) the predictions are improved substantially, as shown in Figure 6.7, which can also been seen from the predictions of the short-circuit resonance and the open-circuit resonance frequency predictions in Table 6.2. Further increase in the number of modes results in uniform convergence to the analytical frequencies as shown in Figures 6.8 and 6.9.

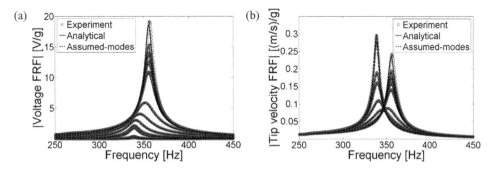

Figure 6.8 Voltage FRFs and (b) tip velocity FRFs of the PZT-5H bimorph cantilever with a tip mass ($N = 5$ in the assumed-modes solution)

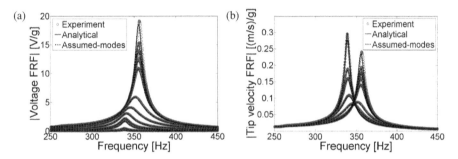

Figure 6.9 Voltage FRFs and (b) tip velocity FRFs of the PZT-5H bimorph cantilever with a tip mass ($N = 10$ in the assumed-modes solution)

6.9 Experimental Validation for a Two-Segment Cantilever

Self-charging structures [22–25] combine flexible piezoceramics and thin-film battery layers with elastic and adhesive layers for multifunctional load-bearing applications. Under dynamic loading conditions, these structures can generate and store electrical energy for use in low-power applications while carrying external loads. A self-charging structure tested under base excitation is shown in Figure 6.10a along with a schematic showing its layers in Figure 6.10b. The battery layers and two of the epoxy layers do not cover the entire overhang length, making the cantilever a two-segment structure. Accurate modeling of this self-charging structure is important in order to predict the fundamental natural frequency for charging the battery layers using the piezoelectric output under resonance excitation. Predicting the voltage FRF with acceptable accuracy is also crucial in order to estimate the level of base acceleration required to reach the battery voltage level and also to predict the maximum AC power output.

The structure shown in Figure 6.10a is thin enough to neglect the shear deformation and rotary inertia effects as far as the fundamental vibration mode is concerned. Furthermore, the configuration is symmetric with respect to the center of the width (therefore the fundamental vibration mode is expected to be pure bending due to symmetric base excitation). If one derives the eigenfunction expression of a two-segment structure and normalizes it appropriately according to the modal analysis procedure (Appendix C.1) by taking into account the

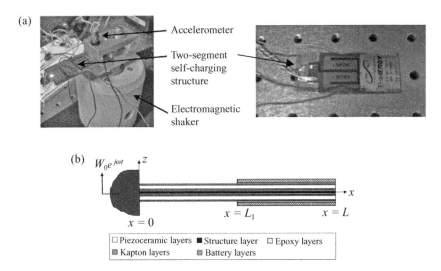

Figure 6.10 (a) Base excitation of a two-segment self-charging cantilever and (b) its schematic

piecewise-defined mass in the base excitation force, the analytical solution given in Chapter 3 can be used; that is, the analytical solution is available for this structural configuration. However, due to the discontinuity at $x = L_1$ in Figure 6.10b, the eigenfunctions need to be piecewise defined and they have to satisfy four more equations at $x = L_1$ (these are called the *matching* conditions) in addition to four boundary conditions (yielding an 8×8 coefficient matrix in the resulting eigenvalue problem). It is worth recalling that the elements of the coefficient matrix will include several hyperbolic functions along with trigonometric functions (often making the matrix ill-conditioned). Compared to a uniform beam, calculation of the eigenvalues is more involved as the transcendental characteristic equation (the determinant of the 8×8 matrix) becomes a lot more complicated than Equation (3.20). One can anticipate that the eigensolution becomes more and more cumbersome with increasing number of segments. The assumed-modes method covered in this chapter is preferred for modeling such configurations with non-uniform geometric and material properties as in the present example. In the following, predictions of the assumed-modes solution are compared against the experimental results for different numbers of modes. The trigonometric admissible function given by Equation (6.234) is used again for its simplicity.

The PZT-5A piezoceramic layers (QP10N, Midé Technology Corporation) of the self-charging structure come from the manufacturer as embedded in kapton layers. Therefore, bonding two of them onto two faces of an aluminum layer (to make a bimorph) using epoxy (3M DP460) results in a nine-layered symmetric structure. Therefore the root region ($0 < x < L_1$) has nine layers consisting of aluminum, piezoceramic, kapton, and epoxy layers. In addition to these layers, bonding two thin-film battery layers (MEC101-7, Infinite Power Solutions) onto two faces toward the tip section of the cantilever makes the tip region ($L_1 < x < L$) thirteen layered. The resulting assembly is symmetric with respect to the neutral surface with fairly even bonding (epoxy) layers and the piezoceramic layers (poled in the thickness direction) are combined in series for the resistor sweep. An experimental setup similar to the

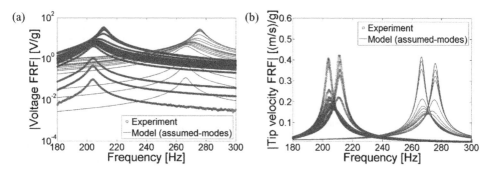

Figure 6.11 (a) Voltage FRFs and (b) tip velocity FRFs of the two-segment cantilever ($N = 1$ in the assumed-modes solution)

one described in Section 4.1 is used for the electromechanical modal analysis of the cantilever. The voltage and tip velocity FRFs are measured for 17 different values of load resistance ranging from 100 Ω to 1 MΩ. Details of the particular self-charging structure shown in Figure 6.10a as well as its electrical storage and mechanical strength aspects can be found in Anton *et al.* [25].

Figure 6.11 shows that using one single trigonometric function in the assumed-modes solution is highly insufficient, as observed in the validation examples of Section 6.8 for the PZT-5H bimorph cantilevers. However, the inaccuracy in Figure 6.11 is much higher than the cases given in Figures 6.2 and 6.6 since the structure tested in this section is more sophisticated than a uniform cantilever. Therefore the admissible function given by Equation (6.234) is less capable of representing its dynamics with a single mode. After the addition of a few more modes, however, the results are improved significantly as depicted in Figure 6.12. For the case with five assumed modes, the error in the short-circuit and open-circuit resonance frequency predictions is less than 1%. As shown in Figures 6.13 and 6.14, along with Table 6.3, including more modes improves the accuracy slightly and the predicted frequencies show slight variations as they tend to converge to the *analytical* values (which are not obtained here).

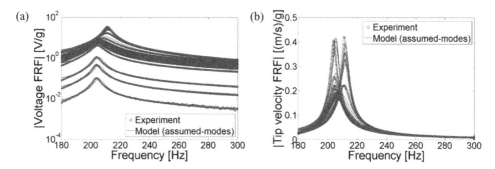

Figure 6.12 (a) Voltage FRFs and (b) tip velocity FRFs of the two-segment cantilever ($N = 5$ in the assumed-modes solution)

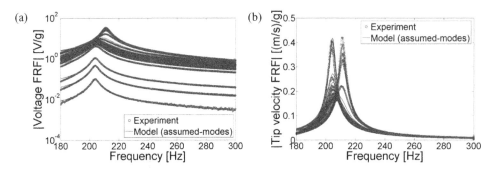

Figure 6.13 (a) Voltage FRFs and (b) tip velocity FRFs of the two-segment cantilever ($N = 10$ in the assumed-modes solution)

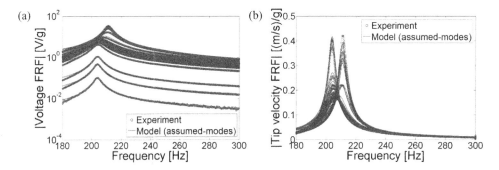

Figure 6.14 (a) Voltage FRFs and (b) tip velocity FRFs of the two-segment cantilever ($N = 20$ in the assumed-modes solution)

Table 6.3 Assumed-modes predictions of the short-circuit and the open-circuit resonance frequencies of the voltage FRF for the two-segment cantilever

	f_1^{sc} (Hz)	f_1^{oc} (Hz)
Experimental	204.0	211.1
Assumed-modes ($N = 1$)	266.4	275.9
Assumed-modes ($N = 5$)	205.4	212.0
Assumed-modes ($N = 10$)	204.5	211.2
Assumed-modes ($N = 20$)	204.1	211.0

The assumed-modes solution derived here is therefore a powerful alternative for the modeling of relatively sophisticated piezoelectric energy harvester configurations.

6.10 Summary

Approximate distributed-parameter modeling of cantilevered piezoelectric energy harvesters is given in this chapter. An electromechanical version of the assumed-modes method of structural dynamics is used to discretize the energy equations into electromechanical Lagrange equations derived from the extended Hamilton's principle. The derivations are given based on the Euler–Bernoulli, Rayleigh, and Timoshenko beam theories. In all cases, an axial displacement variable is defined to account for its coupling with the transverse displacement variable due to structural asymmetry. To demonstrate the modeling of a configuration with asymmetric laminates, the focus is placed on the unimorph configuration. Simplification of the governing equations for symmetric configurations (such as the bimorph configurations covered in Chapters 3 and 4) is also shown and the effect of a tip mass on the resulting formulation is discussed. A short discussion regarding the kinematically admissible functions to be used in the models derived here is also provided. Finally the experimental case studies for a thin bimorph (in the absence and presence of a tip mass) are revisited for validation of the assumed-modes solution using different numbers of admissible trigonometric functions. The predictions of the assumed-modes solution are also compared to the analytical solution and perfect agreement is obtained when a sufficient number of admissible functions is used. Application of the assumed-modes formulation to a two-segment self-charging cantilever with flexible piezoceramics and thin-film batteries is also summarized.

6.11 Chapter Notes

The approximate analytical formulations given in this chapter can be used for predicting the electromechanical response of cantilevers with varying cross-section, asymmetric laminates as well as moderately thick configurations. These are typical problems for which the analytical solution either does not exist or is more involved than the solutions covered in Chapter 3. For instance, multi-segment (or stepped) thin beams (such as the two-segment configuration tested in Section 6.9) can be modeled analytically as long as the segments are geometrically and materially uniform. However, the resulting eigenvalue problem is cumbersome due to the increased dimensions of the differential eigenvalue problem. An n-segment beam results in $4n$-4 matching conditions (in the form of compatibility and continuity equations) at the connection points of the segments in addition to the already existing four boundary conditions. Finding the coefficients of the exact eigenfunctions from the determinant of the $4n \times 4n$ matrix[6] is an involved task that makes the assumed-modes solution preferable. For several other configurations, such as cantilevers with tapered geometry (as modeled by Benasciutti *et al.* [26] and Yang and Tang [27] using finite element software), the analytical solution simply does not exist and the assumed-modes method is a powerful alternative. Since the assumed-modes

---

[6] The *condition number* of the resulting matrix becomes an important problem due to the coexistence of matrix elements (hyperbolic functions and trigonometric functions) with orders-of-magnitude numerical difference.

method is an approximate analytical solution technique, one should use sufficient numbers of modes to ensure convergence as demonstrated in the case studies of this chapter.

The assumed-modes method for distributed-parameter systems is closely related to the Rayleigh–Ritz method as further discussed by Meirovitch [1]. If the modeling assumptions (displacement field etc.) are the same, the discretized equations obtained using both methods are identical for the same admissible functions. The simplest case discussed in this chapter is the Euler–Bernoulli configuration with symmetric laminates, such as the thin and symmetric bimorph configurations. The reduced equations given in Section 6.5.1 for this case are indeed identical to the formerly reported Rayleigh–Ritz type of piezoelectric energy harvester equations [3–5], which follow the derivations given by Hagood *et al.* [2]. As demonstrated by Elvin and Elvin [5] for the Rayleigh–Ritz formulation, lumped-parameters extracted from the assumed-modes solution can be used as transformer parameters in circuit-simulator software for building and analyzing nonlinear energy harvesting circuits (see Chapter 11).

One final remark is that, although modeling based on the Timoshenko beam theory can handle problems with moderate beam thickness, one should formulate the problem using a higher-order shear deformable theory for relatively high thickness levels. An example of such theories is the third-order shear deformable theory introduced by Reddy [28]. Elimination of the shear correction factor is an important advantage of higher-order shear deformable theories, in addition to their improved accuracy in predicting the dynamics of relatively thick structures.

References

1. Meirovitch, L. (2001) *Fundamentals of Vibrations*, McGraw-Hill, New York.
2. Hagood, N.W., Chung, W.H., and Von Flotow, A. (1990) Modelling of piezoelectric actuator dynamics for active structural control. *Journal of Intelligent Material Systems and Structures*, **1**, 327–354.
3. Sodano, H.A., Park, G., and Inman, D.J. (2004) Estimation of electric charge output for piezoelectric energy harvesting. *Strain*, **40**, 49–58.
4. duToit, N.E. and Wardle, B.L. (2007) Experimental verification of models for microfabricated piezoelectric vibration energy harvesters. *AIAA Journal*, **45**, 1126–1137.
5. Elvin, N.G. and Elvin, A.A. (2009) A general equivalent circuit model for piezoelectric generators. *Journal of Intelligent Material Systems and Structures*, **20**, 3–9.
6. Erturk, A. and Inman, D.J. (2008) A distributed parameter electromechanical model for cantilevered piezoelectric energy harvesters. *ASME Journal of Vibration and Acoustics*, **130**, 041002.
7. Dym, C.L. and Shames, I.H. (1973) *Solid Mechanics: A Variational Approach*, McGraw-Hill, New York.
8. Timoshenko, S.P. (1921) On the correction for shear of the differential equation for transverse vibrations of prismatic bars. *Philosophical Magazine*, **41**, 744–746.
9. Timoshenko, S.P. (1922) On the transverse vibrations of bars of uniform cross-section. *Philosophical Magazine*, **43**, 125–131.
10. Mindlin, R.D. (1951) Thickness-shear and flexural vibrations of crystal plates. *Journal of Applied Physics*, **22**, 316–323.
11. Mindlin, R.D. (1952) Forced thickness-shear and flexural vibrations of piezoelectric crystal plates. *Journal of Applied Physics*, **23**, 83–88.
12. Cowper, G.R. (1966) The shear coefficient in Timoshenko beam theory. *ASME Journal of Applied Mechanics*, **33**, 335–340.
13. Kaneko, T. (1975) On Timoshenko's correction for shear in vibrating beams. *Journal of Physics D: Applied Physics*, **8**, 1927–1936.
14. Stephen, N.G. (1978) On the variation of Timoshenko's shear coefficient with frequency. *ASME Journal of Applied Mechanics*, **45**, 695–697.

15. Stephen, N.G. (1980) Timoshenko's shear coefficient from a beam subjected to gravity loading. *ASME Journal of Applied Mechanics*, **47**, 121–127.
16. Stephen, N.G. and Hutchinson, J.R. (2001) Discussion: shear coefficients for Timoshenko beam theory. *ASME Journal of Applied Mechanics*, **68**, 959–961.
17. Hutchinson, J.R. (2001) Shear coefficients for Timoshenko beam theory. *ASME Journal of Applied Mechanics*, **68**, 87–92.
18. Puchegger, S., Bauer, S., Loidl, D., Kromp, K., and Peterlik, H. (2003) Experimental validation of the shear correction factor. *Journal of Sound and Vibration*, **261**, 177–184.
19. Den Hartog, J.P. (1956) *Mechanical Vibrations*, McGraw-Hill, New York.
20. Lee, J. and Schultz, W.W. (2004) Eigenvalue analysis of Timoshenko beams and axisymmetric Mindlin plates by the pseudospectral method. *Journal of Sound and Vibration*, **269**, 609–621.
21. Zhou, D. and Cheung, Y.K. (2001) Vibrations of tapered Timoshenko beams in terms of static Timoshenko beam functions. *ASME Journal of Applied Mechanics*, **68**, 596–602.
22. Erturk, A., Anton, S.R., and Inman, D.J. (2009) Piezoelectric energy harvesting from multifunctional wing spars for UAVs – Part 1: coupled modeling and preliminary analysis. *Proceedings of SPIE*, **7288**, 72880C.
23. Anton, S.R., Erturk, A., and Inman, D.J. (2009) Piezoelectric energy harvesting from multifunctional wing spars for UAVs – Part 2: experiments and storage applications. *Proceedings of SPIE*, **7288**, 72880D.
24. Anton, S.R., Erturk, A., Kong, N., Ha, D.S., and Inman, D.J. (2009) Self-charging structures using piezoceramics and thin-film batteries. Proceedings of the ASME Conference on Smart Materials, Adaptive Structures and Intelligent Systems, Oxnard, CA, September 20–24, 2009.
25. Anton, S.R., Erturk, A., and Inman, D.J. (2010) Multifunctional self-charging structures using piezoceramics and thin-film batteries. *Smart Materials and Structures*, **19**, 115021.
26. Benasciutti, D., Moro, L., Zelenika, S., and Brusa, E. (2009) Vibration energy scavenging via piezoelectric bimorphs of optimized shapes. *Microsystem Technologies*, **16**, 657–668.
27. Yang, Y. and Tang, L. (2009) Equivalent circuit modeling of piezoelectric energy harvesters. *Journal of Intelligent Material Systems and Structures*, **20**, 2223–2235.
28. Reddy, J.N. (1984) A simple higher-order theory for laminated composite plates. *ASME Journal of Applied Mechanics*, **51**, 745–752.

7

Modeling of Piezoelectric Energy Harvesting for Various Forms of Dynamic Loading

This chapter presents derivations for modeling the piezoelectric energy harvesting problem under dynamic loading conditions other than simple harmonic excitation. First the governing electromechanical equations of cantilevered piezoelectric energy harvesters are written by assuming the base acceleration to be an arbitrary function of time. After that, as the first non-harmonic excitation problem of interest, the base excitation is assumed to be periodic in time. The electromechanical frequency response functions (FRFs) derived in Chapter 3 are combined with the Fourier series representation of the base acceleration in order to predict the periodic voltage output. The second problem formulated in this chapter is the case where the base acceleration behaves like ideal white noise, which is a very representative type of random process used in various engineering applications. After deriving the expected value expression for the power output in response to white noise excitation, two approaches of piezoelectric power generation from moving loads are formulated. The first approach considers using a cantilever located at an arbitrary point on a slender bridge while the second approach considers using a piezoceramic patch covering a certain region on the bridge. In both cases, the input is a constant-amplitude load traversing at a certain speed. The local way of formulating piezoelectric power generation from surface patches is also addressed, where the inputs are taken as the dynamic strain components in two orthogonal directions. Finally, after presenting the electromechanical state-space equations for transient base acceleration inputs, two case studies are given for demonstration purpose.

7.1 Governing Electromechanical Equations

Before formulating the piezoelectric energy harvesting problem for different forms of base acceleration, it is worth reviewing the linear electromechanical equations of a symmetric bimorph cantilever under base excitation based on the fundamental results of Chapter 3. For convenience, the base rotation term is dropped as shown in Figure 7.1, and the derivations

Piezoelectric Energy Harvesting, First Edition. Alper Erturk and Daniel J. Inman.

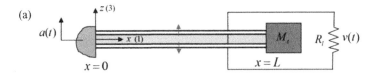

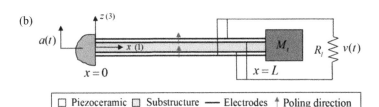

Figure 7.1 Cantilevered bimorph piezoelectric energy harvester configurations under base excitation: (a) series connection; and (b) parallel connection

in this chapter are given for the translational base acceleration $a(t)$, which is arbitrary at this stage (e.g., it might be periodic or random as discussed in Sections 7.2 and 7.3, respectively). The derivations can easily be extended to the case with superimposed base rotation by using the respective FRFs.

Following the analytical derivation steps given in Chapter 3, the dynamics of a bimorph piezoelectric energy harvester under the excitation of base acceleration, $a(t)$, are governed by

$$YI\frac{\partial^4 w_{rel}(x,t)}{\partial x^4} + c_s I\frac{\partial^5 w_{rel}(x,t)}{\partial x^4 \partial t} + c_a\frac{\partial w_{rel}(x,t)}{\partial t} + m\frac{\partial^2 w_{rel}(x,t)}{\partial t^2}$$

$$-\vartheta v(t)\left[\frac{d\delta(x)}{dx} - \frac{d\delta(x-L)}{dx}\right] = -[m + M_t\delta(x-L)]a(t) \tag{7.1}$$

$$C_{\tilde{p}}^{eq}\frac{dv(t)}{dt} + \frac{v(t)}{R_l} + \vartheta\int\limits_0^L\frac{\partial^3 w_{rel}(x,t)}{\partial x^2 \partial t}dx = 0 \tag{7.2}$$

where $w_{rel}(x,t)$ is the vibration response (transverse displacement of the neutral axis relative to the moving base at position x and time t), $v(t)$ is the voltage response (across the external resistive load R_l), YI is the bending stiffness, c_a is the viscous air damping coefficient, c_s is the strain-rate damping coefficient (appears as an effective term $c_s I$ for the composite structure), m is the mass per unit length of the beam, M_t is its tip mass, $C_{\tilde{p}}^{eq}$ is the equivalent capacitance of the piezoceramic layers, ϑ is the electromechanical coupling term in the physical coordinates, and $\delta(x)$ is the Dirac delta function. The electromechanical coupling term is $\vartheta = \bar{e}_{31}bh_{\tilde{p}c}$ if the layers are connected in series and $\vartheta = 2\bar{e}_{31}bh_{\tilde{p}c}$ if the layers are connected in parallel (where $\bar{e}_{31}$ is the plane-stress piezoelectric stress constant, b is the width of the layers, and $h_{\tilde{p}c}$ is the distance from the neutral axis to the center of each piezoceramic layer).

Based on the standard modal analysis procedure (assuming the system to be a normal-mode system),[1] the vibration response is expressed in terms of the modal coordinates $\eta_r(t)$ and the

---

[1] It is worth recalling that combined dynamical systems have certain limitations as far as the normal-mode assumption is concerned [1]. Nevertheless, as an engineering approximation, one can solve the undamped problem for $c_s I = c_a = 0$ (which is a normal-mode system) and introduce modal viscous damping in modal coordinates.

mode shapes $\phi_r(x)$ (mass-normalized eigenfunctions) as

$$w_{rel}(x, t) = \sum_{r=1}^{\infty} \phi_r(x)\eta_r(t) \tag{7.3}$$

The electromechanically coupled ordinary differential equations in modal coordinates are then

$$\frac{d^2\eta_r(t)}{dt^2} + 2\zeta_r\omega_r\frac{d\eta_r(t)}{dt} + \omega_r^2\eta_r(t) - \tilde{\theta}_r v(t) = \sigma_r a(t) \tag{7.4}$$

$$C_p^{eq}\frac{dv(t)}{dt} + \frac{v(t)}{R_l} + \sum_{r=1}^{\infty}\tilde{\theta}_r\frac{d\eta_r(t)}{dt} = 0 \tag{7.5}$$

where ω_r is the undamped natural frequency in short-circuit (i.e., constant electric field) conditions, ζ_r is the modal mechanical damping ratio, σ_r is a modal forcing-related term, and $\tilde{\theta}_r$ is the modal electromechanical coupling ($\tilde{\theta}_r$ and C_p^{eq} are read from Table 3.1 depending on the series or parallel connection of the piezoceramic layers).

For the particular case of harmonic base acceleration, $a(t) = A_0 e^{j\omega t}$, the steady-state analytical solution of the voltage output - to - base acceleration FRF (i.e., the voltage FRF) is given by

$$\alpha(\omega) = \frac{v(t)}{A_0 e^{j\omega t}} = \frac{\sum_{r=1}^{\infty}\dfrac{-j\omega\tilde{\theta}_r\sigma_r}{\omega_r^2 - \omega^2 + j2\zeta_r\omega_r\omega}}{\dfrac{1}{R_l} + j\omega C_p^{eq} + \sum_{r=1}^{\infty}\dfrac{j\omega\tilde{\theta}_r^2}{\omega_r^2 - \omega^2 + j2\zeta_r\omega_r\omega}} \tag{7.6}$$

and the steady-state vibration response - to - base acceleration FRF (i.e., the vibration FRF) is

$$\beta(\omega, x) = \frac{w_{rel}(x, t)}{A_0 e^{j\omega t}}$$

$$= \sum_{r=1}^{\infty}\left[\left(\sigma_r - \tilde{\theta}_r\frac{\sum_{r=1}^{\infty}\dfrac{j\omega\tilde{\theta}_r\sigma_r}{\omega_r^2 - \omega^2 + j2\zeta_r\omega_r\omega}}{\dfrac{1}{R_l} + j\omega C_p^{eq} + \sum_{r=1}^{\infty}\dfrac{j\omega\tilde{\theta}_r^2}{\omega_r^2 - \omega^2 + j2\zeta_r\omega_r\omega}}\right)\frac{\phi_r(x)}{\omega_r^2 - \omega^2 + j2\zeta_r\omega_r\omega}\right] \tag{7.7}$$

For excitations close to a natural frequency, that is, $\omega \approx \omega_r$, the multi-mode equations, Equations (7.6) and (7.7), reduce to the following single-mode equations:

$$\hat{\alpha}(\omega) = \frac{\hat{v}(t)}{A_0 e^{j\omega t}} = \frac{-j\omega R_l\tilde{\theta}_r\sigma_r}{\left(1 + j\omega R_l C_p^{eq}\right)\left(\omega_r^2 - \omega^2 + j2\zeta_r\omega_r\omega\right) + j\omega R_l\tilde{\theta}_r^2} \tag{7.8}$$

$$\hat{\beta}(\omega, x) = \frac{\hat{w}_{rel}(x, t)}{A_0 e^{j\omega t}} = \frac{\left(1 + j\omega R_l C_p^{eq}\right)\sigma_r\phi_r(x)}{\left(1 + j\omega R_l C_p^{eq}\right)\left(\omega_r^2 - \omega^2 + j2\zeta_r\omega_r\omega\right) + j\omega R_l\tilde{\theta}_r^2} \tag{7.9}$$

where $r = 1$ for the frequently investigated case of excitation around the fundamental vibration mode to obtain the largest power output.

Although the foregoing single-mode and multi-mode solutions are derived (Chapter 3) and experimentally validated (Chapter 4) for simple harmonic excitation, the derivations given in the following sections for various dynamic loading conditions show the importance of these equations in finding the response to different types of excitations as well.

7.2 Periodic Excitation

This section formulates the problem of predicting the voltage output across the resistive load given the periodic acceleration input. First the acceleration input is represented using the Fourier series expansion, and then the electromechanical FRFs derived in Chapter 3 are combined with the Fourier series representation of base acceleration for predicting the periodic voltage output. The resulting expressions can be used in problems where the excitation is not simple harmonic motion but it still exhibits periodicity (e.g., machine and mechanism components, parts of human body in walking, etc.).

7.2.1 Fourier Series Representation of Periodic Base Acceleration

The base acceleration in Figure 7.1 is said to be periodic if it satisfies the condition

$$a(t) = a(t + T) \tag{7.10}$$

where T is the period of the motion. The form of the base acceleration is therefore not simple harmonic at a single frequency (unlike the discussion of the previous chapters) but it is periodic. Based on the Fourier series expansion [2], one can express periodic functions as linear combinations of harmonic functions.[2]

The Fourier series expansion of the periodic acceleration can be given by

$$a(t) = p_0 + \sum_{k=1}^{\infty} \left[p_k \cos\left(k\frac{2\pi t}{T}\right) + q_k \sin\left(k\frac{2\pi t}{T}\right) \right] \tag{7.11}$$

where p_0 is the mean value while p_k and q_k ($k = 1, 2, \ldots$ are positive integers) are the Fourier coefficients given by

$$p_0 = \frac{1}{T} \int_0^T a(t)\, dt \tag{7.12}$$

$$p_k = \frac{2}{T} \int_0^T a(t) \cos\left(k\frac{2\pi t}{T}\right) dt \tag{7.13}$$

$$q_k = \frac{2}{T} \int_0^T a(t) \sin\left(k\frac{2\pi t}{T}\right) dt \tag{7.14}$$

---

[2] Complex (exponential) functions can also be used as an alternative to using trigonometric functions in the Fourier series expansion [2].

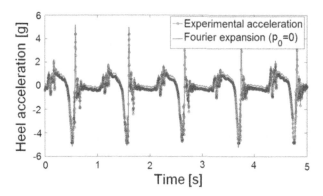

Figure 7.2 Transverse acceleration at the heel during walking (an example of periodic acceleration with negligible mean value)

Two approximations follow the Fourier series expansion of the base acceleration. First, the summation of harmonics will be truncated after taking N harmonic pairs (one should take as many terms as required to ensure convergence). Secondly, the constant part in the expansion will be neglected by setting $p_0 = 0$. Considering typical periodic acceleration fluctuations in most physical systems, this second approximation can be justified for several cases (an illustration follows below). Then the truncated form of the Fourier series expansion becomes

$$a(t) \cong \sum_{k=1}^{N} \left[p_k \cos\left(k\frac{2\pi t}{T} \right) + q_k \sin\left(k\frac{2\pi t}{T} \right) \right] \tag{7.15}$$

where the fluctuating acceleration $a(t)$ is represented by harmonic functions only.

Acceleration fluctuations in several physical systems can indeed be represented in the foregoing form. An example of interest in energy harvesting research is the kinetic energy of a human body. During walking, several parts of the body undergo rigid body motion with acceleration variations. One of the locations with the highest level of acceleration is the *heel*, because it accelerates and decelerates with every step. Figure 7.2 shows the acceleration history measured by locating an accelerometer at the heel to measure the component of the acceleration transverse to the heel in the walking direction. By integrating this time history for five seconds (as if it consists of a single period, although it includes five periods of motion), the Fourier coefficients are obtained for 500 harmonic pairs (the process is easier if one takes only one second of motion in Figure 7.2). The Fourier series expansion is plotted with the original data in Figure 7.2 and it can be seen that the $p_0 = 0$ approximation works very well. The expansion given by Equation (7.15) can therefore be used for representing fluctuating acceleration components with negligible mean value. One should make sure to truncate the expansion only after it converges to the original function.

7.2.2   Periodic Electromechanical Response

After expressing the base acceleration in terms of harmonic functions, the electromechanical response of the system can be estimated based on the linear system assumption. That is, *the electromechanical response of the energy harvester to the summation of a set of harmonics is equal to the summation of its responses to the individual harmonics.*

Using the voltage FRF given by Equation (7.6), the periodic voltage output is given by

$$
v(t) \cong \sum_{k=1}^{N} \left| \alpha \left(k \frac{2\pi t}{T} \right) \right| \left\{ p_k \cos \left[\left(k \frac{2\pi t}{T} \right) + \Phi \left(k \frac{2\pi t}{T} \right) \right] \right.
$$

$$
\left. + q_k \sin \left[\left(k \frac{2\pi t}{T} \right) + \Phi \left(k \frac{2\pi t}{T} \right) \right] \right\}
$$

(7.16)

where $|\alpha(\omega)|$ is the modulus of the voltage FRF and $\Phi(\omega)$ is its phase angle. For the Fourier series representations with frequencies around the higher vibration modes of the harvester, one should use the multi-mode form of the voltage FRF given by Equation (7.6). The highest harmonic of the Fourier expansion $(2N\pi/T)$ can be checked with the natural frequencies of the energy harvester to decide on the number of modes to use in the voltage FRF. If the highest frequency in the Fourier expansion is around the fundamental vibration mode of the energy harvester, the single-mode voltage FRF given by Equation (7.8) can safely be used.

The periodic power output can be obtained from the following expression:

$$
P(t) \cong \frac{1}{R_l} \left\langle \sum_{k=1}^{N} \left| \alpha \left(k \frac{2\pi t}{T} \right) \right| \left\{ p_k \cos \left[\left(k \frac{2\pi t}{T} \right) + \Phi \left(k \frac{2\pi t}{T} \right) \right] \right. \right.
$$

$$
\left. \left. + q_k \sin \left[\left(k \frac{2\pi t}{T} \right) + \Phi \left(k \frac{2\pi t}{T} \right) \right] \right\} \right\rangle^2
$$

(7.17)

Similarly, using the vibration FRF given by Equation (7.7), it is straightforward to express the periodic vibration response of the energy harvester as

$$
w_{rel}(x, t) \cong \sum_{k=1}^{N} \left| \beta \left(k \frac{2\pi t}{T}, x \right) \right| \left\{ p_k \cos \left[\left(k \frac{2\pi t}{T} \right) + \Psi \left(k \frac{2\pi t}{T}, x \right) \right] \right.
$$

$$
\left. + q_k \sin \left[\left(k \frac{2\pi t}{T} \right) + \Psi \left(k \frac{2\pi t}{T}, x \right) \right] \right\}
$$

(7.18)

which can be used for the mechanical design and stress analysis purposes.

7.3 White Noise Excitation

The harmonic and periodic excitations covered in the previous sections are *deterministic* forms of excitation. That is, the characteristics of the excitation can be determined at any future time. Moreover, in general, the response of linear systems to deterministic excitations is also deterministic. Therefore the previous statement regarding the determination of the future behavior is valid for the response as well (assuming that a reliable model exists to express the deterministic response in terms of the deterministic excitation). Several physical phenomena,

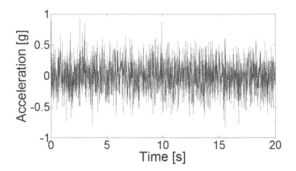

Figure 7.3   Gaussian white noise signal generated in MATLAB

however, do not allow explicit time description in a deterministic sense. If the behavior of a process at an arbitrary future time cannot be predicted, it is called a *non-deterministic process* (or *stochastic process* or *random process*). Several engineering systems and structures undergo random vibrations and the goal of the engineer dealing with random vibrations is to relate the statistical characteristics of the excitation (such as the mean value and standard deviation)[3] to the statistical characteristics of the response (such as the expected value) since a deterministic representation of the response is not possible.

This section derives electromechanical equations for the expected value of the harvested electrical power output by assuming that the base acceleration $a(t)$ is a special type of random process: *ideal white noise*. White noise has a flat *power spectral density* (PSD) over the entire frequency band [3,4]. Moreover, it is a *stationary* random process, meaning that its mean and variance do not change over the time. A sample time history generated using the *white Gaussian noise* command in MATLAB is shown in Figure 7.3 for visualization. The derivation given in this section can therefore be used for estimating the electrical response of piezoelectric energy harvesters excited by random vibration inputs of broad frequency content and flat PSD.

7.3.1   Representation of the Base Acceleration

Since the base acceleration, $a(t)$, is assumed to be ideal white noise, its PSD covers the entire band of frequencies with constant amplitude (i.e., $a(t)$ is made up of all frequencies with equal power contribution). It can be shown that the *autocorrelation function* of ideal white noise is the Dirac delta function. Let the autocorrelation function $R_a(\tau)$ of the white noise type base acceleration be expressed as

$$R_a(\tau) = 2\pi S_0 \delta(\tau) \tag{7.19}$$

---

[3] For standard definitions and derivations in random data analysis and random vibrations used in this section, the reader is referred to Newland [3] and Bendat and Piersol [4], among others [5,6].

where S_0 is the PSD of the base acceleration and $\delta(x)$ is the Dirac delta function. Recalling that the PSD of $a(t)$ is the Fourier transform of its autocorrelation function, one can obtain [3]

$$S_a(\omega) = \frac{1}{2\pi} \int_{-\infty}^{\infty} R_a(\tau) e^{-j\omega\tau} \, d\tau = \frac{1}{2\pi} \int_{-\infty}^{\infty} 2\pi S_0 \delta(\tau) e^{-j\omega\tau} \, d\tau = S_0 \qquad (7.20)$$

which is simply the flat value of the input PSD. In the following derivation, it is assumed that S_0 of the base acceleration is given as the input.

7.3.2 Spectral Density and Autocorrelation Function of the Voltage Response

In a linear stochastic process, it can be shown that [3] the PSD of the voltage output, $S_v(\omega)$, is related to the PSD of the acceleration input through the following equation:

$$S_v(\omega) = |\alpha(\omega)|^2 \, S_a(\omega) \qquad (7.21)$$

where $\alpha(\omega)$ is the FRF that relates the voltage across the load to the base acceleration as given by Equation (7.6) for the full (multi-mode) solution and by Equation (7.8) with the single-mode approximation considering the fundamental vibration mode when $r = 1$.

The inverse Fourier transform of the PSD of the voltage output is its autocorrelation function:

$$R_v(\tau) = \int_{-\infty}^{\infty} S_v(\omega) e^{j\omega\tau} \, d\omega = \int_{-\infty}^{\infty} |\alpha(\omega)|^2 \, S_a(\omega) e^{j\omega\tau} \, d\omega \qquad (7.22)$$

For the case of ideal white noise excitation, using Equation (7.20) in Equation (7.22) leads to

$$R_v(\tau) = S_0 \int_{-\infty}^{\infty} |\alpha(\omega)|^2 \, e^{j\omega\tau} \, d\omega \qquad (7.23)$$

7.3.3 Expected Value of the Power Output

The *mean square value* of the voltage output is related to its autocorrelation function based on the following expression:

$$E\left[v^2(t)\right] = R_v(0) = \int_{-\infty}^{\infty} |\alpha(\omega)|^2 \, S_a(\omega) \, d\omega \qquad (7.24)$$

For the case of white noise excitation,

$$E\left[v^2(t)\right] = R_v(0) = S_0 \int_{-\infty}^{\infty} |\alpha(\omega)|^2 \, d\omega \tag{7.25}$$

It is worth recalling that the electrical power output is simply $v^2(t)/R_l$. Hence the *expected value* of the power output is

$$E\left[P(t)\right] = \frac{S_0}{R_l} \int_{-\infty}^{\infty} |\alpha(\omega)|^2 \, d\omega \tag{7.26}$$

The multi-mode form of the voltage FRF is given by Equation (7.6) but it does not lead to a straightforward expression when used in the above integral. However, considering the fundamental vibration mode in the single-mode voltage FRF (Equation (7.8)) provides a good approximation by covering the most important frequency range as far as the expected value of the power output is concerned:

$$E\left[P(t)\right] \cong \frac{S_0}{R_l} \int_{-\infty}^{\infty} |\hat{\alpha}(\omega)|^2 \, d\omega \tag{7.27}$$

where the single-mode estimate of the voltage output for the fundamental vibration mode is obtained by setting $r = 1$ in Equation (7.8) to give

$$\hat{\alpha}(\omega) = \frac{-j\omega R_l \tilde{\theta}_1 \sigma_1}{\left(1 + j\omega R_l C_p^{eq}\right)\left(\omega_1^2 - \omega^2 + j2\zeta_1\omega_1\omega\right) + j\omega R_l \tilde{\theta}_1^2} \tag{7.28}$$

The following expression can be found in the Appendix of Newland's text [3]:

$$\int_{-\infty}^{\infty} \left|\frac{B_0 + j\omega B_1 - \omega^2 B_2}{A_0 + j\omega A_1 - \omega^2 A_2 - j\omega^3 A_3}\right|^2 \, d\omega = \frac{\pi\left[A_0 A_3 \left(2B_0 B_2 - B_1^2\right) - A_0 A_1 B_2^2 - A_2 A_3 B_0^2\right]}{A_0 A_3 \left(A_0 A_3 - A_1 A_2\right)} \tag{7.29}$$

Putting Equation (7.28) into the form of Equation (7.29) gives the expected power output as

$$E\left[P(t)\right] \cong \frac{\pi S_0 R_l \tilde{\theta}_1^2 \sigma_1^2}{R_l \tilde{\theta}_1^2 + 2\zeta_1\omega_1\left[\left(1 + R_l^2 \tilde{\theta}_1^2 C_p^{eq}\right) + \left(2\zeta_1 + R_l C_p^{eq}\omega_1\right)\left(R_l C_p^{eq}\omega_1\right)\right]} \tag{7.30}$$

Using the dimensionless terms derived in Chapter 5 (mainly Equations (5.13) and (5.14)), one obtains

$$E\left[P(t)\right] \cong \frac{\pi S_0 \upsilon_1 \gamma_1 \sigma_1^2}{2\zeta_1\left[1 + \upsilon_1^2\left(1 + \gamma_1\right)\right] + \upsilon_1\left(\gamma_1 + 4\zeta_1^2\right)} \tag{7.31}$$

where $\upsilon_1 = R_l C_p^{eq} \omega_1$ and $\gamma_1 = \tilde{\theta}_1^2/(C_p^{eq}\omega_1^2)$. Given the PSD level of the white noise (base acceleration), Equation (7.31) gives an estimate for the expected value of the piezoelectric power output for an external load of R_l. Equation (7.31) can be further analyzed for parameter optimization. Similar steps can be followed for estimating the mean square value of the vibration response using the respective FRF.

7.4 Excitation Due to Moving Loads

This section formulates the problem of piezoelectric power generation from moving loads in two subsections. The focus is placed on slender bridge configurations for possible applications to high-span highway bridges. The first case considers the case of using a cantilevered piezoelectric energy harvester by locating it at an arbitrary point on the bridge and the second case investigates the problem of using a thin piezoceramic patch covering an arbitrary region on the slender bridge.

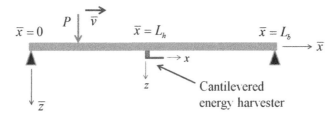

Figure 7.4 Schematic of the moving load problem for a slender bridge with a cantilevered piezoelectric energy harvester (the size of the cantilever is exaggerated)

7.4.1 Cantilevered Piezoelectric Energy Harvester Located on a Bridge

The dynamics of a uniform slender bridge[4] (Figure 7.4) under the excitation of a transversely applied constant amplitude load P (representing a vehicle) moving at a constant speed $\bar{\upsilon}$ are governed by

$$\overline{YI}\frac{\partial^4 \bar{w}(\bar{x}, t)}{\partial \bar{x}^4} + \bar{c}_s \bar{I}\frac{\partial^5 \bar{w}(\bar{x}, t)}{\partial \bar{x}^4 \partial t} + \bar{c}_a \frac{\partial \bar{w}(\bar{x}, t)}{\partial t} + \bar{m}\frac{\partial^2 \bar{w}(\bar{x}, t)}{\partial t^2} = P\delta(\bar{x} - \bar{\upsilon}t) \qquad (7.32)$$

where $\bar{w}(\bar{x}, t)$ is the vibration response of the bridge (transverse displacement of the neutral axis relative to the moving base at position $\bar{x}$ and time t), $\overline{YI}$ is its bending stiffness, $\bar{m}$ is the mass per length, $\bar{c}_s \bar{I}$ and $\bar{c}_a$ represent the stiffness proportional and mass proportional damping components, and $\delta(\bar{x})$ is the Dirac delta function. The cantilevered piezoelectric energy harvester is located at $\bar{x} = L_h$ and it should be noted that its reference frame is the

---

[4] In addition to its physical meaning for civil infrastructure systems (particularly high-span highway bridges), the term *bridge* is used instead of the term *beam* to avoid confusion with the harvester beam (the dynamics of which are governed by a similar equation). The structure can be considered as an arbitrary (but uniform and slender) beam for other possible applications.

xz-frame (as in Figure 7.1) while the reference frame of the bridge is the $\bar{x}\bar{z}$-frame as depicted in Figure 7.4 (therefore x is the axial position on the harvester beam while $\bar{x}$ is the axial position on the bridge). The governing electromechanical equations of the cantilevered energy harvester are as given by Equations (7.1) and (7.2). A reasonable assumption in the following formulation is that the effect of the harvester beam and its dynamics on the bridge is negligible. That is, as the load travels on the bridge, the energy harvester is excited and electrical power is generated but the bridge dynamics are not affected (hence no electrical term exists in Equation (7.32)). The base acceleration of the energy harvester in Equation (7.1) is therefore related to the vibration response of the bridge due to the moving load:

$$a(t) = \left. \frac{\partial^2 \bar{w}(\bar{x}, t)}{\partial t^2} \right|_{\bar{x}=L_h} \tag{7.33}$$

The following derivation therefore provides the solution of $\bar{w}(\bar{x}, t)$ for $0 \le t \le T$ (where $T = L_b/\bar{v}$ is the time of traverse of the moving load over the bridge), which can be used in Equation (7.33) to obtain the input $a(t)$ to Equations (7.1) and (7.2), which govern the dynamics of the energy harvester.

The analytical treatment of the undamped version of Equation (7.32) can be found in many texts and articles [7–10]. The damped problem results in rather lengthy expressions but the modal analysis procedure given in Appendix C.1 is applicable since the system is assumed to be proportionally damped with the form of Equation (7.32).

The vibratory response of the bridge can be expressed as

$$\bar{w}(\bar{x}, t) = \sum_{r=1}^{\infty} \bar{\phi}_r(\bar{x})\bar{\eta}_r(t) \tag{7.34}$$

where $\bar{\eta}_r(t)$ is the modal coordinate and $\bar{\phi}_r(\bar{x})$ is the mass-normalized eigenfunction for the rth vibration mode of the bridge with simple end conditions and given by

$$\bar{\phi}_r(\bar{x}) = \sqrt{\frac{2}{\bar{m}L_b}} \sin\left(\frac{r\pi\bar{x}}{L_b}\right) \tag{7.35}$$

which satisfies the orthogonality conditions

$$\int_0^{L_b} \bar{\phi}_s(\bar{x})\bar{m}\bar{\phi}_r(\bar{x})\,d\bar{x} = \delta_{rs}, \quad \int_0^{L_b} \bar{\phi}_s(\bar{x})\overline{YI}\frac{d^4\bar{\phi}_r(\bar{x})}{d\bar{x}^4}\,d\bar{x} = \bar{\omega}_r^2\delta_{rs} \tag{7.36}$$

where δ_{rs} is the Kronecker delta, defined as being equal to unity for $s = r$ and equal to zero for $s \ne r$, and $\bar{\omega}_r$ is the undamped natural frequency for the rth mode of the bridge:

$$\bar{\omega}_r = (r\pi)^2 \sqrt{\frac{\overline{YI}}{\bar{m}L_b^4}} \tag{7.37}$$

Following the standard modal analysis procedure (Appendix C.1.7), that is, substituting Equation (7.34) into Equation (7.32), multiplying the latter by $\bar{\phi}_s(\bar{x})$, integrating over the length of the beam, and making use of the orthogonality conditions, one obtains

$$\frac{d^2\bar{\eta}_r(t)}{dt^2} + 2\bar{\zeta}_r\bar{\omega}_r\frac{d\bar{\eta}_r(t)}{dt} + \bar{\omega}_r^2\bar{\eta}_r(t) = P\sqrt{\frac{2}{\bar{m}L_b}}\sin\left(\frac{r\pi\bar{v}t}{L_b}\right) \tag{7.38}$$

where $\bar{\zeta}_r$ is the modal mechanical damping ratio ($\bar{\zeta}_r = \bar{c}_s\bar{I}\bar{\omega}_r/2\overline{YI} + \bar{c}_a/2\bar{m}\bar{\omega}_r$). The total solution of this ordinary differential equation is given by

$$\bar{\eta}_r(t) = \bar{\eta}_r^p(t) + \bar{\eta}_r^h(t) \tag{7.39}$$

Here, $\bar{\eta}_r^p(t)$ is the particular solution while $\bar{\eta}_r^h(t)$ is the homogeneous solution.[5] The particular solution can be obtained as

$$\bar{\eta}_r^p(t) = F_r\sin\left(\frac{r\pi\bar{v}t}{L_b} - \varphi_r\right) \tag{7.40}$$

where

$$F_r = P\sqrt{\frac{2}{\bar{m}L_b\left\{[\bar{\omega}_r^2 - (r\pi\bar{v}/L_b)^2]^2 + (2\bar{\zeta}_r\bar{\omega}_r r\pi\bar{v}/L_b)^2\right\}}} \tag{7.41}$$

$$\varphi_r = \tan^{-1}\left(\frac{2\bar{\zeta}_r\bar{\omega}_r r\pi\bar{v}/L_b}{\bar{\omega}_r^2 - (r\pi\bar{v}/L_b)^2}\right) \tag{7.42}$$

The solution of the homogenous problem has the form of

$$\bar{\eta}_r^h(t) = e^{-\bar{\xi}_r\bar{\omega}_r t}\left(A_r\cos\bar{\omega}_r t + B_r\sin\bar{\omega}_r t\right) \tag{7.43}$$

Therefore the total solution becomes

$$\bar{\eta}_r(t) = F_r\sin\left(\frac{r\pi\bar{v}t}{L_b} - \varphi_r\right) + e^{-\bar{\xi}_r\bar{\omega}_r t}\left(A_r\cos\bar{\omega}_r t + B_r\sin\bar{\omega}_r t\right) \tag{7.44}$$

Assuming zero initial conditions for the bridge,

$$\bar{w}(\bar{x}, 0) = 0, \qquad \left.\frac{\partial\bar{w}(\bar{x}, t)}{\partial t}\right|_{t=0} = 0 \tag{7.45}$$

---

[5] Note that one is seeking a solution that is valid for the traverse duration of the vehicle (moving load) and the homogeneous part in Equation (7.39) is significant.

Using Equation (7.34) in Equation (7.45) leads to

$$\left\{\sum_{r=1}^{\infty}\bar{\phi}_r(\bar{x})\left[F_r\sin\left(\frac{r\pi\bar{v}t}{L_b}-\varphi_r\right)+e^{-\bar{\xi}_r\bar{\omega}_r t}\left(A_r\cos\bar{\omega}_r t+B_r\sin\bar{\omega}_r t\right)\right]\right\}_{t=0}=0 \quad (7.46)$$

$$\left\{\sum_{r=1}^{\infty}\bar{\phi}_r(\bar{x})\left[F_r\frac{r\pi\bar{v}}{L_b}\cos\left(\frac{r\pi\bar{v}t}{L_b}-\varphi_r\right)+e^{-\bar{\xi}_r\bar{\omega}_r t}(-A_r\bar{\xi}_r\bar{\omega}_r\cos\bar{\omega}_r t\right.\right.$$
$$\left.\left.-A_r\bar{\omega}_r\sin\bar{\omega}_r t-B_r\bar{\xi}_r\bar{\omega}_r\sin\bar{\omega}_r t+B_r\bar{\omega}_r\cos\bar{\omega}_r t)\right]\right\}_{t=0}=0 \quad (7.47)$$

Using the orthogonality conditions, Equations (7.46) and (7.47) can be reduced to

$$\left[F_r\sin\left(\frac{r\pi\bar{v}t}{L_b}-\varphi_r\right)+e^{-\bar{\xi}_r\bar{\omega}_r t}\left(A_r\cos\bar{\omega}_r t+B_r\sin\bar{\omega}_r t\right)\right]_{t=0}=0 \quad (7.48)$$

$$\left[F_r\frac{r\pi\bar{v}}{L_b}\cos\left(\frac{r\pi\bar{v}t}{L_b}-\varphi_r\right)+e^{-\bar{\xi}_r\bar{\omega}_r t}(-A_r\bar{\xi}_r\bar{\omega}_r\cos\bar{\omega}_r t\right.$$
$$\left.-A_r\bar{\omega}_r\sin\bar{\omega}_r t-B_r\bar{\xi}_r\bar{\omega}_r\sin\bar{\omega}_r t+B_r\bar{\omega}_r\cos\bar{\omega}_r t)\right]_{t=0}=0 \quad (7.49)$$

yielding

$$-F_r\sin\varphi_r+A_r=0 \quad (7.50)$$

$$F_r\frac{r\pi\bar{v}}{L_b}\cos\varphi_r-A_r\bar{\xi}_r\bar{\omega}_r+B_r\bar{\omega}_r=0 \quad (7.51)$$

Hence the coefficients of the homogeneous solution are

$$A_r=F_r\sin\varphi_r \quad (7.52)$$

$$B_r=F_r\left(\bar{\xi}_r\sin\varphi_r-\frac{r\pi\bar{v}}{\bar{\omega}_r L_b}\cos\varphi_r\right) \quad (7.53)$$

The modal response is then

$$\bar{\eta}_r(t)=F_r\left\{\sin\left(\frac{r\pi\bar{v}t}{L_b}-\varphi_r\right)\right.$$
$$\left.+e^{-\bar{\xi}_r\bar{\omega}_r t}\left[\sin\varphi_r\cos\bar{\omega}_r t+\left(\bar{\xi}_r\sin\varphi_r-\frac{r\pi\bar{v}}{\bar{\omega}_r L_b}\cos\varphi_r\right)\sin\bar{\omega}_r t\right]\right\} \quad (7.54)$$

yielding[6]

$$\bar{w}(\bar{x}, t) = \sum_{r=1}^{\infty} \bar{\phi}_r(\bar{x}) F_r \left\{ \sin \left(\frac{r\pi \bar{v} t}{L_b} - \varphi_r \right) \right.$$
$$\left. + e^{-\bar{\xi}_r \bar{\omega}_r t} \left[\sin \varphi_r \cos \bar{\omega}_r t + \left(\bar{\xi}_r \sin \varphi_r - \frac{r\pi \bar{v}}{\bar{\omega}_r L_b} \cos \varphi_r \right) \sin \bar{\omega}_r t \right] \right\} \quad (7.55)$$

Therefore, the acceleration input to the energy harvester is

$$a(t) = \frac{\partial^2}{\partial t^2} \left\langle \sum_{r=1}^{\infty} \bar{\phi}_r(L_h) F_r \left\{ \sin \left(\frac{r\pi \bar{v} t}{L_b} - \varphi_r \right) \right. \right.$$
$$\left. \left. + e^{-\bar{\xi}_r \bar{\omega}_r t} \left[\sin \varphi_r \cos \bar{\omega}_r t + \left(\bar{\xi}_r \sin \varphi_r - \frac{r\pi \bar{v}}{\bar{\omega}_r L_b} \cos \varphi_r \right) \sin \bar{\omega}_r t \right] \right\} \right\rangle \quad (7.56)$$

which can be further expanded after the application of the differentiation. The resulting expression is quite lengthy and numerical solution can be used to solve for the voltage response (see Section 7.6). Recall that Equation (7.55) is valid for $0 \leq t \leq T$ where $T = L_b/\bar{v}$ is the time of traverse of the moving load over the bridge. Therefore the base acceleration input to the energy harvester is also valid for $0 \leq t \leq T$.

The response history depends very much on the speed of the moving load. Frýba [9] normalizes the $\pi \bar{v}/L_b$ term (see the way it appears in Equations (7.41) and (7.56)) with respect to the fundamental natural frequency of the bridge and defines the dimensionless parameter $\alpha = \pi \bar{v}/\bar{\omega}_1 L_b$. Olsson [10] reports that $\alpha = 1$ typically corresponds to a vehicle speed of 400–1500 km/h depending on the structural flexibility of the bridge, defining a conservative upper limit for most practical purposes. The dependence of the bridge response on the vehicle speed is such that, for the fast vehicle speed of $\alpha = 1$, the mid-span ($\bar{x} = L_b/2$) reaches its maximum deflection (implying a quarter cycle of vibration) after the vehicle leaves the bridge [10]. Simulations given in Frýba [9] show that the maximum dynamic deflection is obtained for $0.5 \leq \alpha \leq 0.7$. For large vehicle speeds, the deflection rapidly tends to zero but converges to the static deflection for small vehicle speeds. The reader is referred to Frýba [9] and Olsson [10] for further discussion.

7.4.2   Thin Piezoelectric Layer Covering a Region on the Bridge

For the cases with very low oscillation frequencies that cannot excite a given cantilever effectively, an alternative approach is presented in this section. Figure 7.5 shows a slender bridge with a piezoceramic patch covering the region $L_{h1} \leq \bar{x} \leq L_{h2}$ for harvesting energy from the vibrations induced by the moving load. For the typical dimensions of commercially available piezoceramics, obviously $(L_{h2} - L_{h1}) \ll L_b$ (therefore the schematic in Figure 7.5 is an exaggerated view). However, several of such patches can be combined to generate usable

---

[6] Following his solution based on a different approach (Laplace–Carson integral transformation), Frýba [9] studies special cases regarding the speed of the load and the damping in the structure. Here, the response expression is left in its general form and the reader is referred to Frýba [9] for a detailed discussion.

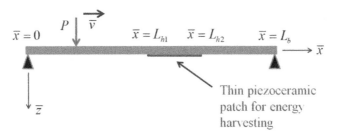

Figure 7.5 Schematic of the moving load problem for a slender bridge with a thin piezoceramic patch covering a region for energy harvesting (the size of the patch is exaggerated)

electrical power output from the dynamic strain induced on the surface by the moving load. Therefore the purpose here is to relate the surface strain of the bridge to the voltage output of the piezoceramic. The dynamics of the bridge are governed by Equation (7.32) and it is assumed that the effect of the patch and piezoelectric power generation on the bridge dynamics is negligible. Therefore the vibratory response of the bridge to the moving load for $0 \leq t \leq T$ is given by Equation (7.55).

Assuming that the electrodes of the piezoceramic patch are connected to a resistive electrical load R_l, the integral form of Gauss's law can be recalled from Chapter 3 as

$$\frac{d}{dt}\left(\int_A \mathbf{D} \cdot \mathbf{n}\, dA\right) = \frac{v(t)}{R_l} \tag{7.57}$$

where $v(t)$ is the voltage across the resistive load, $\mathbf{D}$ is the vector of electric displacement components, $\mathbf{n}$ is the unit outward normal, and the integration is performed over the electrode area A of the piezoceramic patch. Following steps similar to those described in Section 3.1.3, Equation (7.57) leads to

$$C_p \frac{dv(t)}{dt} + \frac{v(t)}{R_l} = -\bar{e}_{31} h_{pc} b_p \int_{L_{h1}}^{L_{h2}} \frac{\partial^3 \bar{w}(\bar{x}, t)}{\partial \bar{x}^2 \partial t}\, d\bar{x} \tag{7.58}$$

where b_p is the width of the piezoceramic (equal to the width of the electrodes), h_{pc} is the distance from the neutral axis of the bridge to the center of the piezoceramic patch, $\bar{e}_{31}$ is the plane-stress piezoelectric stress constant, and C_p is the capacitance of the piezoceramic patch ($C_p = \bar{\varepsilon}_{33}^S b_p (L_{h2} - L_{h1})/h_p$ where h_p is the thickness of the piezoceramic).

Substituting the response form given by Equation (7.34) into Equation (7.58) gives

$$\frac{dv(t)}{dt} + \frac{v(t)}{\tau} = \sum_{r=1}^{\infty} \psi_r \frac{d\bar{\eta}_r(t)}{dt} \tag{7.59}$$

where $\bar{\eta}_r(t)$ is expressed by Equation (7.54) and the modal electromechanical coupling term is

$$\psi_r = -\frac{\bar{e}_{31}h_{pc}b_p}{C_p}\int_{L_{h1}}^{L_{h2}}\frac{d^2\bar{\phi}_r(\bar{x})}{d\bar{x}^2}d\bar{x} = -\frac{\bar{e}_{31}h_{pc}b_p}{C_p}\frac{d\bar{\phi}_r(\bar{x})}{d\bar{x}}\Big|_{\bar{x}=L_{h1}}^{\bar{x}=L_{h2}} \tag{7.60}$$

In Equation (7.59), τ is the time constant of the circuit given by

$$\tau = R_l C_p \tag{7.61}$$

Equation (7.59) is a first-order ordinary differential equation that can be solved by using the following integrating factor:

$$\gamma(t) = e^{t/\tau} \tag{7.62}$$

yielding

$$v(t) = e^{-t/\tau}\int e^{t/\tau}\sum_{r=1}^{\infty}\psi_r\frac{d\bar{\eta}_r(t)}{dt}dt \tag{7.63}$$

where the initial voltage is assumed to be zero and the modal velocity response of the bridge is obtained from Equation (7.54) as

$$\frac{d\bar{\eta}_r(t)}{dt} = \frac{d}{dt}\left\langle F_r\left\{\sin\left(\frac{r\pi\bar{v}t}{L_b}-\varphi_r\right)\right.\right.$$
$$+e^{-\bar{\xi}_r\bar{\omega}_rt}\left[\sin\varphi_r\cos\bar{\omega}_rt + \left(\bar{\xi}_r\sin\varphi_r - \frac{r\pi\bar{v}}{\bar{\omega}_rL_b}\cos\varphi_r\right)\sin\bar{\omega}_rt\right]\right\}\right\rangle \tag{7.64}$$

The electrical power history is therefore

$$P(t) = \frac{1}{R_l}\left[e^{-t/\tau}\int e^{t/\tau}\sum_{r=1}^{\infty}\psi_r\frac{d\bar{\eta}_r(t)}{dt}dt\right]^2 \tag{7.65}$$

which is valid for $0 \leq t \leq T$.

7.5   Local Strain Fluctuations on Large Structures

The problem of piezoelectric power generation from surface strain fluctuations on large structures is covered in this section. The formulation aims to predict the power output extracted from the dynamic strain induced in a small rectangular piezoceramic patch attached to a vibrating host structure (Figure 7.6). The discussion given here can therefore be considered as a *local* way of treating the problem of Section 7.4.2. The input in the following derivation is considered as the dynamic strain components in two orthogonal directions rather than the original source of the excitation (which might be a moving load, wind, or machinery-induced vibrations among other possible sources). The thin piezoceramic patch is considered as a Kirchhoff plate (so that the transverse shear stress components and the normal stress component in the thickness direction are negligible).

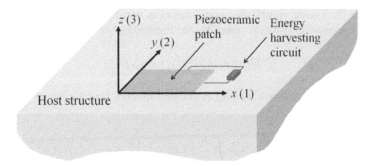

Figure 7.6 Piezoceramic patch attached to a large structure for power generation from surface strain fluctuations

7.5.1   Power Output to General Strain Fluctuations

The schematic of a thin piezoceramic patch bonded onto the surface of a large structure (e.g., a civil engineering structure, such as a bridge) is shown in Figure 7.6, where the loading (not shown) is such that the two orthogonal directions are the *principal strain* [11,12] directions of the structure at the location of interest at an arbitrary instant of time. Hence the two-dimensional strain state can be represented by the strain components in these two directions. As in the discussion of the moving load problem, it is assumed that the effect of piezoelectric power generation on the dynamics of the large structure is negligible. Therefore the local strain components measured in the absence of the piezoceramic patch are the same as those measured in its presence. Moreover, the piezoceramic patch is very thin compared to the thickness of the large structure so that the strain distribution on a plane at half its thickness is equal to the local surface strain distribution on the large structure.

For the rectangular thin piezoceramic patch undergoing strain fluctuations in the 1- and 2-directions (Figure 7.6), the electric displacement is expressed as (Appendix A.4)

$$D_3 = \bar{e}_{31} S_1 + \bar{e}_{32} S_2 + \bar{\varepsilon}_{33}^S E_3 \tag{7.66}$$

where S_1 and S_2 are the strain components,[7] E_3 is the electric field component while the plane-stress piezoelectric stress constants and the permittivity component are given by

$$\bar{e}_{31} = \bar{e}_{32} = \frac{d_{31}}{s_{11}^E + s_{12}^E} \tag{7.67}$$

$$\bar{\varepsilon}_{33}^S = \bar{\varepsilon}_{33}^T - \frac{2d_{31}^2}{s_{11}^E + s_{12}^E} \tag{7.68}$$

Here, a poled piezoceramic plate is considered (so the symmetry of transverse isotropy about the z-axis is applied). In Equations (7.67) and (7.68), d_{31} is the piezoelectric strain constant,

---

[7] Note that the derivations given here are for homogeneous piezoceramics with conventional electrodes. Fiber-based orthotropic piezoceramics (such as macro-fiber composites) do not give any output to strain fluctuations in the direction that is orthogonal to the fiber alignment (which may or may not be favorable depending on the signs of S_1 and S_2).

s_{11}^E and s_{12}^E are the elastic compliance components at constant electric field, and $\bar{\varepsilon}_{33}^T$ is the permittivity component at constant stress. Hence Equation (7.66) can be rewritten as

$$D_3 = \bar{e}_{31} (S_1 + S_2) + \bar{\varepsilon}_{33}^S E_3 \tag{7.69}$$

If the voltage across the resistive load R_l is denoted by $v(t)$, Gauss's law given by Equation (7.57) leads to

$$\frac{dv(t)}{dt} + \frac{v(t)}{\tau} = \frac{\bar{e}_{31}}{C_p} \int_A \frac{\partial}{\partial t} [S_1(x, y, t) + S_2(x, y, t)] \, dA \tag{7.70}$$

Here, the time constant τ and the capacitance of the piezoceramic C_p are given by

$$\tau = R_l C_p, \quad C_p = \frac{\bar{\varepsilon}_{33}^S A}{h_p} \tag{7.71}$$

where A is the electrode area. If the strain field is homogeneous in both directions under the piezoceramic such that S_1 and S_2 are space independent:

$$\frac{dv(t)}{dt} + \frac{v(t)}{\tau} = \frac{\bar{e}_{31} A}{C_p} \frac{d}{dt} [S_1(t) + S_2(t)] \tag{7.72}$$

The voltage response can be obtained from Equation (7.72) as

$$v(t) = \frac{\bar{e}_{31} A}{C_p} e^{-t/\tau} \int e^{t/\tau} \frac{d}{dt} [S_1(t) + S_2(t)] \, dt \tag{7.73}$$

where the initial voltage and the initial strain components are assumed to be zero.
   Eventually the time history of the power output is given by

$$P(t) = \frac{1}{R_l} \left\{ \frac{\bar{e}_{31} A}{C_p} e^{-t/\tau} \int e^{t/\tau} \frac{d}{dt} [S_1(t) + S_2(t)] \, dt \right\}^2 \tag{7.74}$$

where the inputs are orthogonal strain components $S_1(t)$ and $S_2(t)$.

7.5.2 Steady-State Power Output to Harmonic Strain Fluctuations

Let the dynamic strain components be harmonic at the same frequency for simplicity:

$$S_1(t) = \tilde{S}_1 e^{j\omega t}, \quad S_2(t) = \tilde{S}_2 e^{j\omega t} \tag{7.75}$$

where ω is the frequency, j is the unit imaginary number, $\tilde{S}_1$ and $\tilde{S}_2$ are the strain values. In practice, each of the strain components could also have a DC component, which has no contribution to the alternating voltage output of the piezoceramic patch at steady state. If the steady-state voltage output is $v(t) = Ve^{j\omega t}$ (where V is the complex voltage), Equation (7.72) reduces to

$$\left(j\omega C_p + \frac{1}{R_l}\right) V = j\omega \bar{e}_{31} A \left(\tilde{S}_1 + \tilde{S}_2\right)$$ (7.76)

Therefore, the steady-state voltage output is

$$v(t) = j\omega \bar{e}_{31} A \left(\tilde{S}_1 + \tilde{S}_2\right) \left(j\omega C_p + \frac{1}{R_l}\right)^{-1} e^{j\omega t}$$ (7.77)

The steady-state power amplitude ($|P| = |V|^2 / R_l$) is obtained from Equation (7.77) as

$$P = \frac{\omega^2 \bar{e}_{31}^2 A^2 \left(\tilde{S}_1 + \tilde{S}_2\right)^2 R_l}{1 + \omega^2 R_l^2 C_p^2}$$ (7.78)

which can be used to find the optimum electrical load as

$$\left.\frac{\partial P}{\partial R_l}\right|_{R_l = R_l^{opt}} = 0 \rightarrow R_l^{opt} = \frac{1}{\omega C_p}$$ (7.79)

Finally, Equation (7.79) can be used in Equation (7.78) to give the maximum power amplitude:

$$P_{max} = \frac{\omega \bar{e}_{31}^2 A^2 \left(\tilde{S}_1 + \tilde{S}_2\right)^2}{2C_p}$$ (7.80)

7.5.3 Strain Gage Measurements and Strain Transformations

In practice, it is useful to estimate the amount of power that can be harvested before bonding a piezoceramic patch onto the host structure. Moreover, one should check the directions of principal strain components and how they change with time (if they do). Strain gages can be used for measuring the strain fluctuation on the surface of large structures, such as bridges. Typically, *strain gage rosettes* are employed, which consist of multiple gages oriented at a fixed angle with respect to each other. Figure 7.7 shows the schematic of a commonly configuration: a *rectangular rosette* (where the angle between the gages is 45°). In general, at least three independent strain readings are required to define the two-dimensional state of strain assuming that no other information is available.[8] The rectangular rosette configuration shown in Figure 7.7 is gives three simultaneous strain measurements S_A, S_B, and S_C in the A-, B-, and C-directions, respectively. The directions of the principal strain components (S_1 and S_2) are denoted by 1 and 2.

---

[8] The discussion given here is related to *Mohr's circle* and the reader is referred to any elementary book on the mechanics of materials for details (e.g., Gere and Timoshenko [11], Beer *et al.* [12]).

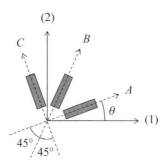

Figure 7.7   Schematic of a rectangular rosette configuration with gages A, B, and C

At an arbitrary instant of time, the principal strain components are obtained from the following strain transformation [11,12]:

$$S_{1,2} = \frac{S_A + S_C}{2} \pm \frac{1}{\sqrt{2}}\sqrt{(S_A - S_B)^2 + (S_B - S_C)^2} \tag{7.81}$$

and the angle between gage A and the direction of positive S_1 is

$$\theta = \frac{1}{2}\tan^{-1}\left(\frac{S_A - 2S_B + S_C}{S_A - S_C}\right) \tag{7.82}$$

Therefore, Equations (7.81) and (7.82) give an idea regarding how the strain levels and the direction of principal strains change in time if the loading and structural conditions result in such variations. It is worth noticing from both power expressions (for general strain input and harmonic strain input) given by Equations (7.74) and (7.78) that the power output is quadratically proportional to the *strain resultant* $S_1 + S_2$ (which is the key dynamic parameter). It is useful also to note from Equation (7.81) that

$$S_1 + S_2 = \frac{S_A + S_C}{2} + \frac{S_A + S_C}{2} = S_A + S_C \tag{7.83}$$

According to Equation (7.83), the strain resultant of two arbitrary but orthogonal directions $(S_A + S_C)$ is equal to that of the principal strain directions $(S_1 + S_2)$. Although determining the two-dimensional strain state on the surface requires three independent strain measurements, two orthogonal strain measurements are sufficient to estimate the power output in response to the resultant of the principal strain components. However, calculation of other parameters (such as the directions of principal strain components) requires the strain measured in the third direction (S_B in the present case) as well.

7.6   Numerical Solution for General Transient Excitation

For the problems of energy harvesting with transient base acceleration or other types of general excitation, it might become necessary to use numerical solution. This section provides the set of first-order ordinary differential equations with electromechanical coupling for time-domain simulations using an appropriate ordinary differential equation solver (such as the *ode45* command of MATLAB).

7.6.1 Initial Conditions in Modal Coordinates

Let the initial displacement and velocity conditions of an energy harvester beam (Figure 7.1) be given in physical coordinates by

$$w_{rel}(x, 0) = \kappa(x), \quad \left.\frac{\partial w_{rel}(x, t)}{\partial t}\right|_{t=0} = \mu(x) \tag{7.84}$$

The orthogonality relation given by Equation (C.31) in Appendix C can be used to give the initial conditions in modal coordinates as

$$\eta_r(0) = \int_0^L \kappa(x)m\phi_r(x)\,dx + \kappa(L)M_t\phi_r(L) + \left[\frac{d\kappa(x)}{dx}I_t\frac{d\phi_r(x)}{dx}\right]_{x=L} \tag{7.85}$$

$$\left.\frac{d\eta_r(t)}{dt}\right|_{t=0} = \int_0^L \mu(x)m\phi_r(x)\,dx + \mu(L)M_t\phi_r(L) + \left[\frac{d\mu(x)}{dx}I_t\frac{d\phi_r(x)}{dx}\right]_{x=L} \tag{7.86}$$

It is convenient to denote the initial condition for the voltage across the load by

$$v(0) = v_0 \tag{7.87}$$

7.6.2 State-Space Representation of the Electromechanical Equations

The modal electromechanical equations, Equations (7.4) and (7.5), can be expanded to give the following infinite set of second-order ordinary differential equations along with a first-order ordinary differential equation:

$$\frac{d^2\eta_1(t)}{dt^2} + 2\zeta_1\omega_1\frac{d\eta_1(t)}{dt} + \omega_1^2\eta_1(t) - \tilde{\theta}_1 v(t) = \sigma_1 a(t)$$

$$\frac{d^2\eta_2(t)}{dt^2} + 2\zeta_2\omega_2\frac{d\eta_2(t)}{dt} + \omega_2^2\eta_2(t) - \tilde{\theta}_2 v(t) = \sigma_2 a(t)$$

$$\vdots \tag{7.88}$$

$$C_{\tilde{p}}^{eq}\frac{dv(t)}{dt} + \frac{v(t)}{R_l} + \tilde{\theta}_1\frac{d\eta_1(t)}{dt} + \tilde{\theta}_2\frac{d\eta_2(t)}{dt} + \ldots = 0$$

In general, for a given acceleration history $a(t)$, it is sufficient to take a finite number of vibration modes so that

$$\frac{d^2\eta_1(t)}{dt^2} + 2\zeta_1\omega_1\frac{d\eta_1(t)}{dt} + \omega_1^2\eta_1(t) - \tilde{\theta}_1 v(t) = \sigma_1 a(t)$$

$$\frac{d^2\eta_2(t)}{dt^2} + 2\zeta_2\omega_2\frac{d\eta_2(t)}{dt} + \omega_2^2\eta_2(t) - \tilde{\theta}_2 v(t) = \sigma_2 a(t)$$

$$\vdots \tag{7.89}$$

$$\frac{d^2\eta_M(t)}{dt^2} + 2\zeta_M\omega_M\frac{d\eta_M(t)}{dt} + \omega_M^2\eta_M(t) - \tilde{\theta}_M v(t) = \sigma_M a(t)$$

$$C_{\tilde{p}}^{eq}\frac{dv(t)}{dt} + \frac{v(t)}{R_l} + \tilde{\theta}_1\frac{d\eta_1(t)}{dt} + \tilde{\theta}_2\frac{d\eta_2(t)}{dt} + \cdots + \tilde{\theta}_M\frac{d\eta_M(t)}{dt} = 0$$

where M is the total number of vibration modes considered in the solution. Hence the vibration response given by Equation (7.3) is approximated as a finite series of the eigenfunctions weighted by the modal coordinates:

$$w_{rel}(x, t) \cong \sum_{r=1}^{M} \phi_r(x)\eta_r(t) = \phi_1(x)\eta_1(t) + \phi_2(x)\eta_2(t) + \cdots + \phi_M(x)\eta_M(t) \tag{7.90}$$

It is convenient to define the following state variables for $r = 1, \ldots, M$:

$$u_r^{(1)} = \eta_r(t), \quad u_r^{(2)} = \frac{d\eta_r(t)}{dt}, \quad u^{(3)} = v(t) \tag{7.91}$$

Then, Equations (7.89) become

$$\frac{du_1^{(1)}}{dt} = u_1^{(2)}$$

$$\frac{du_1^{(2)}}{dt} = -2\zeta_1\omega_1 u_1^{(2)} - \omega_1^2 u_1^{(1)} + \tilde{\theta}_1 u^{(3)} + \sigma_1 a(t)$$

$$\frac{du_2^{(1)}}{dt} = u_2^{(2)}$$

$$\frac{du_2^{(2)}}{dt} = -2\zeta_2\omega_2 u_2^{(2)} - \omega_2^2 u_2^{(1)} + \tilde{\theta}_2 u^{(3)} + \sigma_2 a(t)$$

$$\vdots \tag{7.92}$$

$$\frac{du_M^{(1)}}{dt} = u_M^{(2)}$$

$$\frac{du_M^{(2)}}{dt} = -2\zeta_M\omega_M u_M^{(2)} - \omega_M^2 u_M^{(1)} + \tilde{\theta}_M u^{(3)} + \sigma_M a(t)$$

$$\frac{du^{(3)}}{dt} = -\frac{1}{C_{\tilde{p}}^{eq} R_l}u^{(3)} - \frac{\tilde{\theta}_1}{C_{\tilde{p}}^{eq}}u_1^{(2)} - \frac{\tilde{\theta}_2}{C_{\tilde{p}}^{eq}}u_2^{(2)} - \cdots - \frac{\tilde{\theta}_M}{C_{\tilde{p}}^{eq}}u_M^{(2)}$$

Rearranging these equations and using the shorthand overdot for the time differentiation leads to

$$
\begin{bmatrix}
\dot{u}_1^{(1)} \\
\dot{u}_2^{(1)} \\
\vdots \\
\dot{u}_M^{(1)} \\
\dot{u}_1^{(2)} \\
\dot{u}_2^{(2)} \\
\vdots \\
\dot{u}_M^{(2)} \\
\dot{u}^{(3)}
\end{bmatrix}
=
\begin{bmatrix}
u_1^{(2)} \\
u_2^{(2)} \\
\vdots \\
u_M^{(2)} \\
-2\zeta_1\omega_1 u_1^{(2)} - \omega_1^2 u_1^{(1)} + \tilde{\theta}_1 u^{(3)} + \sigma_1 a(t) \\
-2\zeta_2\omega_2 u_2^{(2)} - \omega_2^2 u_2^{(1)} + \tilde{\theta}_2 u^{(3)} + \sigma_2 a(t) \\
\vdots \\
-2\zeta_M\omega_M u_M^{(2)} - \omega_M^2 u_M^{(1)} + \tilde{\theta}_M u^{(3)} + \sigma_M a(t) \\
-u^{(3)}/R_l C_{\tilde{p}}^{eq} - \tilde{\theta}_1 u_1^{(2)}/C_{\tilde{p}}^{eq} - \tilde{\theta}_2 u_2^{(2)}/C_{\tilde{p}}^{eq} - \cdots - \tilde{\theta}_M u_M^{(2)}/C_{\tilde{p}}^{eq}
\end{bmatrix}
\tag{7.93}
$$

Equations (7.93) are the $2M + 1$ first-order electromechanical (state-space) equations for the $2M + 1$ unknowns of the truncated distributed-parameter system. The initial conditions are expressed using Equations (7.85)–(7.87):

$$
u_r^{(1)}(0) = \int_0^L \kappa(x)m\phi_r(x)\,dx + \kappa(L)M_t\phi_r(L) + \left[\frac{d\kappa(x)}{dx}I_t\frac{d\phi_r(x)}{dx}\right]_{x=L}
\tag{7.94}
$$

$$
u_r^{(2)}(0) = \int_0^L \mu(x)m\phi_r(x)\,dx + \mu(L)M_t\phi_r(L) + \left[\frac{d\mu(x)}{dx}I_t\frac{d\phi_r(x)}{dx}\right]_{x=L}
\tag{7.95}
$$

$$
u^{(3)}(0) = v_0
\tag{7.96}
$$

For an arbitrary base acceleration $a(t)$ and with the initial conditions given by Equations (7.94)–(7.96), Equations (7.93) can be employed to solve for the $2M + 1$ state variables $u_r^{(1)}$, $u_r^{(2)}$, and $u^{(3)}$ ($r = 1, \ldots, M$) which are defined by Equations (7.91). Depending on the frequency content of $a(t)$, in general, it is sufficient to take only a few vibration modes in the solution (even $M = 1$ can be sufficient if the response is strongly dominated by the fundamental vibration mode). Note that one of the state variables directly gives the voltage output, though it is necessary to use the mechanical state variables (modal coordinates) as the coefficients of the eigenfunctions in Equation (7.90) if one is interested in the vibration response as well.

7.7   Case Studies

Two case studies are presented to demonstrate the use of some of the derivations given in this chapter. The problems considered in the case studies are related to Sections 7.2 and 7.5.

The first case study investigates piezoelectric power generation from a bimorph cantilever located on the crank-to-rocker connecting link of a four-bar mechanism. The periodic motion of the mechanism creates a periodic base acceleration for the energy harvester, which is then used in the electromechanical equations derived in Section 7.2. The second case study estimates the piezoelectric power generation potential of a piezoceramic patch from surface strain fluctuations. After a theoretical analysis, strain measurements taken on a steel multi-girder bridge are investigated for calculating the strain resultant so that the power output of the piezoceramic patch could be estimated.

7.7.1 Periodic Excitation of a Bimorph Energy Harvester on a Mechanism Link

Figure 7.8a shows a four-bar mechanism with a cantilevered bimorph energy harvester attached to the link that connects the crank (link 1) to the rocker (link 3). It is assumed that the links are rigid and the presence of the cantilever does not affect the kinematics of the mechanism. The coupler point C of the mechanism follows the closed path shown in Figure 7.8b in one revolution of the crank link. The crank link rotates with an angular velocity of 120 rpm ($\omega = 4\pi$ rad/s in Figure 7.8a). The dimensions of the links are given in Table 7.1 and the coupler segment makes an angle (γ) of $10°$ with link 2. Expectedly, as the coupler point C follows the path given in Figure 7.8b with a period of 0.5 seconds, the acceleration at the base of the cantilever fluctuates periodically (considering the steady-state motion of the mechanism). Although the problem discussed here can find applications in robot manipulators and constitutes an example of *energy harvesting from rigid body motions*[9] rather than structural vibrations, the primary goal here is to create an arbitrary periodic acceleration to demonstrate the use of the equations derived in Section 7.2. The properties of the cantilevered bimorph with a tip mass located on the mechanism are given in Table 7.2.

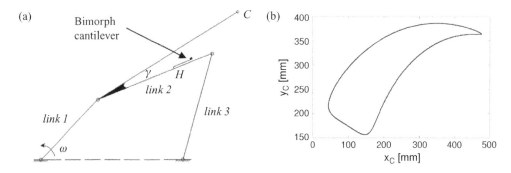

Figure 7.8 (a) Schematic of a four-bar mechanism with a cantilevered energy harvester located on link 2 and (b) the coupler curve of point C due to one revolution of link 1

Various methods are available for the steady-state kinematic analysis of mechanisms. Raven's method [13] is used here in order to obtain the acceleration at the base of the

---

[9] Another typical example of rigid body kinetic energy potential is the human body with several links undergoing periodic motion while walking at a constant speed (see Figure 7.2).

Table 7.1 Properties of the four-bar mechanism

Link/segment	Length
Link 1 (mm)	200
Link 2 (mm)	300
Link 3 (mm)	270
Ground link (mm)	350
Coupler segment (on link 2) (mm)	400
Harvester position on link 2 (mm)	200

cantilever. Since the longitudinal vibration modes are usually at very high frequencies, the transverse component of the acceleration at point H in Figure 7.8a is assumed to be the only source of excitation input for the cantilever. Figure 7.9a shows the periodic acceleration history ($a(t)$ in Equations (7.1) and (7.2)) in the transverse direction at the base of the cantilever. Since the rotational speed of the crank is 120 rpm, the time history shown in Figure 7.9a covers two periods of motion. The analytical base acceleration history is used in Equations (7.12)–(7.14) in order to obtain the Fourier coefficients (by taking $T = 1$ s although one could consider only $T = 0.5$ s). The Fourier series expansion of the acceleration history obtained from Equation (7.11) is compared to the exact acceleration history in Figure 7.9a (by using 30 harmonic pairs in the summation). The approximate expansion (that neglects the mean value of the acceleration) obtained using Equation (7.15) is also shown in the same figure. Clearly, the acceleration fluctuation is such that its mean value is indeed negligible and the harmonic components of the Fourier series expansion can successfully represent $a(t)$. The Fourier coefficients associated with these harmonics and their frequencies are plotted in Figure 7.9b.

For the parameters of the bimorph cantilever with a tip mass listed in Table 7.2, the FRFs for the voltage to base acceleration and the transverse tip displacement to base acceleration are shown for a set of resistors in Figures 7.10a and 7.10b, respectively. Note that the layers of the piezoceramic are combined in series and a modal mechanical damping ratio of 1% is assumed. For this cantilever, the fundamental short-circuit and open-circuit resonance frequencies are read from the voltage FRF as 56.4 Hz and 59.3 Hz, respectively. Since the purpose is to demonstrate the formulation, the harvester is not optimized for the input given in Figure 7.9a. For the purpose of optimizing the cantilever, the graph of the Fourier coefficients can be

Table 7.2 Geometric and material properties of the bimorph cantilever

	Piezoceramic (PZT-5H)		Substructure (brass)
Length (L) (mm)	40		40
Width (b) (mm)	5		5
Thickness (h_p, h_s) (mm)	0.2 (each)		0.1
Tip mass (M_t) (kg)		0.001	
Mass density (ρ_p, ρ_s) (kg/m^3)	7500		9000
Elastic modulus ($\bar{c}_{11}^E$, Y_s) (GPa)	60.6		105
Piezoelectric constant ($\bar{e}_{31}$) (C/m^2)	−16.6		—
Permittivity constant ($\bar{\varepsilon}_{33}^S$) (nF/m)	25.55		—

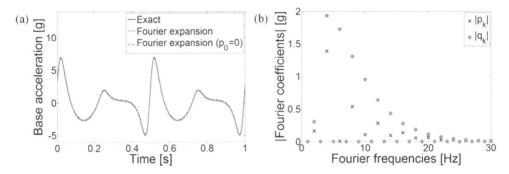

Figure 7.9 (a) Lateral acceleration at the base of the cantilever for two periods of motion and (b) Fourier coefficients of the harmonics with their frequencies

utilized so that the fundamental resonance frequency of the harvester is around the frequencies of the highest energy input (less than 10 Hz in this case). However, it is worth mentioning that the acceleration amplitudes in the present case are very high. One should check the periodic vibration response of the cantilever if it is to be designed to have a resonance less than 10 Hz.

Equations (7.16) and (7.18) are used to simulate the voltage and the vibration histories for different values of load resistance and the results are given in Figures 7.11–7.13. Note that the voltage amplitude increases monotonically with increasing load resistance as in the case of simple harmonic excitation (Chapters 3 and 4). The tip displacement amplitude of the cantilever is less than 0.6 mm, which is around 1.5% of the overhang length (it can be considered as a linear response). Considering Figures 7.11–7.13, it is worth highlighting that the shunt damping effect of piezoelectric power generation on the periodic response of the cantilever is not significant. This is simply because the excitation frequencies (Fourier frequencies in Figure 7.9b) are considerably below the resonance frequency (close to the quasi-static region in Figure 7.10b). For the Fourier frequencies with non-zero energy input, which is roughly the range below 20 Hz in Figure 7.9b, the effect of changing load resistance on the vibration response is indeed negligible as can be seen from the tip displacement FRF of Figure 7.10b, in agreement with the time histories of periodic motion.

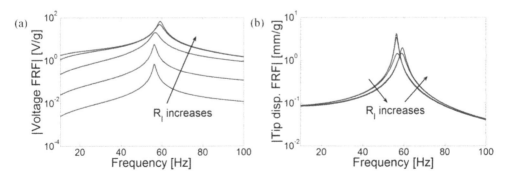

Figure 7.10 (a) Voltage FRFs and (b) tip displacement FRFs (relative to the base) of the bimorph energy harvester for a set of resistors ($R_l = 1\,\text{k}\Omega, 10\,\text{k}\Omega, 100\,\text{k}\Omega, 1\,\text{M}\Omega, 10\,\text{M}\Omega$)

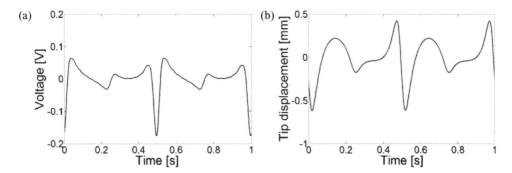

Figure 7.11   (a) Voltage history and (b) tip displacement history of the cantilever for two periods of motion ($R_l = 10\,\mathrm{k\Omega}$)

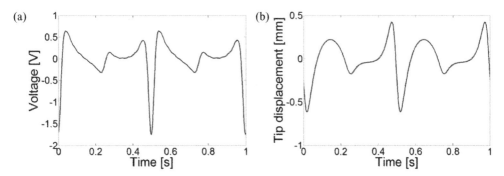

Figure 7.12   (a) Voltage history and (b) tip displacement history of the cantilever for two periods of motion ($R_l = 100\,\mathrm{k\Omega}$)

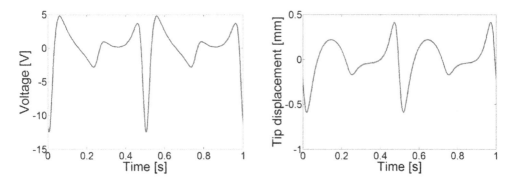

Figure 7.13   (a) Voltage history and (b) tip displacement history of the cantilever for two periods of motion ($R_l = 1\,\mathrm{M\Omega}$)

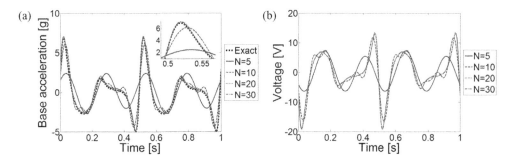

Figure 7.14 (a) Acceleration input and (b) voltage output histories of the cantilever for two periods of motion with a different number of harmonic pairs in the Fourier expansion (open-circuit conditions)

The final part of this case study concerns the convergence of the response with the increasing number of harmonics. It was shown previously with Figure 7.9a that the Fourier series expansion with 30 harmonics was able to represent the base acceleration with good accuracy. Figure 7.14a shows the Fourier series expansion of the base acceleration input (with the zero main value assumption) for a different number of harmonic pairs along with the exact acceleration input. The Fourier series representation converges to the exact acceleration with the increasing number of modes. The open-circuit voltage response is plotted in Figure 7.14b and, as expected, the voltage response converges to a certain pattern with the same rate as the acceleration converges to the exact input. Therefore, the number of harmonic pairs that successfully represents the base acceleration is sufficient to represent the voltage response in the linear electromechanical problem.

7.7.2 Analysis of a Piezoceramic Patch for Surface Strain Fluctuations of a Bridge

The piezoceramic patch (QP10N, Midé Technology Corporation) shown in Figure 7.15 is analyzed here for piezoelectric power generation from surface strain fluctuations following the derivations given in Section 7.5. This commercially available patch uses PZT-5A (embedded in kapton); therefore, the piezoelectric stress constant is obtained as

Figure 7.15 The piezoceramic patch analyzed for power generation from surface strain fluctuations

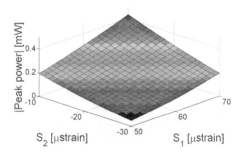

Figure 7.16 Variation of the maximum power amplitude for different combinations of the principal strain components at 10 Hz

$\bar{e}_{31} = -16\,\text{C/m}^2$ from Table E.4 in Appendix E. The electrode area is $A = 9.46\,\text{cm}^2$ and the published capacitance of $C_p = 60\,\text{nF}$ is used in the calculations. It is assumed that the strain fluctuation is harmonic at the same frequency in both principal strain directions. The input parameters to predict the maximum power amplitude are then the principal strain amplitudes and the frequency of oscillations (Equation (7.80)).

Suppose that the strain state at the point of interest on the host structure is such that the principal strain components have the opposite sign at an arbitrary instant of time (so they are 180° out of phase) and they oscillate at 10 Hz. Figure 7.16 shows the peak power amplitude obtained from Equation (7.80) for different combinations of $\tilde{S}_1$ and $\tilde{S}_2$. It is arbitrarily assumed for simulation purposes that $\tilde{S}_1$ takes values between 50 and 70 microstrain while $\tilde{S}_2$ takes values between -30 and -10 microstrain. Recall from Equation (7.80) that the important parameter is the resultant of these strain components given by $\tilde{S}_1 + \tilde{S}_2$. Therefore, the maximum power output obtained for the resultant of 70 and -10 microstrain is 0.43 mW, while the minimum power output of 0.0477 mW is obtained for the resultant of 50 and -30 microstrain. As expected from Equation (7.80), the power output depends quadratically on the strain resultant. Therefore, as the strain resultant is changed by a factor of three, the power output changes by a factor of nine.

The peak power amplitude is plotted in Figure 7.17 for various strain resultant and frequency combinations. For a fixed excitation frequency, the power output depends quadratically on

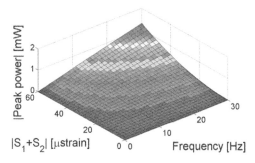

Figure 7.17 Variation of the maximum power amplitude for different combinations of the strain resultant and the frequency

Figure 7.18 Three-span continuous curved multi-girder steel bridge in Roanoke, VA (span length: 33 m). [14]

the strain resultant, whereas, for a fixed strain resultant, the power output changes linearly with frequency. The maximum power output of 1.3 mW is obtained in Figure 7.16 for a 60 microstrain resultant and 30 Hz frequency using this 9.46 cm^2 piezoceramic patch. The optimum electrical load of this maximum power output can be calculated from Equation (7.79) as 88.4 kΩ.

Next, the strain measurements taken on the steel multi-girder bridge shown in Figure 7.18 [14] are considered to estimate the piezoelectric power that can be generated using the patch discussed in this section. Dynamic strain measurements were taken at a location on the bridge by the MISTRAS Group Inc. Products and Systems Division using a rectangular rosette strain gage configuration (Figure 7.7). Strain measurements recorded at the point of interest with a sampling frequency of 82.6 Hz are shown in Figure 7.19, where the orientation of the gages

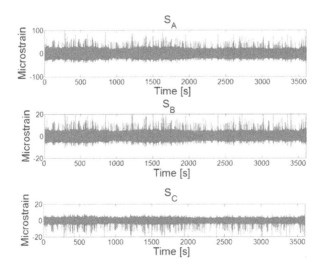

Figure 7.19 Dynamic strain components measured using a rectangular rosette strain gage configuration (see Figure 7.7 for the arrangement of gages A, B, and C)

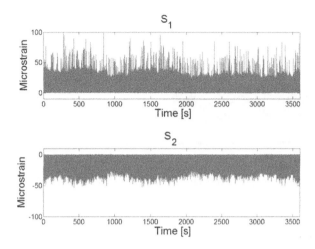

Figure 7.20   Dynamic principal strain components

A, B, and C can be found in Figure 7.7. Therefore the measurements A and C are taken in two orthogonal directions which are not necessarily the principal strain directions. The data shown in Figure 7.19 are then used in Equation (7.81) and the dynamic strain components shown in Figure 7.20 are obtained.

A close-up view of the principal strain components is shown in Figure 7.21 at an arbitrary instant of time. Note that one component oscillates around a positive static value whereas the other one oscillates around a negative static value. However, the dynamic (oscillatory) components exhibit a favorable behavior as their main signal patterns do not cancel each other drastically. It is observed through FFT (Fast Fourier Transform) analysis that a major harmonic exists at approximately 22.6 Hz in both components.

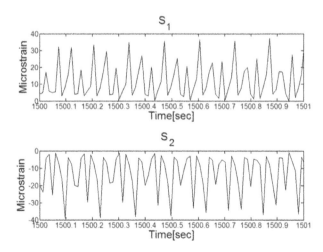

Figure 7.21   Close-up views of the dynamic principal strain components at an arbitrary instant of time

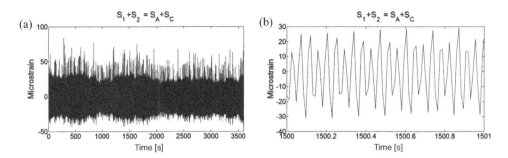

Figure 7.22 Resultant of the principal strain components: (a) entire time history; and (b) a close-up view at an arbitrary instant of time

Recall from Equations (7.74) and (7.80) that the key parameter is the resultant of the principal strain components $S_1 + S_2$, which is identical to $S_1 + S_2 = S_A + S_C$ from Equation (7.83). Figures 7.22a and 7.22b show this strain resultant for the entire time history and for an arbitrary instant of time, respectively. If the dynamic behavior of the strain resultant shown in Figure 7.22b can be approximated as a harmonic strain fluctuation of ± 25 microstrain at 22.6 Hz, a peak power amplitude of 0.16 mW is estimated from Figure 7.17. Therefore, assuming homogeneous strain behavior in the small region, by combining six of such patches, one can reach the peak power amplitude of 1 mW based on this first approximation.

Depending on the structure and the loading conditions, the dynamic strain behavior as well as the principal strain directions can be strongly time variant. In fact, the problem is completely stochastic (see Section 7.3) in most civil infrastructure systems, such as the bridge example considered here. If one is interested in the power output for a particular strain resultant history, such as the one shown in Figure 7.22a (although it will not be repeatable), it can be used in the integrand of Equation (7.74) and numerical solution can be performed. If the statistical characteristics of the strain components are known, however, random data analysis [3,4] can be performed by implementing the existing harmonic equations appropriately (as done for the base excitation problem in Section 7.3). From the stochastic analysis point of view, therefore, representing the strain components in the form of a random process (such as the white noise shown in Figure 7.3) can be a better approximation than taking a single harmonic and assuming steady-state behavior.

7.8 Summary

Derivations for modeling the piezoelectric energy harvesting problem under dynamic loading conditions other than simple harmonic excitation are presented. First the governing electromechanical equations of cantilevered bimorph piezoelectric energy harvesters are written in physical and modal coordinates by assuming the base acceleration to be an arbitrary function of time. After that, the problem of periodic base excitation of cantilevers is visited first. The base acceleration is represented as a Fourier series expansion and the electromechanical frequency response functions derived in Chapter 3 are combined with the Fourier series representation of the base acceleration in order to predict the periodic voltage output. The second problem considered in this chapter is the case where the base acceleration has the form of ideal white noise, which is a stationary random process with a very broad frequency content of uniform power distribution. After expressing the expected value of the piezoelectric power output in terms of the power spectral density of the white noise input, two approaches of piezoelectric power generation from moving loads are formulated. The form of excitation in both cases is a transversely applied load moving at a constant speed. The first approach considers using a cantilever located at an arbitrary point on the bridge while the second approach considers using a piezoceramic patch covering a certain region on the bridge. Formulating the piezoelectric power generation problem from surface patches is also addressed, where the inputs are taken as the dynamic strain components in two orthogonal directions. Finally, for transient and general base excitations, the electromechanical state-space equations are presented in modal coordinates for numerical solution. Two illustrative case studies are also presented to demonstrate the use of the equations derived in the periodic excitation and the strain fluctuation sections.

7.9   Chapter Notes

The derivations given in this chapter provide the mathematical background for predicting the electromechanical response in various linear problems other than the case of simple harmonic excitation. It should be noted that different sections of this chapter can be combined to formulate a given problem accurately. For instance, the base acceleration obtained in Section 7.4 can be used in Section 7.6 for the numerical solution of the electromechanical response by using the transient acceleration caused by the moving load. Likewise, Section 7.5 provides both time-domain and frequency-domain solutions for the problem of energy harvesting from local strain fluctuations. If the strain input is better represented by white noise, the frequency-domain solution can be used in Section 7.3 for Fourier transformation to obtain an expression for the expected value of the power output in terms of the input power spectral density of the strain-rate resultant.

In the derivations of this chapter, some of the expressions are given in their most general form. The reader should consider the problem at hand to make simplifications as required. For instance, in the case study of periodic excitation given in Section 7.7, the Fourier coefficients with significant energy input are observed to be at frequencies less than the fundamental natural frequency of the energy harvester (compare Figures 7.9b and 7.10). Based on such preliminary analysis, one can decide on how many harmonic pairs to take in the Fourier series expansion. Moreover, the range of frequency input in this particular case shows that using the single-mode voltage FRFs (instead of the multi-mode FRFs) in the Fourier expansion of the voltage response gives a good approximation. In a similar fashion, if time-domain numerical solution is inevitable (Section 7.6), the frequency spectrum of the input acceleration should be checked to see how many modes are required to represent the electromechanical response with good accuracy. Therefore the designer or analyst can make simplifications with such straightforward considerations.

It is worth directing the reader to some of the recent work on the modeling and analysis of non-harmonic excitation. Scruggs [15] investigates the optimal control of a linear energy harvester network for increased power flow to a storage system under stochastic excitation. Seuaciuc-Osorio and Daqaq [16] formulate the problem of piezoelectric energy harvesting under time-varying frequency by focusing on a piezo-stack configuration for lumped-parameter modeling. Adhikari *et al.* [17] consider the Gaussian white noise excitation of the lumped-parameter problem for a piezo-stack configuration and obtain an expression analogous to the expected value of the power derived in this chapter (Section 7.3). In addition, rigorous theoretical analyses of white Gaussian and colored random excitations are presented by Daqaq [18] for a lumped-parameter electromagnetic energy harvester with stiffness nonlinearity.

References

1. Banks, H.T., Luo, Z.H., Bergman, L.A., and Inman, D.J. (1998) On the existence of normal modes of damped discrete-continuous systems. *ASME Journal of Applied Mechanics*, **65**, 980–989.
2. Greenberg, M.D. (1998) *Advanced Engineering Mathematics*, Prentice Hall, Englewood Cliffs, NJ.
3. Newland, D.E. (1993) *Random Vibrations, Spectral and Wavelet Analysis*, John Wiley & Sons, Inc., New York.
4. Bendat, J.S. and Piersol, A.G. (1986) *Random Data Analysis and Measurement Procedures*, John Wiley & Sons, Inc., New York.
5. Meirovitch, L. (2001) *Fundamentals of Vibrations*, McGraw-Hill, New York.
6. Inman, D.J. (2007) *Engineering Vibration*, Prentice Hall, Englewood Cliffs, NJ.
7. Timoshenko, S., Young, D.H., and Weaver, W. (1974) *Vibration Problems in Engineering*, John Wiley & Sons, Inc., New York.
8. Rao, S.S. (2007) *Vibration of Continuous Systems*, John Wiley & Sons, Inc., Hoboken, NJ.
9. Frýba, L. (1972) *Vibration of Solids and Structures under Moving Loads*, Noordhoff, Groningen.
10. Olsson, M. (1991) On the fundamental moving load problem. *Journal of Sound and Vibration*, **145**, 299–307.
11. Gere, J.M. and Timoshenko, S.P. (1990) *Mechanics of Materials*, PWS-Kent, Boston, MA.
12. Beer, F.P., Johnston, E.R., and DeWolf, J.T. (2001) *Mechanics of Materials*, McGraw-Hill, New York.
13. Uicker, J.J., Pennock, G.R., and Shigley, J.E. (2003) *Theory of Machines and Mechanisms*, Oxford University Press, Oxford.
14. Erturk, A. and Inman, D.J. (2011) Piezoelectric power generation for civil infrastructure systems. Proceedings of the 18th SPIE Annual International Symposium on Smart Structures and Materials & Nondestructive Evaluation and Health Monitoring, San Diego, CA, March 6–10, 2011.
15. Scruggs, J.T. (2009) An optimal stochastic control theory for distributed energy harvesting networks. *Journal of Sound and Vibration*, **320**, 707–725.
16. Seuaciuc-Osorio, T. and Daqaq, M.F. (2010) Energy harvesting under excitations of time-varying frequency. *Journal of Sound and Vibration*, **329**, 2497–2515.
17. Adhikari, S., Friswell, M.I., and Inman, D.J. (2009) Piezoelectric energy harvesting from broadband random vibrations. *Smart Materials and Structures*, **18**, 115005.
18. Daqaq, M.F. (2010) Response of uni-modal Duffing-type harvesters to random forced excitations. *Journal of Sound and Vibration*, **329**, 3621–3631.

8

Modeling and Exploiting Mechanical Nonlinearities in Piezoelectric Energy Harvesting

This chapter focuses on exploiting mechanical nonlinearities in piezoelectric energy harvesting. Enhancement of the frequency bandwidth using the monostable[1] form of the piezoelectrically coupled Duffing equation with softening and hardening stiffness effects is discussed. A perturbation-based electromechanical formulation is given for the weakly non-linear monostable problem with analytical expressions for the electromechanically coupled nonlinear vibration response and voltage response. A theoretical case study is then presented to demonstrate the use of the nonlinear electromechanical frequency response expressions and comparisons are given against time-domain numerical simulations. After the monostable case, the outstanding broadband power generation performance of the strongly nonlinear bistable Duffing oscillator with piezoelectric coupling is investigated theoretically and experimentally through the piezomagnetoelastic energy harvester. It is experimentally verified that the broadband high-energy orbits of the piezomagnetoelastic energy harvester can give an order-of-magnitude larger power compared to the conventional piezoelastic configuration. Chaotic vibration of the bistable piezomagnetoelastic energy harvester configuration is also discussed. Experimental performance results of a bistable carbon-fiber-epoxy plate with piezoceramics are summarized for nonlinear broadband piezoelectric energy harvesting using an alternative structural configuration.

---

[1] The terms *monostable* and *bistable* are used here for the *static* equilibrium of the respective configuration (i.e., statically monostable and statically bistable). In this context, for instance, the dynamic response of a monostable Duffing oscillator can have multiple stable solutions but it is statically monostable.

Piezoelectric Energy Harvesting, First Edition. Alper Erturk and Daniel J. Inman.
© 2011 John Wiley & Sons, Ltd. Published 2011 by John Wiley & Sons, Ltd.

8.1   Perturbation Solution of the Piezoelectric Energy Harvesting Problem: the Method of Multiple Scales

This section reviews the linear single-mode piezoelectric energy harvesting problem to compare the resonance approximation of the exact governing equations to the exact solution of the perturbation-based approximate governing equations in order to show that they are identical (so that the weakly nonlinear electromechanical problem of the next section can be investigated directly using the perturbation approach). The perturbation technique used here is the method of multiple scales [1]; however, other techniques such as the Lindstedt–Poincaré method or the method of harmonic balance can also be used.

8.1.1   Linear Single-Mode Equations of a Piezoelectric Energy Harvester

The linear lumped-parameter equations of a piezoelectric energy harvester under harmonic excitation can be given for the fundamental vibration mode as

$$\ddot{x} + 2\varepsilon\mu\omega_n\dot{x} + \omega_n^2 x - \varepsilon\chi v = Fe^{j\omega t} \tag{8.1}$$

$$\dot{v} + \lambda v + \kappa\dot{x} = 0 \tag{8.2}$$

where x is the displacement response (at the tip of the cantilever), v is the voltage response across the external electrical load, ω_n is the undamped fundamental natural frequency (in short-circuit conditions), F is the harmonic forcing amplitude at frequency ω, j is the unit imaginary number, λ is the reciprocal of the time constant of the resistive–capacitive circuit (hence is inversely proportional to the load resistance), χ is the piezoelectric coupling term in the mechanical equation, κ is the piezoelectric coupling term in the electrical equation, and the overdot stands for differentiation with respect to time.[2] Furthermore, ε is a small bookkeeping parameter [1] and μ is a mechanical damping term ($\zeta = \varepsilon\mu$ where ζ is the mechanical damping ratio). The bookkeeping parameter (which can be considered to be in the order of the mechanical damping ratio ζ) is introduced for the purpose of perturbation analysis, therefore Equations (8.1) and (8.2) are rearranged (scaled) forms of typical single-mode equations derived in Chapter 3.

8.1.2   Exact Solution

For the linear electromechanical problem described by Equations (8.1) and (8.2), the steady-state response expressions can be given by $x = Xe^{j\omega t}$ and $v = Ve^{j\omega t}$ where X and V are the complex displacement and voltage terms, respectively. Then,

$$\left(-\omega^2 + j2\varepsilon\mu\omega\omega_n + \omega_n^2\right) X - \varepsilon\chi V = F \tag{8.3}$$

$$(j\omega + \lambda) V + j\omega\kappa X = 0 \tag{8.4}$$

---

[2] The backward and forward coupling terms are denoted by different letters for the general case (since different normalizations are used by different authors [2–4]) considering that the mechanical equation is given for physical coordinates rather than modal coordinates ($x(t) \cong \phi_1(L)\eta_1(t)$ for the derivations given in Chapter 3). Equations (8.1) and (8.2) are given in the form such that the coefficients of the highest-order mechanical and electrical variables, respectively, in the mechanical and electrical equations, are unity.

The complex voltage is

$$V = \frac{j\omega\kappa F}{\left(\omega_n^2 - \omega^2 + j2\varepsilon\mu\omega\omega_n\right)(\lambda + j\omega) + j\varepsilon\omega\kappa\chi} \tag{8.5}$$

or alternatively

$$V = \frac{j\omega\kappa\left(\lambda - j\omega\right)F}{\left(\omega_n^2 - \omega^2 + j2\varepsilon\mu\omega\omega_n\right)\left(\lambda^2 + \omega^2\right) + j\varepsilon\omega\kappa\chi\left(\lambda - j\omega\right)} \tag{8.6}$$

The complex displacement is

$$X = \frac{(\lambda + j\omega)F}{\left(\omega_n^2 - \omega^2 + j2\varepsilon\mu\omega\omega_n\right)(\lambda + j\omega) + j\varepsilon\omega\kappa\chi} \tag{8.7}$$

or alternatively

$$X = \frac{\left(\lambda^2 + \omega^2\right)F}{\left(\omega_n^2 - \omega^2 + j2\varepsilon\mu\omega\omega_n\right)\left(\lambda^2 + \omega^2\right) + j\varepsilon\omega\kappa\chi\left(\lambda - j\omega\right)} \tag{8.8}$$

Thus, the moduli of Equations (8.6) and (8.8) are then

$$|V| = \frac{\left(\omega^2 + \lambda^2\right)^{1/2}\kappa\omega F}{\left\{\left[\left(\omega_n^2 - \omega^2\right)\left(\lambda^2 + \omega^2\right) + \varepsilon\kappa\chi\omega^2\right]^2 + \left[2\varepsilon\mu\omega\omega_n\left(\lambda^2 + \omega^2\right) + \varepsilon\lambda\kappa\chi\omega\right]^2\right\}^{1/2}} \tag{8.9}$$

$$|X| = \frac{\left(\lambda^2 + \omega^2\right)F}{\left\{\left[\left(\omega_n^2 - \omega^2\right)\left(\lambda^2 + \omega^2\right) + \varepsilon\kappa\chi\omega^2\right]^2 + \left[2\varepsilon\mu\omega\omega_n\left(\lambda^2 + \omega^2\right) + \varepsilon\lambda\kappa\chi\omega\right]^2\right\}^{1/2}} \tag{8.10}$$

which are the exact voltage and displacement amplitudes obtained from Equations (8.1) and (8.2).

8.1.3   Resonance Approximation of the Exact Solution

If the excitation frequency is very close to the natural frequency (which is the desired case in energy harvesting), one can use

$$\omega = \omega_n + \varepsilon\sigma \tag{8.11}$$

where σ is the *detuning parameter* and is order 1, i.e., $\sigma = O(1)$. Since ε is small, the following approximation can be made:

$$\omega^2 = \omega_n^2 + 2\varepsilon\omega_n\sigma + \varepsilon^2\sigma^2 \cong \omega_n^2 + 2\varepsilon\omega_n\sigma \tag{8.12}$$

Then, substituting Equations (8.11) and (8.12) into Equations (8.9) and (8.10) and making simplifications yields

$$|V| = \frac{F}{\varepsilon} \frac{\kappa \omega_n \left(\omega_n^2 + \lambda^2\right)^{1/2}}{\left\{\left[\kappa \chi \omega_n^2 - 2\sigma \omega_n \left(\lambda^2 + \omega_n^2\right)\right]^2 + \left[2\mu \omega_n^2 \left(\lambda^2 + \omega_n^2\right) + \lambda \kappa \chi \omega_n\right]^2\right\}^{1/2}} \tag{8.13}$$

$$|X| = \frac{F}{\varepsilon} \frac{\left(\lambda^2 + \omega_n^2\right)}{\left\{\left[\kappa \chi \omega_n^2 - 2\sigma \omega_n \left(\lambda^2 + \omega_n^2\right)\right]^2 + \left[2\mu \omega_n^2 \left(\lambda^2 + \omega_n^2\right) + \lambda \kappa \chi \omega_n\right]^2\right\}^{1/2}} \tag{8.14}$$

which are the resonance approximations to the exact solution of the exact form of the governing equations (exact in the lumped-parameter or single-mode sense). In both expressions, the first term (F/ε) represents the resonance behavior, that is, the order difference between the forcing amplitude and the response amplitude.

8.1.4 Perturbation Solution

The basic idea of the method of multiple scales is to represent the response as a function of multiple variables, i.e., scales, rather than a single variable [1]. In order to obtain a first approximation to the response, the time scales up to order ε are defined as

$$T_0 = t, \quad T_1 = \varepsilon t \tag{8.15}$$

The response expressions are then

$$x(t;\varepsilon) = X_0(T_0, T_1) + \varepsilon X_1(T_0, T_1) + \dots \tag{8.16}$$

$$v(t;\varepsilon) = V_0(T_0, T_1) + \varepsilon V_1(T_0, T_1) + \dots \tag{8.17}$$

The derivatives with respect to time are also expansions as partial derivatives with respect to the time scales:

$$\frac{d}{dt} = D_0 + \varepsilon D_1 + \dots \tag{8.18}$$

$$\frac{d^2}{dt^2} = D_0^2 + 2\varepsilon D_0 D_1 + \dots \tag{8.19}$$

where D_n is shorthand for the partial differential operator $\partial/\partial T_n$.

Substituting Equations (8.16)–(8.19) into Equations (8.1) and (8.2) gives

$$\left(D_0^2 + 2\varepsilon D_0 D_1\right)(X_0 + \varepsilon X_1) + 2\varepsilon \mu \omega_n (D_0 + \varepsilon D_1)(X_0 + \varepsilon X_1) + \omega_n^2 (X_0 + \varepsilon X_1)$$
$$- \varepsilon \chi (V_0 + \varepsilon V_1) = \varepsilon f e^{j\omega T_0} \tag{8.20}$$

$$(D_0 + \varepsilon D_1)(V_0 + \varepsilon V_1) + \lambda (V_0 + \varepsilon V_1) + \kappa (D_0 + \varepsilon D_1)(X_0 + \varepsilon X_1) = 0 \tag{8.21}$$

where soft excitation is assumed as common practice in primary resonance analysis (which is the only case in the linear problem) so that the forcing will appear in the same perturbation equation as damping:

$$F = \varepsilon f \tag{8.22}$$

Equations (8.20) and (8.21) are the *approximate* governing equations obtained from the exact governing equations using the method of multiple scales.

Expanding Equations (8.20) and (8.21), and then equating the coefficients of the ε^0 terms, gives

$$D_0^2 X_0 + \omega_n^2 X_0 = 0 \tag{8.23}$$
$$D_0 V_0 + \lambda V_0 = -\kappa D_0 X_0 \tag{8.24}$$

Similarly, equating the coefficients of the ε terms in Equations (8.20) and (8.21) yields

$$D_0^2 X_1 + \omega_n^2 X_1 = -2 D_0 D_1 X_0 - 2 \mu \omega_n D_0 X_0 + \chi V_0 + f e^{j \omega T_0} \tag{8.25}$$
$$D_1 V_1 + \lambda V_1 = -D_1 V_0 - \kappa (D_0 X_1 + D_1 X_0) \tag{8.26}$$

From Equations (8.23) and (8.24),

$$X_0 = A (T_1) e^{j \omega_n T_0} + cc \tag{8.27}$$
$$V_0 = B (T_1) e^{-\lambda T_0} - \frac{j \omega_n \kappa A (T_1) e^{j \omega_n T_0}}{\lambda + j \omega_n} + cc \tag{8.28}$$

where $A (T_1)$ and $B (T_1)$ are unknown functions at this point and cc stands for the complex conjugate of the preceding terms.

Substituting Equations (8.27) and (8.28) into Equation (8.25) and eliminating the secular terms [1] gives

$$-2 j \omega_n D_1 \left(A e^{j \omega_n T_0} \right) - 2 j \mu \omega_n^2 A e^{j \omega_n T_0} - \frac{j \kappa \chi \omega_n}{\lambda + j \omega_n} A e^{j \omega_n T_0} + f e^{j \omega T_0} = 0 \tag{8.29}$$

Using Equation (8.11) in the power of the forcing term, Equation (8.29) reduces to

$$-2 j \omega_n (D_1 A + \mu \omega_n A) - \frac{j \kappa \chi \omega_n A}{\lambda + j \omega_n} + f e^{j \sigma T_1} = 0 \tag{8.30}$$

The unknown function A can be expressed in polar form as

$$A = a e^{j \beta} \tag{8.31}$$

where a and β are real functions of T_1 and it is worth recalling from Equations (8.16) and (8.27) that the first approximation to the vibration response is $x = a e^{j(\omega_n t + \beta)} + O (\varepsilon)$ (hence a is the vibration response amplitude one is seeking).

Substituting Equation (8.31) into Equation (8.30) and introducing

$$\gamma = \sigma T_1 - \beta \tag{8.32}$$

yields the following two equations obtained from the real and the imaginary parts:

$$-2\omega_n a' - 2\mu\omega_n^2 a - \frac{\kappa\chi\lambda\omega_n a}{\lambda^2 + \omega_n^2} + f\sin\gamma = 0 \tag{8.33}$$

$$2\omega_n a\sigma - 2\omega_n a\gamma' - \frac{\kappa\chi\omega_n^2 a}{\lambda^2 + \omega_n^2} + f\cos\gamma = 0 \tag{8.34}$$

where the prime denotes differentiation with respect to T_1. Equations (8.33) and (8.34) are the electromechanical *modulation equations*.

Steady-state response implies $a' = \gamma' = 0$, reducing the modulation equations to

$$2\mu\omega_n^2 a + \frac{\kappa\chi\lambda\omega_n a}{\lambda^2 + \omega_n^2} = f\sin\gamma \tag{8.35}$$

$$-2\omega_n a\sigma + \frac{\kappa\chi\omega_n^2 a}{\lambda^2 + \omega_n^2} = f\cos\gamma \tag{8.36}$$

or alternatively

$$\left[2\mu\omega_n^2\left(\lambda^2 + \omega_n^2\right) + \lambda\kappa\chi\omega_n\right]a = f\left(\lambda^2 + \omega_n^2\right)\sin\gamma \tag{8.37}$$

$$\left[-2\sigma\omega_n\left(\lambda^2 + \omega_n^2\right) + \kappa\chi\omega_n^2\right]a = f\left(\lambda^2 + \omega_n^2\right)\cos\gamma \tag{8.38}$$

Equations (8.37) and (8.38) can be combined to give the steady-state vibration amplitude as

$$a = \frac{\left(\lambda^2 + \omega_n^2\right)f}{\left\{\left[\kappa\chi\omega_n^2 - 2\sigma\omega_n\left(\lambda^2 + \omega_n^2\right)\right]^2 + \left[2\mu\omega_n^2\left(\lambda^2 + \omega_n^2\right) + \lambda\kappa\chi\omega_n\right]^2\right\}^{1/2}} \tag{8.39}$$

which is identical to Equation (8.14).

At steady state, the homogeneous part of Equation (8.28) vanishes (i.e., $B\left(T_1\right)e^{-\lambda T_0} \to 0$ as $T_0 = t \to \infty$ since $B\left(T_1\right)$ is bounded and $\lambda > 0$), yielding the following expression for the steady-state voltage amplitude in terms of the vibration amplitude:

$$b = \frac{\kappa\omega_n a}{\left(\lambda^2 + \omega_n^2\right)^{1/2}} \tag{8.40}$$

Substituting Equation (8.39) into Equation (8.40) gives

$$b = \frac{\kappa\omega_n\left(\omega_n^2 + \lambda^2\right)^{1/2}f}{\left\{\left[\kappa\chi\omega_n^2 - 2\sigma\omega_n\left(\lambda^2 + \omega_n^2\right)\right]^2 + \left[2\mu\omega_n^2\left(\lambda^2 + \omega_n^2\right) + \lambda\kappa\chi\omega_n\right]^2\right\}^{1/2}} \tag{8.41}$$

which is nothing but Equation (8.13). Therefore, *the exact solution of the perturbation-based approximation to the exact governing equations is identical to the resonance approximation to the exact solution of the exact governing equations.* The foregoing discussion shows the validity of the perturbation solution around the resonance in the particular electromechanical problem of linear piezoelectric energy harvesting. In nonlinear problems, however, the analytical solution of the exact governing equations often does not exist. Nevertheless, one can compare the perturbation solution against time-domain simulations to check its completeness and validity.

8.2 Monostable Duffing Oscillator with Piezoelectric Coupling

This section uses the perturbation-based approach to obtain analytical expressions for the electromechanical frequency response curves in a typical weakly nonlinear energy harvester: the monostable Duffing oscillator with piezoelectric coupling. Enhancement of frequency bandwidth using hardening and softening stiffness effects is discussed along with the effects of other system parameters and perturbation-based analytical solutions are compared against time-domain numerical solutions. The effects of system parameters on the nonlinear electromechanical response are discussed.

8.2.1 Analytical Expressions Based on the Perturbation Solution

The lumped-parameter equations of a piezoelectric energy harvester with cubic nonlinearity in the displacement term (often results from *geometric nonlinearities* [5]) under harmonic excitation can be given as

$$\ddot{x} + 2\varepsilon\mu\omega_n\dot{x} + \omega_n^2 x + \varepsilon\alpha x^3 - \varepsilon\chi v = \varepsilon f \cos\omega t \tag{8.42}$$

$$\dot{v} + \lambda v + \kappa\dot{x} = 0 \tag{8.43}$$

where α is the coefficient of the nonlinear term: $\alpha > 0$ is the *hardening* stiffness case whereas $\alpha < 0$ is the *softening* stiffness case ($\alpha = 0$ is the linear case given by Equations (8.1) and (8.2)).[3] The mechanical forcing term is in the same order as the mechanical damping and piezoelectric coupling (soft forcing is considered directly as previously defined by Equation (8.22)), therefore the primary resonance [5] behavior is of interest ($\omega \approx \omega_n$).

Substituting Equations (8.16)–(8.19) into Equations (8.42) and (8.43) and using Equation (8.11) gives

$$\left(D_0^2 + 2\varepsilon D_0 D_1\right)(X_0 + \varepsilon X_1) + 2\varepsilon\mu\omega_n(D_0 + \varepsilon D_1)(X_0 + \varepsilon X_1) + \omega_n^2(X_0 + \varepsilon X_1) \\ + \varepsilon\alpha(X_0 + \varepsilon X_1)^3 - \varepsilon\chi(V_0 + \varepsilon V_1) = \varepsilon f \cos(\omega_n T_0 + \sigma T_1) \tag{8.44}$$

$$(D_0 + \varepsilon D_1)(V_0 + \varepsilon V_1) + \lambda(V_0 + \varepsilon V_1) + \kappa(D_0 + \varepsilon D_1)(X_0 + \varepsilon X_1) = 0 \tag{8.45}$$

which can be expanded to equate the coefficients of the ε^0 and ε terms separately, yielding

$$D_0^2 X_0 + \omega_n^2 X_0 = 0 \tag{8.46}$$

$$D_0 V_0 + \lambda V_0 = -\kappa D_0 X_0 \tag{8.47}$$

---

[3] Another form of nonlinearity that can be pronounced for relatively high excitation levels is piezoelectric nonlinearity [2,6–8]. In the simplest terms for the lumped-parameter problem, the electromechanical coupling terms become functions of x as analyzed by Triplett and Quinn [2] and more sophisticated approaches require electroelasticity-based distributed-parameter analysis as in Stanton *et al.* [7,8]. In the absence of other nonlinearities, the effect of piezoelastic nonlinearities [8] is similar to the softening stiffness effect covered in this section.

and

$$D_0^2 X_1 + \omega_n^2 X_1 = -2D_0 D_1 X_0 - 2\mu\omega_n D_0 X_0 - \alpha X_0^3 + \chi V_0 + f\cos{(\omega_n T_0 + \sigma T_1)} \quad (8.48)$$

$$D_1 V_1 + \lambda V_1 = -D_1 V_0 - \kappa(D_0 X_1 + D_1 X_0) \quad (8.49)$$

The solutions of Equations (8.46) and (8.47) are the same as those given by Equations (8.27) and (8.28), respectively. Therefore, Equations (8.27) and (8.28) can be used in Equation (8.48), and eliminating the secular terms leads to[4]

$$-2j\omega_n D_1\left(Ae^{j\omega_n T_0}\right) - 2j\mu\omega_n^2 Ae^{j\omega_n T_0} - 3\alpha A^2 \bar{A} e^{j\omega_n T_0}$$
$$-\frac{j\kappa\chi\omega_n}{\lambda + j\omega_n} Ae^{j\omega_n T_0} + \frac{1}{2} f e^{j(\omega_n T_0 + \sigma T_1)} = 0 \quad (8.50)$$

or alternatively

$$-2j\omega_n\left(A' + \mu\omega_n A\right) - 3\alpha A^2 \bar{A} - \frac{j\kappa\chi\omega_n}{\lambda^2 + \omega_n^2}(\lambda - j\omega_n)A + \frac{1}{2} f e^{j\sigma T_1} = 0 \quad (8.51)$$

where $\bar{A}$ is the complex conjugate of A.

To solve Equation (8.51), A is expressed in polar form as

$$A = \frac{1}{2} a e^{j\beta} \quad (8.52)$$

where a and β are real functions of T_1 and the factor 1/2 is again due to Euler's formula since the first approximation to the vibration response is a cosine function:

$$x = a\cos{(\omega_n t + \beta)} + O(\varepsilon) \quad (8.53)$$

Substituting Equation (8.52) into Equation (8.51) along with the use of Equation (8.32) gives the electromechanical modulation equations:[5]

$$-2\omega_n a' - 2\mu\omega_n^2 a - \frac{\kappa\chi\lambda\omega_n a}{\lambda^2 + \omega_n^2} + f\sin\gamma = 0 \quad (8.54)$$

$$2\omega_n\sigma a - 2\omega_n a\gamma' - \frac{3}{4}\alpha a^3 - \frac{\kappa\chi\omega_n^2 a}{\lambda^2 + \omega_n^2} + f\cos\gamma = 0 \quad (8.55)$$

At steady state, $a' = \gamma' = 0$, yielding

$$2\mu\omega_n^2 a + \frac{\kappa\chi\lambda\omega_n a}{\lambda^2 + \omega_n^2} = f\sin\gamma \quad (8.56)$$

$$-2\omega_n\sigma a + \frac{3}{4}\alpha a^3 + \frac{\kappa\chi\omega_n^2 a}{\lambda^2 + \omega_n^2} = f\cos\gamma \quad (8.57)$$

---

[4] Note that the factor 1/2 multiplying the mechanical forcing term is due to Euler's formula ($e^{j\theta} = \cos\theta + j\sin\theta$).

[5] Stability analysis of the electromechanical response can be performed using the modulation equations [5].

Therefore the frequency response equation that relates the vibration response amplitude a and the detuning parameter σ for a given forcing level f can be given by

$$\left[\left(2\mu\omega_n^2 + \frac{\kappa\chi\lambda\omega_n}{\lambda^2 + \omega_n^2}\right)^2 + \left(\frac{3}{4}\alpha a^2 + \frac{\kappa\chi\omega_n^2}{\lambda^2 + \omega_n^2} - 2\omega_n\sigma\right)^2\right]a^2 = f^2 \tag{8.58}$$

which gives a sixth-order equation in a (which is cubic in a^2) similar to the one derived using the using the Lindstedt–Poincaré method [1] in the absence of piezoelectric nonlinearities [2]. To plot the frequency response (a vs. σ), one can either solve the cubic equation for a^2 or simply express σ explicitly in terms of a as suggested by Nayfeh and Mook [5] (which is preferred here):

$$\sigma = \frac{1}{2\omega_n}\left\{\frac{3}{4}\alpha a^2 + \frac{\kappa\chi\omega_n^2}{\lambda^2 + \omega_n^2} \pm \left[\frac{f^2}{a^2} - \left(2\mu\omega_n^2 + \frac{\kappa\chi\lambda\omega_n}{\lambda^2 + \omega_n^2}\right)^2\right]^{1/2}\right\} \tag{8.59}$$

Using Equation (8.59), the frequency response curve of the vibration response can be constructed for a given excitation level.

The steady-state relation between the voltage amplitude b and the vibration amplitude a given by Equation (8.40) can be rearranged to give

$$a = \frac{\left(\lambda^2 + \omega_n^2\right)^{1/2}}{\kappa\omega_n}b \tag{8.60}$$

Therefore the frequency response expression for the voltage amplitude is

$$\sigma = \frac{1}{2\omega_n}\left\{\frac{3}{4}\alpha\frac{\lambda^2 + \omega_n^2}{\kappa^2\omega_n^2}b^2 + \frac{\kappa\chi\omega_n^2}{\lambda^2 + \omega_n^2} \pm \left[\frac{\kappa^2\omega_n^2 f^2}{\left(\lambda^2 + \omega_n^2\right)b^2} - \left(2\mu\omega_n^2 + \frac{\kappa\chi\lambda\omega_n}{\lambda^2 + \omega_n^2}\right)^2\right]^{1/2}\right\} \tag{8.61}$$

which can be used to construct the b vs. σ curve.

8.2.2 State-Space Representation of the Governing Equations for Numerical Solution

Time-domain numerical solution using computer code is an alternative way of solving the exact governing nonlinear equations. The first-order (or state-space) representation of Equations (8.42) and (8.43) can be given by

$$\begin{bmatrix} \dot{u}_1 \\ \dot{u}_2 \\ \dot{u}_3 \end{bmatrix} = \begin{bmatrix} u_2 \\ -2\varepsilon\mu\omega_n u_2 - \omega_n^2 u_1 - \alpha u_1^3 + \varepsilon\chi u_3 + \varepsilon f \cos\Omega t \\ -\lambda u_3 - \kappa u_2 \end{bmatrix} \tag{8.62}$$

where the state variables are $u_1 = x$, $u_2 = \dot{x}$, and $u_3 = v$. The electromechanically coupled equations given by Equation (8.62) can be used in an ordinary differential equation solver for numerical simulations (e.g., the *ode45* command of MATLAB).

8.2.3 Theoretical Case Study

This section considers a theoretical case study to demonstrate the nonlinear electromechanical frequency response expressions given by Equations (8.59) and (8.61) with comparisons against the time-domain simulations of Equation (8.62) for certain frequencies. The aim is to investigate the effects of hardening and softening stiffness (along with other parameters) on the nonlinear frequency response curves as well as to validate the analytical perturbation equations.

The system parameters used in the first simulations of this case study are $\omega_n = 1$, $\mu = 0.5$, $\chi = 5$, $\kappa = 0.5$, and $\lambda = 0.05$. It is important to note that the small bookkeeping parameter ε is not embedded in these terms, in agreement with the representation given by Equation (8.42). For instance, $\mu = O(1)$ but the physical damping ratio is expectedly order ε, that is, $\zeta = \varepsilon\mu = O(\varepsilon)$. For a mechanical damping ratio of $\zeta = 0.5\%$, $\varepsilon = 0.01$ becomes the small parameter to have $\mu = 0.5$ in this case. Therefore, the detuning parameter range of $-5 \le \sigma \le 5$ covers $-1.05 \le \omega_n + \varepsilon\sigma \le 1.05$ which is a reasonably small range around the undamped natural frequency $\omega_n = 1$. Note that the system is close to open-circuit conditions for $\lambda = 0.05$. Consequently, the resonance of the linear problem occurs close to the open-circuit resonance frequency for these system parameters (a value larger than unity in this case – see Chapter 5).

It is well known that the nature of the forced response in nonlinear dynamical systems strongly depends on the excitation amplitude [1,5,9, 10]. Therefore the forcing amplitude (f) and the coefficient of cubic nonlinearity (α) are the primary parameters of interest to investigate the resulting nonlinear phenomena. In the first set of simulations, the nonlinear stiffness parameter is taken as $\alpha = 5$ (therefore *hardening stiffness* is expected). Figure 8.1 shows the vibration and the voltage frequency response curves for an excitation amplitude of $f = 0.1$ (which is a relatively low excitation amplitude since $f = O(1)$). Note that the horizontal axis is not the frequency but the detuning parameter (a measure of the amount of perturbation

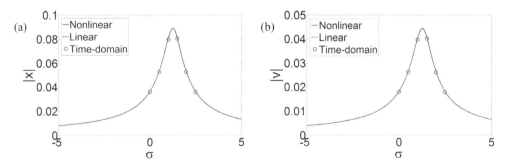

Figure 8.1 (a) Displacement and (b) voltage frequency response curves for an excitation amplitude of $f = 0.1$ obtained by the nonlinear perturbation solution ($\alpha = 5$), linear perturbation solution ($\alpha = 0$), and time-domain numerical solutions ($\alpha = 5$, $x(0) = 1$, $\dot{x}(0) = 1$, $v(0) = 0$)

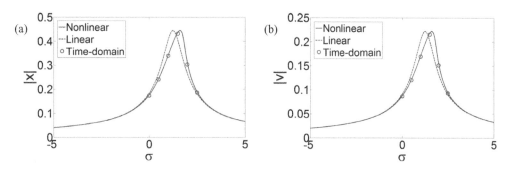

Figure 8.2 (a) Displacement and (b) voltage frequency response curves for an excitation amplitude of $f = 0.5$ obtained by the nonlinear perturbation solution ($\alpha = 5$), linear perturbation solution ($\alpha = 0$), and time-domain numerical solutions ($\alpha = 5$, $x(0) = 1$, $\dot{x}(0) = 1$, $v(0) = 0$)

relative to the linear short-circuit natural frequency). The solid and dashed lines, respectively, are obtained from the analytical expressions of the perturbation solution (Equations (8.59) and (8.61)) for $\alpha = 5$ and $\alpha = 0$. Therefore the dashed line in each graph is nothing but the linear analytical solution reviewed in Section 8.1. It is clear from Figure 8.1 that the effect of nonlinearity (the cubic part of the stiffness) is negligible for this low excitation amplitude. Steady-state response amplitudes obtained from the time-domain numerical simulations using Equation (8.62) (given for certain frequencies in Figure 8.1) agree perfectly with the perturbation solutions. Recall that the peak values do not correspond to $\sigma = 0$ because the system is close to open-circuit conditions for $\lambda = 0.05$.

If the excitation amplitude is increased to $f = 0.5$, as shown in Figure 8.2, the nonlinear perturbation solution deviates considerably from the linear solution and the cubic component of the stiffness is no longer negligible. Since the coefficient of the nonlinear term is positive, the resulting effect is bending of the frequency response curve toward the right as a result of the hardening stiffness. The data points obtained for certain frequencies by numerical solution of the exact governing equations agree very well with the nonlinear frequency response curves obtained from the perturbation solution.

Having observed that the nonlinear term becomes more effective with increasing excitation amplitude, the displacement and the voltage frequency response curves are plotted for different excitation levels in Figures 8.3a and 8.3b, respectively. As the excitation level increases, a high-energy attractor develops over a broad range of frequencies (along with an unstable solution branch shown with the dashed lines in Figure 8.3). Therefore, compared to the bandwidth of large-amplitude response of energy harvesters with linear stiffness (Chapters 3 and 4), that of a configuration with nonlinear stiffness can be much broader, which is preferred in vibration-based energy harvesting to cover a wider range of excitation frequencies. The results of time-domain numerical solution (given for six different σ values in Figure 8.3) perfectly agree with the perturbation solution.

It should be noted from Figure 8.3 that multiple stable solutions may coexist over a range of frequencies depending on the excitation amplitude. For instance, for an excitation amplitude of $f = 1.5$ at $\sigma = 4$ (i.e., $\omega = 1.04$), there is approximately a factor of 4.6 between the response amplitudes of the stable high-energy and low-energy attractors. Obviously the high-energy response is preferred for larger electrical output and the resulting response strongly depends

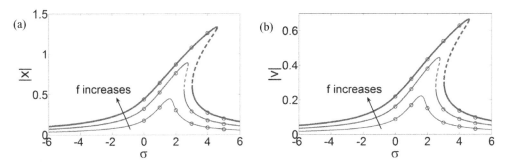

Figure 8.3 (a) Displacement and (b) voltage frequency response curves for different excitation levels ($f = 0.5$, $f = 1.0$, $f = 1.5$). Solid lines are stable and dashed lines are unstable analytical solutions (perturbation-based) and circles are time-domain numerical solutions ($\mu = 0.5$, $\alpha = 5$, $\chi = 5$, $\kappa = 0.5$, $\lambda = 0.05$)

on the initial conditions. In the time-domain solutions (circles in Figure 8.3), the coexisting data points of $f = 1.5$ at $\sigma = 4$ are obtained for two different sets of initial conditions. Among other possible combinations, the initial condition set $x(0) = 1$, $\dot{x}(0) = 1$, $v(0) = 0$ results in a high-energy response (the data point on the upper branch) while the initial conditions of $x(0) = 1$, $\dot{x}(0) = 0$, $v(0) = 0$ yield a low-energy response (the data point on the lower branch). Figure 8.4 shows the time-domain response histories for these two sets of initial conditions with

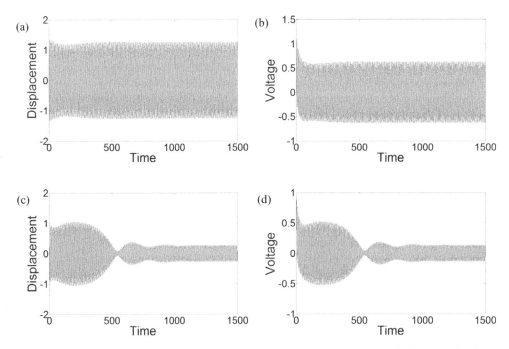

Figure 8.4 Time history examples of coexisting high-energy and low-energy solutions: (a) displacement and (b) voltage histories for $x(0) = 1$, $\dot{x}(0) = 1$, $v(0) = 0$; (c) displacement and (d) voltage histories for $x(0) = 1$, $\dot{x}(0) = 0$, $v(0) = 0$ (both cases have $f = 1.5$ at $\sigma = 4$)

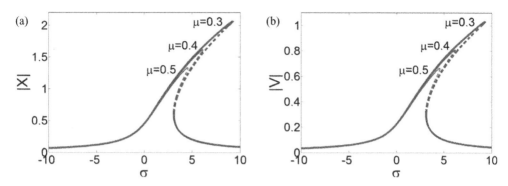

Figure 8.5 (a) Displacement and (b) voltage frequency response curves for different values of mechanical damping. Solid lines are stable and dashed lines are unstable perturbation-based analytical solutions ($f = 1.5, \alpha = 5, \chi = 5, \kappa = 0.5, \lambda = 0.05$)

the same excitation amplitude and frequency. Unlike linear systems with positive damping, the steady-state forced response of nonlinear systems usually depends on the initial conditions. As in the example given here, when more than one stable steady-state solution exists, the initial conditions determine the steady-state response physically realized by the system. This concept is related to the *domains of attraction* and the reader can find details in the nonlinear dynamics literature [5].

Having observed the favorable effect of nonlinear stiffness on the frequency bandwidth for relatively high forcing levels, the excitation level of $f = 1.5$ for different values of mechanical damping is considered in Figure 8.5: $\mu = 0.5$, $\mu = 0.4$, and $\mu = 0.3$ (which correspond to $\zeta = 0.005$, $\zeta = 0.004$, and $\zeta = 0.003$, respectively, since $\varepsilon = 0.01$). Both the displacement and the voltage frequency response curves show that the effect of nonlinearity becomes more pronounced with reduced mechanical damping. In addition to the increase in the peak response amplitude (as in the linear problem), the bandwidth of high-energy stable solutions also increases as the mechanical damping is decreased. Therefore, light mechanical damping is preferred for enhanced bandwidth of effective operation for an energy harvester with nonlinear stiffness.

Another interesting parameter that has been kept constant in the foregoing simulations is the reciprocal of the time constant denoted by λ. As mentioned previously, for $\lambda = 0.05$ used in the above discussion, the system is close to open-circuit conditions (upper limit of the voltage output). Since λ is inversely proportional to the external load resistance, the system is closer to short-circuit conditions for its larger values. Figure 8.6 shows the displacement and voltage frequency response curves for $\lambda = 0.05$, $\lambda = 0.5$, and $\lambda = 5$. Similar to the case of the linear problem (Chapters 3 and 4), moderate values of load resistance result in considerable vibration attenuation according to Figure 8.6a and the voltage output increases monotonically with increasing load resistance (i.e., decreasing λ) in Figure 8.6b. It is worth highlighting the fact that, for the moderate value of λ among the values used here (which results in the largest shunt damping compared to others), the unstable solution branch denoted by the dashed lines disappears and the response is no longer multi-valued. From the mathematical point of view, the effects of piezoelectric coupling and nonlinear stiffness are in the same order in the mechanical equation. Therefore, the effect of piezoelectric power generation can indeed affect the resulting nonlinear phenomena depending on the external load used. The matched

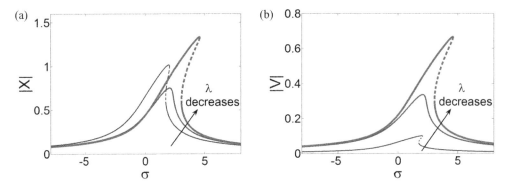

Figure 8.6 (a) Displacement and (b) voltage frequency response curves for different values of the reciprocal of time constant ($\lambda = 0.05, \lambda = 0.5, \lambda = 5$). Solid lines are stable and dashed lines are unstable perturbation-based analytical solutions ($f = 1.5, \alpha = 5, \mu = 0.5, \chi = 5, \kappa = 0.5$)

resistance of maximum power transfer from the mechanical domain to the electrical domain results in strong vibration attenuation as a result of the principle of conservation of energy. Therefore, the dilemma in exploiting the nonlinear dynamic response using a *weakly* nonlinear configuration (such that the piezoelectric coupling and the nonlinearity are in the same order) is that the external load of maximum power might remove the desired consequences of the nonlinear dynamic phenomena, which should be considered by the designer. The presence of an external load (which is essential in energy harvesting) can shorten the high-energy solution branch dramatically (compare the $\lambda = 0.05$ and $\lambda = 0.5$ cases in Figure 8.6). Other than open-circuit conditions ($\lambda \rightarrow 0$), coexisting solutions appear close to short-circuit conditions ($\lambda \rightarrow \infty$) due to the reduced shunt damping effect (see $\lambda = 5$ in Figure 8.6a); however, the electrical response is very low (Figure 8.6b) and short-circuit conditions have no practical value in energy harvesting.

So far the discussion has been given for the hardening stiffness case considering a fixed positive value of the coefficient of nonlinearity. Figure 8.7 shows the displacement and voltage

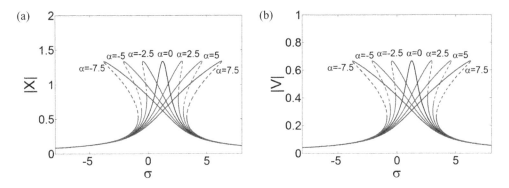

Figure 8.7 (a) Displacement and (b) voltage frequency response curves for different values of the cubic nonlinearity coefficient. Solid lines are stable and dashed lines are unstable perturbation-based analytical solutions ($f = 1.5, \mu = 0.5, \chi = 5, \kappa = 0.5, \lambda = 0.05$)

frequency response curves for a range of the coefficient of cubic nonlinearity. The negative values of α result in the softening stiffness effect (bending the frequency response curves toward the left), which is the opposite of the hardening stiffness effect obtained for the positive values of α. The previous discussion regarding the effects of other system parameters is valid for the softening stiffness case as well.

8.3 Bistable Duffing Oscillator with Piezoelectric Coupling: the Piezomagnetoelastic Energy Harvester

In this section, the bistable form of the Duffing oscillator with piezoelectric coupling is investigated by theoretical simulations following Erturk *et al.* [11,12], prior to the experimental results of the next section. The bistable piezomagnetoelastic structure is introduced along with its governing lumped-parameter equations [11,12] exhibiting strongly nonlinear behavior. The focus is placed on high-energy periodic orbits of this configuration for broadband energy harvesting and the chaotic electromechanical response is also discussed.

8.3.1 Lumped-Parameter Electromechanical Equations

The magnetoelastic structure that forms the basis of this section was first investigated by Moon and Holmes [13] as a mechanical structure that exhibits strange attractor motions. As shown in Figure 8.8a, the magnetoelastic device consists of a ferromagnetic cantilever with two permanent magnets located symmetrically near the free end and is subjected to harmonic base excitation. The bifurcations of the static problem are described by a butterfly catastrophe [14] with a sixth-order magnetoelastic potential. Therefore, depending on the magnet spacing, the ferromagnetic beam may have five (with three stable), three (with two stable), or one (stable) equilibrium positions [13]. For the case with three equilibrium positions (statically *bistable* configuration), the governing lumped-parameter equation of motion for the fundamental vibration mode has the well-known form of the bistable Duffing equation [9, 10,13], the nonlinear dynamics of which were extensively analyzed by Holmes [15].

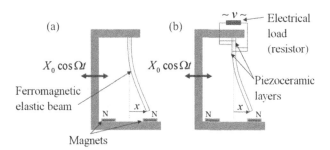

Figure 8.8   Schematics of (a) the magnetoelastic structure investigated by Moon and Holmes [13] and (b) the piezomagnetoelastic power generator investigated by Erturk *et al.* [11,12]

In order to use the magnetoelastic configuration as a piezoelectric energy harvester, Erturk *et al.* [11,12] attached two piezoceramic layers onto the root of the cantilever and obtained a bimorph as depicted in Figure 8.8b, where the electrical outputs are connected to a resistive electrical load. Introducing piezoelectric coupling to the bistable Duffing equation and applying Kirchhoff's laws to the circuit with a resistor leads to the following electromechanical equations describing the nonlinear system dynamics for the fundamental vibration mode [11,12]:[6]

$$\ddot{x} + 2\zeta\dot{x} - \frac{1}{2}x\left(1 - x^2\right) - \chi v = F\cos\omega t \tag{8.63}$$

$$\dot{v} + \lambda v + \kappa\dot{x} = 0 \tag{8.64}$$

where x is the tip displacement of the beam in the transverse direction, v is the voltage across the load resistance, ζ is the mechanical damping ratio, ω is the excitation frequency, F is the excitation force due to base acceleration ($F \propto \omega^2 X_0$ where X_0 is the base displacement amplitude), χ is the piezoelectric coupling term in the mechanical equation, κ is the piezoelectric coupling term in the electrical circuit equation, and λ is the reciprocal of the time constant ($\lambda \propto 1/R_l C_p^{eq}$ where R_l is the load resistance and C_p^{eq} is the equivalent capacitance of the piezoceramic layers). The dimensionless Equations (8.63) and (8.64) are not necessarily in the same form as Equations (8.42) and (8.43), although the same letters are assigned to some of the terms. The problem investigated in this section is *strongly* nonlinear, unlike the *weakly* nonlinear problem of Section 8.2 (i.e., in the perturbation sense, mechanical nonlinearity and piezoelectric coupling are not of the same order). The three static equilibrium positions obtained from Equation (8.63) are $(x, \dot{x}) = (0, 0)$ (a saddle) and $(x, \dot{x}) = (\pm 1, 0)$ (two sinks). Note that the inherent piezoelastic nonlinearities [2, 6–8] are ignored in Equations (8.63) and (8.64) by assuming the standard form of the linear piezoelectric constitutive equations (Appendix A).

8.3.2 Time-Domain Simulations of the Electromechanical Response

The first-order form of Equations (8.63) and (8.64) can be expressed as

$$\begin{bmatrix} \dot{u}_1 \\ \dot{u}_2 \\ \dot{u}_3 \end{bmatrix} = \begin{bmatrix} u_2 \\ -2\zeta u_2 + \frac{1}{2}u_1\left(1 - u_1^2\right) + \chi u_3 + F\cos\omega t \\ -\lambda u_3 - \kappa u_2 \end{bmatrix} \tag{8.65}$$

where the state variables are $u_1 = x$, $u_2 = \dot{x}$, and $u_3 = v$.

The time-domain voltage simulations shown in Figures 8.9 and 8.10 are obtained using Equation (8.65) with $\omega = 0.8$, $\zeta = 0.01$, $\chi = 0.05$, $\kappa = 0.5$, and $\lambda = 0.05$ (close to open-circuit conditions). In the first case (Figure 8.9), the forcing term is $F = 0.08$ and the motion starts with an initial deflection at one of the stable equilibrium positions ($x(0) = 1$ with zero initial velocity and voltage: $\dot{x}(0) = v(0) = 0$). The resulting vibratory motion is on a chaotic

---

[6] The factor 1/2 in front of the magnetoelastic restoring force ensures that the undamped ($\zeta = 0$) short-circuit ($v \rightarrow 0$) natural frequency of small oscillations around either magnet is *unity* (which can be shown by substituting $x = 1 + \bar{x}$ and linearizing in $\bar{x}$). This is the convention preferred because the relevant phenomena discussed herein are for excitation frequencies below this *post-buckled* natural frequency (thus for $\omega < 1$).

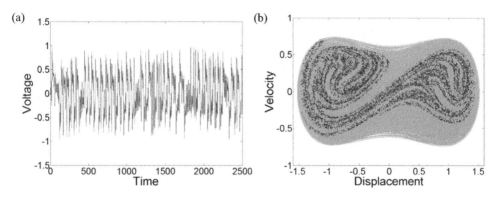

Figure 8.9 (a) Theoretical voltage history obtained from the strange attractor motion and (b) the Poincaré map of the motion on its phase portrait ($x(0) = 1$, $\dot{x}(0) = 0$, $v(0) = 0$, $F = 0.08$, $\omega = 0.8$)

strange attractor (yielding the chaotic voltage history shown in Figure 8.9a) and the Poincaré map of this strange attractor motion is shown in Figure 8.9b on its phase portrait.

If the excitation amplitude is increased by 50% (to $F = 0.12$) by keeping the same initial conditions, the transient chaotic response is followed by large-amplitude periodic oscillations on a high-energy orbit with a substantially improved voltage response (Figure 8.10a). That is, the attractor for this excitation amplitude is a large-amplitude periodic attractor, no longer a strange attractor. Remarkably, Figure 8.10b shows that the response on a very similar large-amplitude periodic attractor can be obtained with the original excitation amplitude ($F = 0.08$) but different initial conditions (simply by imposing an initial velocity condition

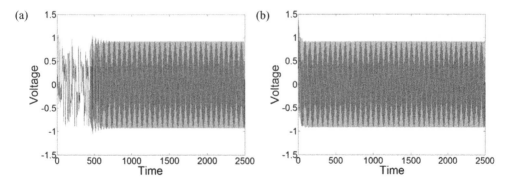

Figure 8.10 Theoretical voltage histories: (a) large-amplitude response due to the excitation amplitude ($x(0) = 1$, $\dot{x}(0) = 0$, $v(0) = 0$, $F = 0.12$, $\omega = 0.8$); (b) large-amplitude response due to the initial conditions for a lower excitation amplitude ($x(0) = 1$, $\dot{x}(0) = 1.3$, $v(0) = 0$, $F = 0.08$, $\omega = 0.8$)

so that $x(0) = 1$, $\dot{x}(0) = 1.3$, $v(0) = 0$.[7] This second case corresponds to the coexistence of a strange attractor with a large-amplitude periodic attractor for the same forcing level as formerly discussed by Guckenheimer and Holmes [9]. For the same amount of forcing, depending on the initial conditions, the system can be attracted by the high-energy periodic orbit (Figure 8.10b) instead of the strange attractor (Figure 8.9a). Obviously the large-amplitude periodic orbit is preferred for energy harvesting and therefore the focus is mainly placed on the response form of Figure 8.10b in the following discussion.

8.3.3 Performance Comparison of the Piezomagnetoelastic and the Piezoelastic Configurations in the Phase Space

Having observed the large-amplitude electromechanical response of the piezomagnetoelastic energy harvester configuration described by Equation (8.65), comparisons can be made against the conventional piezoelastic configuration (which is the commonly employed cantilever configuration without the magnets causing the bistability – see Chapters 3 and 4).

The lumped-parameter electromechanical equations of the linear piezoelastic configuration for the fundamental vibration mode are[8]

$$\ddot{x} + 2\zeta\dot{x} + \frac{1}{2}x - \chi v = F \cos \omega t \qquad (8.66)$$

$$\dot{v} + \lambda v + \kappa \dot{x} = 0 \qquad (8.67)$$

which can be given in the first-order form as

$$\begin{bmatrix} \dot{u}_1 \\ \dot{u}_2 \\ \dot{u}_3 \end{bmatrix} = \begin{bmatrix} u_2 \\ -2\zeta u_2 - \frac{1}{2}u_1 + \chi u_3 + F \cos \omega t \\ -\lambda u_3 - \kappa u_2 \end{bmatrix} \qquad (8.68)$$

For the same numerical input ($\omega = 0.8$, $\zeta = 0.01$, $\chi = 0.05$, $\kappa = 0.5$, and $\lambda = 0.05$), initial conditions, and the forcing amplitude of Figure 8.10b ($x(0) = 1$, $\dot{x}(0) = 1.3$, $v(0) = 0$, $F = 0.08$), one can simulate the voltage response of the piezoelastic configuration using Equation (8.68).

Figure 8.11a shows the velocity vs. displacement trajectories of the piezomagnetoelastic and the piezoelastic configurations. As can be seen from the steady-state periodic orbits appearing in this figure after the transients, for the same excitation amplitude, system parameters,

---

[7] Holmes [15] presented Lyapunov functions showing the stability of these limit cycles obtained under large forcing (as well as the stability of the limit cycles around the focus points for small forcing). The presence of piezoelectric coupling does not alter the positive definiteness of the Lyapunov function, nor does it alter the negative definiteness of its time derivative.

[8] In Equation (8.66), the factor 1/2 is preserved so the undamped short-circuit natural frequency of the piezoelastic configuration is $1/\sqrt{2}$ (hence the theoretical discussion is close to the experimental setup discussed in the next section). As will be seen in Section 8.4, the fundamental post-buckled resonance frequency of the piezomagnetoelastic configuration is 10.6 Hz, whereas the fundamental resonance frequency of the piezoelastic configuration is 7.4 Hz (these frequencies are roughly proportional to 1 and $1/\sqrt{2}$ obtained from Equations (8.63) and (8.66), respectively).

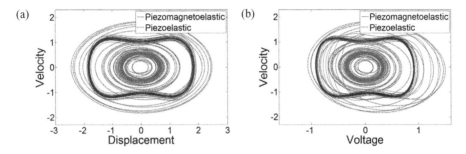

Figure 8.11   Comparison of the (a) velocity vs. displacement and (b) velocity vs. voltage phase portraits of the piezomagnetoelastic and piezoelastic configurations ($x(0) = 1$, $\dot{x}(0) = 1.3$, $v(0) = 0$, $F = 0.08$, $\omega = 0.8$)

and forcing amplitude, the steady-state vibration amplitude of the piezomagnetoelastic configuration can be much larger than that of the piezoelastic configuration. As expected, the large-amplitude periodic response – also called the limit-cycle oscillation (LCO) – on the high-energy orbit of the piezomagnetoelastic configuration is also observed in the velocity vs. voltage trajectory shown in Figure 8.11b.[9]

The superiority of the piezomagnetoelastic configuration over the piezoelastic configuration can be shown by plotting these trajectories at several other frequencies. The three-dimensional voltage vs. velocity trajectories are given for the frequency range of $\omega = 0.5 - 1$ in Figure 8.12. In all cases, the system parameters, initial conditions, and forcing amplitude[10] are identical. In Figure 8.12a ($\omega = 0.5$), the electrical output of the piezomagnetoelastic configuration is not considerably larger because the elastic beam oscillates around $x(0) = 1$ (small-amplitude LCO [9,15]). That is, the forcing amplitude is insufficient to overcome the attraction of the magnetic force at the respective focus. As a result, the piezomagnetoelastic configuration oscillates on a low-energy orbit and its electrical response amplitude is indeed comparable to that of the piezoelastic configuration. In Figures 8.12b, 8.12d, 8.12e, and 8.12f (for $\omega = 0.6$, $\omega = 0.8$, $\omega = 0.9$, and $\omega = 1$), however, the piezomagnetoelastic configuration shows a very large-amplitude electromechanical response on high-energy orbits compared to the orbits of the piezoelastic configuration. Only near the resonance frequency (approximately $1/\sqrt{2}$) of the piezoelastic configuration is the response amplitude of the piezoelastic configuration larger (Figure 8.12c). Nevertheless, at this particular frequency where the *resonant* (piezoelastic) configuration generates more voltage, the difference in the response amplitudes is not as dramatic as at other frequencies where the *non-resonant* (piezomagnetoelastic) configuration is much superior.

---

[9] For the system parameters used in these simulations, the system is very close to open-circuit conditions and the phase between the voltage and the velocity is approximately 90°. Therefore, in open-circuit conditions, it is reasonable to plot the velocity vs. voltage output as the *electromechanical phase portrait* (as an alternative to the conventional velocity vs. displacement phase portrait). From the experimental point of view, it is advantageous to plot these two independent measurements (voltage output of the piezoceramic vs. the velocity signal from the laser vibrometer) rather than integrate the experimental velocity history (which often results in a non-uniform drift).

[10] Note that the forcing amplitude in the base excitation problem is proportional to the square of the frequency. Keeping the forcing amplitude F constant at different frequencies implies keeping the base acceleration amplitude the same. Hence the base displacement amplitudes are different.

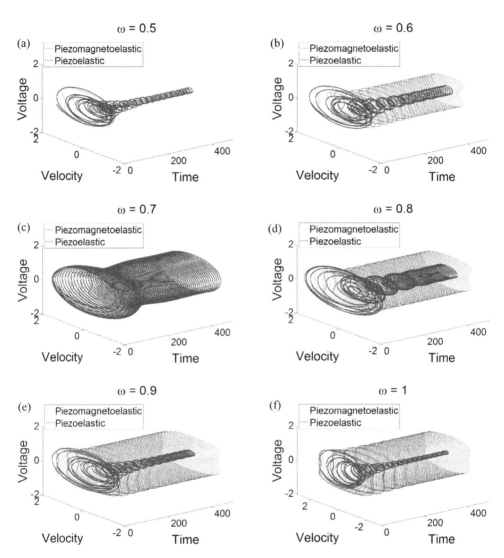

Figure 8.12 Comparison of the voltage vs. velocity phase trajectories of the piezomagnetoelastic and piezoelastic configurations for (a) $\omega = 0.5$, (b) $\omega = 0.6$, (c) $\omega = 0.7$, (d) $\omega = 0.8$, (e) $\omega = 0.9$, (f) $\omega = 1$ ($x(0) = 1$, $\dot{x}(0) = 1.3$, $v(0) = 0$, $F = 0.08$)

8.3.4 Comparison of the Chaotic Response and the Large-Amplitude Periodic Response

As mentioned previously, the large-amplitude attracting orbits of the piezomagnetoelastic configuration might coexist with strange attractors [9,15]. The advantage of the large-amplitude periodic response of the piezomagnetoelastic configuration shown in Figure 8.10b over the piezoelastic configuration is evident in Figures 8.11 and 8.12d. For the same excitation amplitude, if the initial velocity condition is removed, it is known from

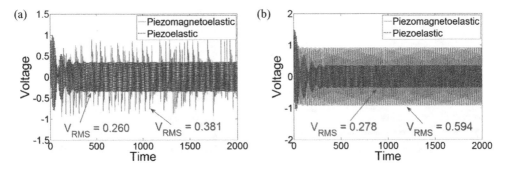

Figure 8.13 Comparisons of (a) the chaotic response ($x(0) = 1$, $\dot{x}(0) = 0$, $v(0) = 0$, $F = 0.08$, $\omega = 0.8$) and (b) the large-amplitude periodic response in the piezomagnetoelastic configuration ($x(0) = 1$, $\dot{x}(0) = 1.3$, $v(0) = 0$, $F = 0.08$, $\omega = 0.8$) against the piezoelastic configuration

Figure 8.9 that the response will be on a strange attractor. It is useful to compare the chaotic response in the piezomagnetoelastic configuration to the periodic response in the piezoelastic configuration (for $x(0) = 1$, $\dot{x}(0) = 0$, $v(0) = 0$, $F = 0.08$, $\omega = 0.8$). For the time interval shown in Figure 8.13a, the chaotic response of the piezomagnetoelastic configuration generates 46.5% larger RMS (Root Mean Square) voltage than the periodic response of the conventional piezoelastic configuration. When the initial velocity condition ($\dot{x}(0) = 1.3$) is imposed to catch the large-amplitude orbit (as in Figure 8.10b), the RMS voltage output of the piezomagnetoelastic configuration is 113.7% larger than that of the conventional piezoelastic configuration for the same time interval (Figure 8.13b). Note that the only difference between the simulations in Figures 8.13a and 8.13b is the initial velocity condition, which strongly affects the steady-state response of the piezomagnetoelastic configuration. The large-amplitude periodic response of the piezomagnetoelastic configuration is preferred to chaos not just because of its 2.5 times larger gain compared to that of the chaotic response, but also because the periodic response is preferred to chaotic response for processing the voltage output using a nonlinear energy harvesting circuit for charging a battery or a capacitor efficiently [16–21].

8.4   Experimental Performance Results of the Bistable Piezomagnetoelastic Energy Harvester

8.4.1   Experimental Setup

The piezomagnetoelastic energy harvester and the setup used for the experimental verifications are shown in Figure 8.14a [11,12]. The piezomagnetoelastic configuration (elastic cantilever with the magnets causing the bistability) and the piezoelastic configuration (conventional elastic cantilever obtained when the magnets are removed) are shown in Figures 8.14b and 8.14c, respectively. Harmonic base excitation is provided by a seismic shaker (Acoustic Power Systems APS-113), the acceleration at the base of the cantilever is measured by a small accelerometer (PCB Piezotronics Model U352C67), and the velocity response of the cantilever is recorded by a laser vibrometer (Polytec PDV-100). The time history of the base acceleration, voltage, and vibration responses are recorded by a National Instruments NI-cDAQ data acquisition system (with a sampling frequency of 2000 Hz). The ferromagnetic beam

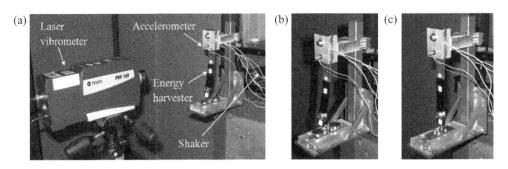

Figure 8.14 (a) Experimental setup with the piezoelectric energy harvester, seismic shaker, accelerometer, and laser vibrometer; (b) piezomagnetoelastic configuration; (c) piezoelastic configuration

(made of tempered blue steel) is 145 mm long (overhang length), 26 mm wide, and 0.26 mm thick. A lumped mass of 14 g is attached close to the tip for improved dynamic flexibility. Two PZT-5A piezoceramic layers (QP16N, Midé Technology Corporation) are attached to both faces of the beam at the root using a high-shear-strength epoxy and their electrodes are connected in parallel. The spacing between the symmetrically located circular rare earth magnets is 50 mm (center to center) and this distance is selected to realize the three equilibrium cases described by Equations (8.63) and (8.64). The tip deflection of the magnetically buckled beam in the static case to either side is approximately 15 mm relative to the statically unstable equilibrium position. The post-buckled fundamental resonance frequency of the beam is 10.6 Hz (around both focus points),[11] whereas the fundamental resonance frequency of the unbuckled beam (when the magnets are removed) is 7.4 Hz (both under the open-circuit conditions of piezoceramics, i.e., at constant electric displacement). Before the comparison of the piezomagnetoelastic (Figure 8.14b) and the piezoelastic (Figure 8.14c) configurations, it is worth reviewing the individual performance of the piezomagnetoelastic configuration [11,12].

8.4.2 Performance Results of the Piezomagnetoelastic Configuration

For a base acceleration amplitude of $0.5g$ (where g is the gravitational acceleration: $g = 9.81 \, \text{m/s}^2$) at 8 Hz, with an initial deflection at one of the stable equilibrium positions (15 mm to the shaker side), zero initial velocity, and voltage, the chaotic open-circuit voltage response shown in Figure 8.15a is obtained [11,12]. The Poincaré map of the strange attractor motion is displayed in Figure 8.15b on its phase portrait. These figures are obtained from a measurement taken for about 15 minutes (1784400 data points due to a sampling frequency of 2000 Hz) and they exhibit very good qualitative agreement with the numerical simulations given by Figure 8.9.

If the excitation amplitude is increased to $0.8g$ (similar to the amount of increase in the theoretical simulation) by keeping the frequency the same, the response turns from transient

---

[11] Since the frequency of natural oscillations around each magnet is around 10.6 Hz, the bistable configuration is indeed fairly symmetric (though small imperfections always exist). Therefore the experimental prototype agrees well with the symmetric double-well potential assumption of Equation (8.63).

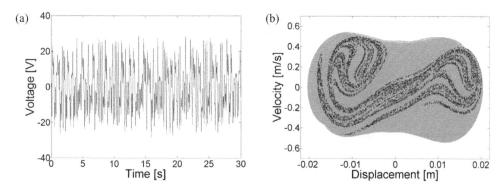

Figure 8.15   (a) Experimental voltage history obtained from the strange attractor motion and (b) the Poincaré map of the motion on its phase portrait (excitation: 0.5*g* at 8 Hz)

chaos into large-amplitude periodic motion with a strong improvement in the open-circuit voltage response as shown in Figure 8.16a (analogous to Figure 8.10a). Therefore, the strange attractor of Figure 8.15a is replaced by a large-amplitude periodic attractor in Figure 8.16a due to the increased forcing amplitude. A similar improvement is obtained in Figure 8.16b, where the excitation amplitude is kept as the original one (0.5*g*) and a disturbance (hand impulse) is applied at $t = 11$ s (as a simple alternative to creating an initial velocity condition, analogous to Figure 8.10b). Therefore this second case corresponds to the coexistence of a strange attractor and a large-amplitude periodic attractor for the same level of forcing [9,15]. The system stays on the strange attractor (Figure 8.15) in the absence of any disturbance and the aforementioned disturbance results in a large-amplitude periodic response (Figure 8.16b) with a substantial gain as discussed in the following section. These experimental results (Figure 8.16) exhibit good agreement with the theoretical discussion (Figure 8.10). The next step is to compare the piezomagnetoelastic and the piezoelastic configurations experimentally.

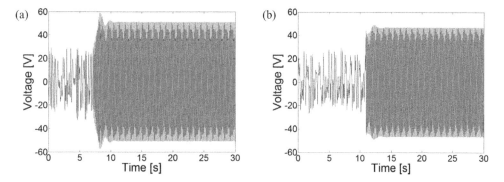

Figure 8.16   Experimental voltage histories: (a) large-amplitude response due to the excitation amplitude (excitation: 0.8*g* at 8 Hz); (b) large-amplitude response due to a disturbance at $t = 11$ s for a lower excitation amplitude (excitation: 0.5*g* at 8 Hz)

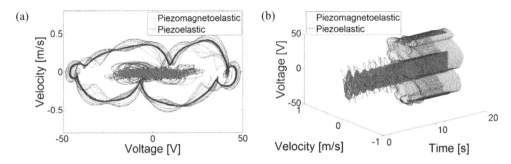

Figure 8.17 (a) Two-dimensional and (b) three-dimensional comparison of the electromechanical trajectories of the piezomagnetoelastic and piezoelastic configurations (excitation: 0.5g at 8 Hz with a disturbance at $t = 11$ s)

8.4.3 Comparison of the Piezomagnetoelastic and the Piezoelastic Configurations for Voltage Generation

Figure 8.17a compares the velocity vs. open-circuit voltage phase portraits of the piezomagnetoelastic and the piezoelastic[12] configurations for a base excitation amplitude of 0.5g at 8 Hz. The chaotic part of response in the piezomagnetoelastic configuration belongs to the time interval before the disturbance is applied. This figure is therefore analogous to the theoretical demonstration given by Figure 8.11b. The three-dimensional view of the electromechanical trajectory in the phase space is shown in Figure 8.17b and it exhibits good qualitative agreement with its simplified theoretical counterpart based on lumped-parameter modeling (e.g., Figure 8.12d).

8.4.4 On the Chaotic and the Large-Amplitude Periodic Regions of the Response

Before undertaking the broadband comparisons of the piezomagnetoelastic and piezoelastic configurations, the voltage history of Figure 8.17b is reconsidered in two parts. The time history until the instant of the disturbance is chaotic, which would yield a strange attractor motion similar to Figure 8.15 if no disturbance were applied. After the disturbance is applied at $t = 11$ s, the large-amplitude response on a high-energy orbit is obtained. In order to visualize the advantage of the second region in the response history of Figure 8.17b, the open-circuit voltage histories of the piezomagnetoelastic and piezoelastic configurations are compared for the same harmonic input (0.5g at 8 Hz). Figure 8.18a shows the acceleration input to the piezomagnetoelastic and piezoelastic configurations at an arbitrary instant of time. The voltage input to the seismic shaker is identical for both configurations, yielding very similar base acceleration amplitudes for a fair comparison. Figure 8.18b displays the comparison of the piezomagnetoelastic and piezoelastic configurations, where the former exhibits chaotic response while the latter has already reached its periodic steady-state response at the input frequency. As a rough

---

[12] In the experiments, the piezoelastic configuration is obtained simply by removing the magnets of the piezomagnetoelastic configuration after the experiments of the latter are completed.

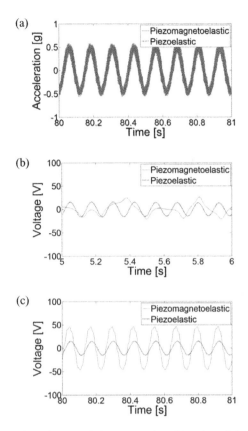

Figure 8.18 Comparison of the input and the output time histories of the piezomagnetoelastic and piezoelastic configurations: (a) input acceleration histories; (b) voltage outputs in the chaotic response region of the piezomagnetoelastic configuration; (c) voltage outputs in the large-amplitude response region of the piezomagnetoelastic configuration (excitation: $0.5g$ at 8 Hz with a disturbance at $t = 11$ s)

comparison, from Figure 8.18b, it is not possible to claim that the chaotic response of the piezomagnetoelastic configuration has any substantial advantage over the harmonic response of the piezoelastic configuration as their amplitudes look very similar (this is further discussed in the following paragraph). Figure 8.18c shows the voltage histories of these configurations some time after the disturbance is applied to the piezomagnetoelastic configuration and the large-amplitude periodic response is obtained. Obviously, if the same disturbance is applied to the piezoelastic configuration, the trajectory (in the phase space) returns to the same low-amplitude orbit after some transients since no such high-energy attractor exists in the piezoelastic configuration. Therefore the response amplitude of the piezoelastic configuration is identical in Figures 8.18b and 8.18c. However, the large-amplitude response of the piezomagnetoelastic configuration can give more than three times larger RMS voltage output according to Figure 8.18c.

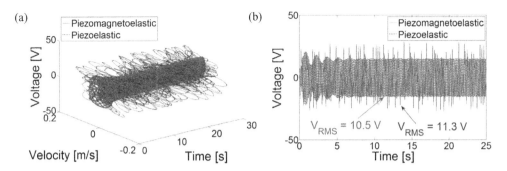

Figure 8.19 Comparison of the chaotic response in the piezomagnetoelastic configuration and the periodic response in the piezoelastic configuration (excitation: 0.5g at 8 Hz, no disturbance is applied): (a) voltage vs. velocity trajectory; and (b) voltage history showing the RMS values

In order to verify the claim that the chaotic response region of the piezomagnetoelastic configuration may not have a substantial advantage over the periodic response of the piezoelastic configuration, the time histories of both configurations are compared in the absence of any disturbance (so the piezomagnetoelastic configuration stays on the strange attractor as in Figure 8.15 rather than being attracted by the high-energy orbit as in Figure 8.17). For a base acceleration amplitude of 0.5g at 8 Hz, the electromechanical trajectories of the piezomagnetoelastic and piezoelastic configurations are shown in Figure 8.19. The three-dimensional chaotic trajectory of the piezomagnetoelastic configuration shows larger instantaneous voltage and velocity amplitudes in Figure 8.19a. However, as shown in Figure 8.19b, the RMS voltage output of the piezomagnetoelastic configuration (obtained for the interval of 0–25 seconds) is only 7.6% larger than that of the piezoelastic configuration (it is observed that these RMS values do not change considerably with increasing time interval). Consequently, no substantial improvement is obtained in the chaotic response region. Moreover, one would prefer a periodic signal to a chaotic signal for processing the harvested energy using an efficient energy harvesting circuit [16–21]. Along with the theoretical discussion given in Section 8.3.4, this observation justifies the efforts of focusing on the large-amplitude periodic oscillations rather than chaos in the bistable structure. The results for the broadband power generation performance of both configurations are presented next.

8.4.5 Broadband Performance Comparison

A harmonic base excitation of 0.5g amplitude (yielding an RMS value of approximately 0.35g) is applied at frequencies of 5 Hz, 6 Hz, 7 Hz, and 8 Hz to both configurations (piezoelastic and piezomagnetoelastic). Figure 8.20 shows the comparison of the average steady-state power vs. load resistance graphs of the piezomagnetoelastic and piezoelastic configurations at these frequencies. Note that the excitation amplitudes (i.e., the base acceleration levels) of both configurations are very similar in all cases. The piezomagnetoelastic energy harvester gives larger power at 5 Hz, 6 Hz, and 8 Hz by an order of magnitude, while the piezoelastic configuration gives larger power only at 7 Hz (by a factor of 2.3), which is close to its resonance

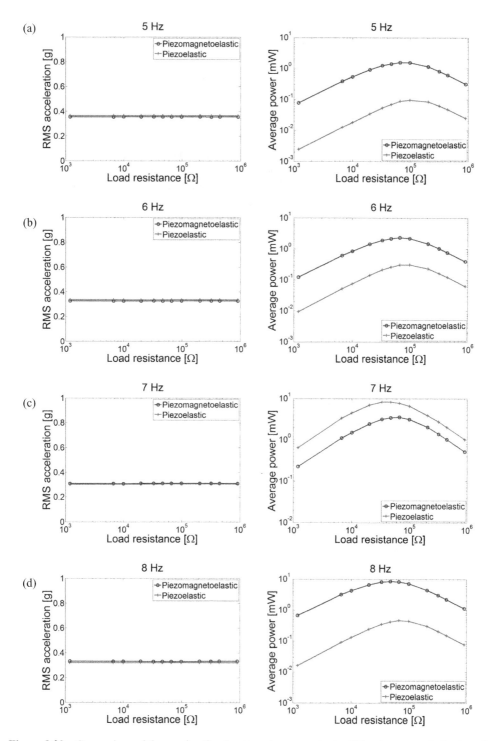

Figure 8.20 Comparison of the acceleration inputs and power outputs of the piezomagnetoelastic and the piezoelastic configurations at steady state for a range of excitation frequencies: (a) 5 Hz; (b) 6 Hz; (c) 7 Hz; (d) 8 Hz

Table 8.1 Comparison of the average power outputs of the piezomagnetoelastic and piezoelastic energy harvester configurations (RMS acceleration input: $0.35g$)

Excitation frequency (Hz)	5	6	7	8
Piezomagnetoelastic configuration (mW)	1.57	2.33	3.54	8.45
Piezoelastic configuration (mW)	0.10	0.31	8.23	0.46

frequency.[13] The average power outputs read from these graphs for the optimum values of load resistance are listed in Table 8.1. Note that not all the response forms in the piezomagnetoelastic configuration start with chaotic motion. For instance, for excitation at 6 Hz, the motion of the piezomagnetoelastic configuration starts with a small-amplitude LCO around one of the focus points. The disturbance applied at $t = 11$ s results in a large-amplitude LCO as summarized in Figure 8.21. Therefore, not only chaotic vibrations but also small-amplitude LCOs around one of the magnets can be turned into large-amplitude LCOs (i.e., small-amplitude and large-amplitude LCOs may coexist). It is worth adding that not only for open-circuit conditions (which do not cause any dissipation in the mechanical domain) but even for the load of maximum generation, the system can stay on the high-energy orbit. That is, the effect of piezoelectric power generation does not create significant enough dissipation through Joule heating to suppress the resulting nonlinear phenomena (large-amplitude LCO in this case), which is a favorable situation. The reason is that the system is strongly nonlinear; that is, in the mathematical (perturbation) sense, mechanical nonlinearity and piezoelectric coupling are not of the same order (e.g., if one scaled the terms in Equation (8.63), the piezoelectric coupling would be $O(\varepsilon)$ while the mechanical nonlinearity is $O(1)$).

Variation of the optimum average electrical power outputs of both configurations with the excitation frequency is plotted in Figure 8.22. It is important to notice in this figure that, at several frequencies, the non-resonant piezomagnetoelastic energy harvester can indeed generate an order-of-magnitude more power for the same input. The resonant piezoelastic energy harvester can generate larger power only within a narrow band around its fundamental resonance frequency. However, this power output is not an order of magnitude larger than that of the piezomagnetoelastic configuration (in qualitative agreement with Figure 8.12c). It can be concluded that the piezomagnetoelastic configuration exhibits a substantially better broadband power generation performance provided that the input excitation results in oscillations on its high-energy orbits in the frequency range of interest. Given the frequency range and the amplitude of harmonic base excitation at these frequencies, the piezomagnetoelastic energy harvester should be designed to be attracted by these high-energy orbits.

8.4.6 Vertical Excitation of the Piezomagnetoelastic Energy Harvester

In practice, the direction of vibratory motion is often vertical as depicted in Figure 8.23. In such a case, the gravitational force field might distort the static equilibrium of the flexible cantilever considerably, which might affect the condition of static bistability. The sag of the

---

[13] For the same excitation amplitude, the piezomagnetoelastic configuration cannot escape the attraction of the low-energy orbit when excited at 4 Hz (analogous to the theoretical case in Figure 8.12a), whereas the attractor is chaotic at 9 Hz.

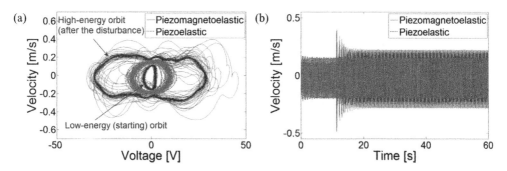

Figure 8.21   (a) Comparison of the electromechanical trajectories for the piezomagnetoelastic and the piezoelastic configurations showing that the starting motion of the piezomagnetoelastic configuration is a low-energy LCO until the disturbance at 11s. (b) Velocity histories of both configurations showing the instant of disturbance for the piezomagnetoelastic configuration (excitation: 0.5g at 6 Hz)

elastic beam due to its own weight can cause the beam to become biased toward the lower magnet (i.e., the magnet that is at the ground side). As a result, the statically bistable beam can become statically monostable since the combined effect of the attraction of the lower magnet and gravity overcomes the attraction of the upper magnet. From the dynamic point of view, the beam cannot escape the attraction of the lower magnet and the only possible steady-state response becomes small-amplitude oscillations around the lower magnet (low-energy orbit). A simple adjustment of the magnet spacing (to balance the transverse force resultant at the tip) can resolve the problem and the broadband phenomena discussed herein can be preserved for vertical excitation as well (not necessarily for the exact same frequency range).

Figure 8.24a shows the static deflection of the piezoelastic cantilever used in the experiments (in the absence of magnets) when it is located horizontally for vertical excitation. Since the sag of the flexible beam due to its weight is considerable, the piezomagnetoelastic configuration is no longer statically bistable when located horizontally. In order to reduce the attraction of the

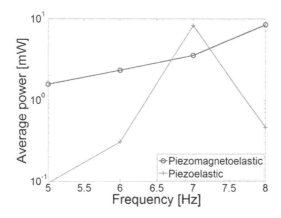

Figure 8.22   Comparison of the average power outputs of the piezomagnetoelastic and piezoelastic energy harvester configurations (RMS acceleration input: 0.35g)

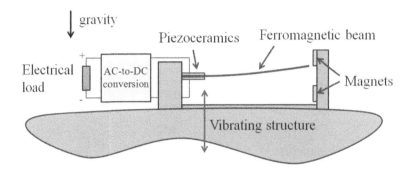

Figure 8.23   Schematic of a piezomagnetoelastic energy harvester under vertical excitation

lower magnet, it is moved downward by an empirical amount (Figure 8.24b). Once the force balance at the tip of the beam is achieved, it is observed that similar large-amplitude LCOs can be obtained for the vertical excitation of the piezomagnetoelastic configuration as well (the symmetry of the *potential wells* [10], however, is distorted, which introduces quadratic nonlinearity into Equation (8.63) in simplest terms). As in the case of horizontal excitation, LCOs of the piezomagnetoelastic configuration have much larger amplitude compared to those of the piezoelectric configuration (Figure 8.24c).

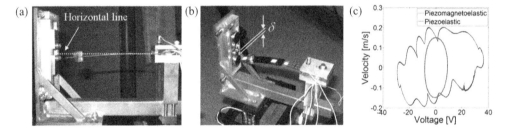

Figure 8.24   (a) Static deflection of the piezoelastic beam due to gravity (in the absence of magnets); (b) piezomagnetoelastic configuration with the lower magnet moved downward to balance the magnetic attraction; (c) comparison of the steady-state periodic orbits in the piezomagnetoelastic and piezoelastic configurations after the adjustment (for a vertical excitation amplitude of $0.5g$ at 5.5 Hz)

8.5   A Bistable Plate for Piezoelectric Energy Harvesting

8.5.1   *Nonlinear Phenomena in the Bistable Plate*

Having discussed the bistable piezomagnetoelastic energy harvester both theoretically and experimentally, the performance results of an alternative bistable energy harvester configuration are discussed in this section: a bistable carbon-fiber-epoxy plate with piezoceramics [22]. An advantage of this configuration is that the bistability is a result of the curing process and the unsymmetric laminate characteristics of the plate [23] and it does not require external magnets. Moreover, having two length dimensions is another advantage of the plate configuration (compared to the beam configuration) that offers flexibility to create multiple resonance frequencies (at least two instead of one) over a given frequency range. The main

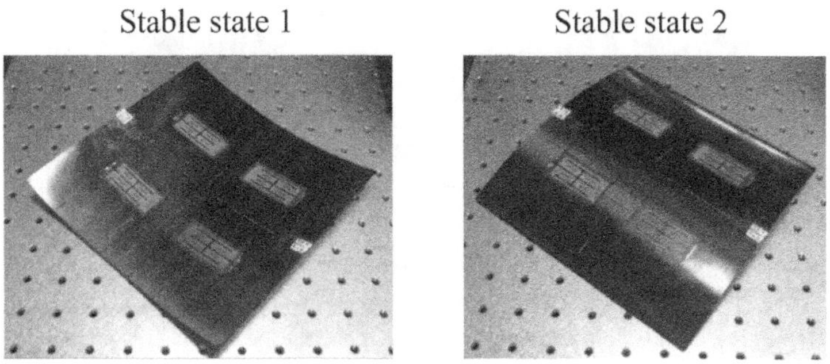

Stable state 1   Stable state 2

Figure 8.25   Statically stable configurations of the bistable piezoelectric energy harvester plate

disadvantage of the particular carbon-fiber-epoxy sample discussed in this section (which was not originally designed or fabricated for energy harvesting) is its high stiffness, requiring larger acceleration levels to realize dynamic snap-through and other nonlinear phenomena compared to the piezomagnetoelastic energy harvester that uses flexible spring steel as the substructure material.

Figure 8.25 shows the statically stable configurations of the $[90_2/0_2]$ 200 mm $\times$ 200 mm experimental sample investigated in this section. Four PZT-5A piezoceramic layers (QP16N, Midé Technology Corporation) are bonded onto the smooth face of the plate at the locations of large strain fluctuation (as the plate snaps through one stable state to the other).[14] It is reasonable to expect that the two potential wells [10] of this configuration are not symmetric, hence the stable static equilibrium positions are not symmetric with respect to the unstable static equilibrium position (which is the flat plate configuration). Nevertheless, the form of Equation (8.63), which assumes symmetry for the stable static equilibrium positions ($x = \pm 1$), can still describe the basic nonlinear phenomena (e.g., chaos and large-amplitude LCO) of the bistable plate configuration, while this section focuses on the experimental results only.[15]

For the base excitation tests, the plate is clamped from its center to the armature of the seismic shaker (Acoustic Power Systems APS-113) used in the experiments of Section 8.4. A small accelerometer (PCB Piezotronics Model U352C67) is located on the moving armature of the shaker and the velocity response close to the center of the plate is recorded by a laser vibrometer (Polytec OFV303 laser head and OFV3001 vibrometer). The electrical outputs of the piezoceramic patches are combined in parallel for the experimental resistor sweep to investigate the power generation performance of the plate for different nonlinear phenomena. In order to increase the dynamic flexibility of the plate so that the excitation level for achieving snap-through motion is reduced, identical lumped masses (of 15.6 g) are attached at the corners of the plate (Figure 8.26).

---

[14] In simplest terms, the electromechanical coupling is proportional to the integral of the curvature of the piezo-ceramic (Chapter 3). If the change of curvature (in one direction) is large during the snap-through motion, then the electrical output is also large.

[15] Alternatively, one can modify the static part of Equation (8.63) to have non-symmetric stable equilibrium positions, which creates a quadratic term in addition to the cubic term.

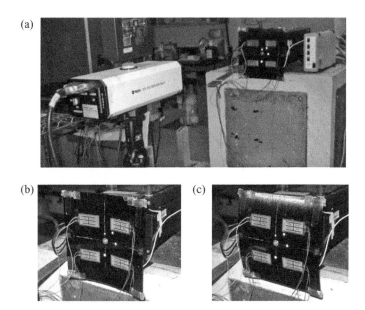

Figure 8.26 (a) Experimental setup and close-up views of the plate on the shaker with (b) stable state 1 and (b) stable state 2

For small oscillations around the two stable states, the linear velocity and voltage FRFs of the plate are measured (per base acceleration input) as shown in Figures 8.27a and 8.27b, respectively. The first two linear modes observed around state 1 are $\omega_1^{s1} = 10.9\,\text{Hz}$ and $\omega_2^{s1} = 15.8\,\text{Hz}$, whereas those around state 2 are $\omega_1^{s2} = 9.4\,\text{Hz}$ and $\omega_2^{s2} = 12.9\,\text{Hz}$. The first mode around each stable state (ω_1^{s1}, ω_1^{s2}) is related more to the out-of-plane rotation about the joint between the plate and the shaker (like a rigid body rotation) and is associated with much less elastic deformation compared to the second mode (ω_2^{s1}, ω_2^{s2}). These FRFs verify the previous discussion that the potential wells of the bistable plate are indeed not symmetric and the plate is stiffer for small oscillations around state 1. Considering the dominant elastic modes in Figure 8.27, it can be observed that small-amplitude oscillations around state 1 give larger voltage output even though the respective modal frequency is higher than that of state 2

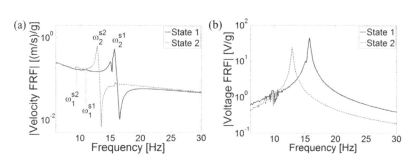

Figure 8.27 FRFs for the (a) velocity to base acceleration and (b) voltage to base acceleration of the plate for small oscillations around the two stable configurations

(implying a stiffer mode). Therefore the arrangement of the piezoceramics is more favorable for small oscillations around state 1.

Various nonlinear dynamic phenomena can be realized with the bistable plate. Figure 8.28 shows examples of nonlinear open-circuit voltage histories observed at different frequencies along with the resistor sweep curves of the RMS voltage output for each case. The response histories in Figures 8.28a–8.28c are obtained with the static initial condition at state 2, while the response history shown in Figure 8.28d is obtained with the static initial condition at stable state 1. In all four cases shown in Figure 8.28, the excitation amplitude is $2g$ since the plate is considerably stiffer than need be as it was not originally designed and fabricated for energy harvesting. The large-amplitude LCO observed for harmonic excitation at 8.6 Hz (Figure 8.28a) is analogous to the large-amplitude LCO of the bistable piezomagnetoelastic energy harvester beam; therefore, it is a very favorable response type for energy harvesting (it coexists with a low-energy LCO around the respective potential well). When the excitation frequency is increased to 9.8 Hz, an intermittent chaotic response of very large voltage output is observed as shown in Figure 8.28b (this bursting response of the plate needs further investigation as it tends to exhibit transient characteristics). Persistent chaotic response (Figure 8.28c) is obtained when the excitation frequency is 12.5 Hz and the voltage output is still very large as in the previous two nonlinear phenomena. The last response sample is the subharmonic resonance case shown in Figure 8.28d. The excitation frequency in this case is 20.2 Hz and the response has a slower component around $\omega_1^{s1} = 10.9$ Hz (since this last case is for oscillations around state 1). The resulting steady-state response is not associated with considerable elastic deformation (dominated by the out-of-plane rotation of the plate about the joint between the shaker and the plate), hence the electrical output is much less than the previous three cases. Nevertheless, subharmonic resonance is a nonlinear phenomenon that can make it possible to obtain a response at a frequency other than the excitation frequency.[16] Note that the bistable piezoelectric energy harvester plate exhibits the foregoing nonlinear phenomena over a fairly wide frequency range (8–21 Hz).

8.5.2 Broadband Power Generation Performance

The average electrical power output vs. load resistance curves for the aforementioned four nonlinear phenomena are plotted in Figure 8.29. For the chaotic response case, it is ensured that a sufficiently long time history is averaged until the power output converges to a certain value for each resistor. Remarkably, it is observed in Figure 8.29 that the maximum power output of all cases is for the load resistance of 22 kΩ. As shown in Table 8.2, the average power outputs of the energetic intermittent chaos and large-amplitude LCO cases are 34 mW and 27 mW, respectively. These are followed by the average power outputs of the persistent chaos and the subharmonic resonance cases, which are 11 mW and 0.67 mW, respectively.

The backward and forward frequency sweep diagrams [22] of the bistable piezoelectric energy harvester plate are summarized in Figure 8.30, exhibiting rich nonlinear dynamic phenomena over a broad excitation frequency range. The velocity and voltage graphs vs. frequency in Figure 8.30 are similar to *bifurcation diagrams* [9,10] (excitation frequency

---

[16] In this respect, superharmonic resonances [5] (not discussed here) can be more interesting than subharmonic resonances to obtain a response at a higher frequency by exciting a given nonlinear energy harvester at a lower frequency (which might find applications in energy harvesting with micro-scale devices). Barton *et al.* [24] recently reported superharmonic resonances down to 1:5 for an electromagnetic energy harvester.

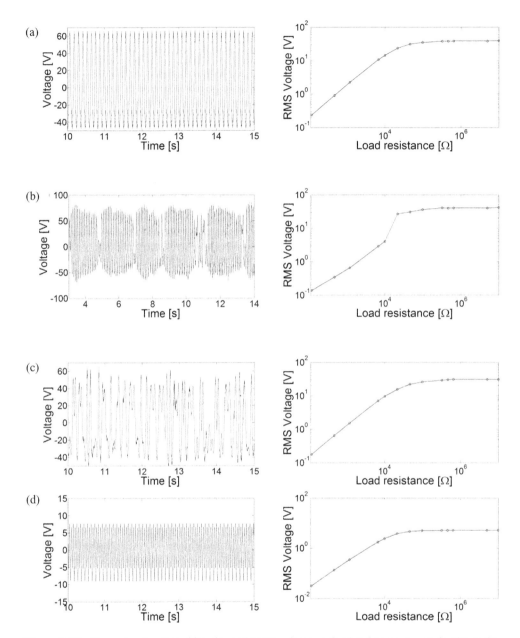

Figure 8.28 Open-circuit voltage histories and RMS voltage vs. load resistance curves for the excitation amplitude of 2g at different frequencies: (a) large-amplitude LCO (8.6 Hz); (b) intermittent chaos (9.8 Hz); (c) persistent chaos (12.5 Hz); (d) subharmonic resonance (20.2 Hz)

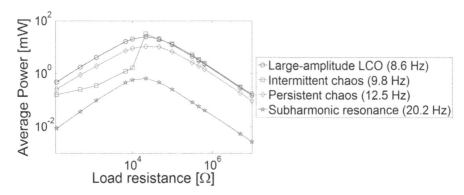

Figure 8.29 Average power vs. load resistance curves for different types of nonlinear response (excitation amplitude: 2g)

being the *bifurcation parameter* under constant excitation amplitude) but they do not necessarily include all coexisting solutions (except for the coexisting large-amplitude and small-amplitude LCOs shown in Figure 8.30a). Figure 8.30a is a forward sweep starting with the static initial condition at state 2, while Figure 8.30b is a backward sweep starting with the static initial condition at state 1. Note that the plate undergoes snap-through and changes in the equilibrium states as a result of the experimental frequency sweep. Figure 8.30 is therefore a summary of the rich broadband nonlinear dynamic response characteristics of the bistable piezoelectric energy harvester plate.

Table 8.2 Average power outputs for different nonlinear phenomena

	Average power (mW)	Excitation frequency (Hz)	Optimum resistance (kΩ)
Intermittent chaos	34	9.8	22
Large-amplitude LCO	27	8.6	22
Persistent chaos	11	12.5	22
Subharmonic resonance	0.67	20.2	22

8.6 Summary

This chapter investigates the utilization of mechanical nonlinearities for performance improvement in piezoelectric energy harvesting. First, the focus is placed on the (statically) monostable Duffing oscillator with piezoelectric coupling for bandwidth enhancement through the hardening and softening stiffness effects causing cubic nonlinearity. A perturbation-based electromechanical formulation is presented for the weakly nonlinear monostable problem with analytical expressions for the electromechanically coupled nonlinear vibration response and voltage response. The use of the nonlinear electromechanical frequency response expressions is demonstrated with a theoretical case study and the results of the perturbation simulations are verified against the time-domain numerical simulations. After the weakly nonlinear monostable piezoelectric Duffing oscillator with hardening/softening stiffness, the strongly nonlinear

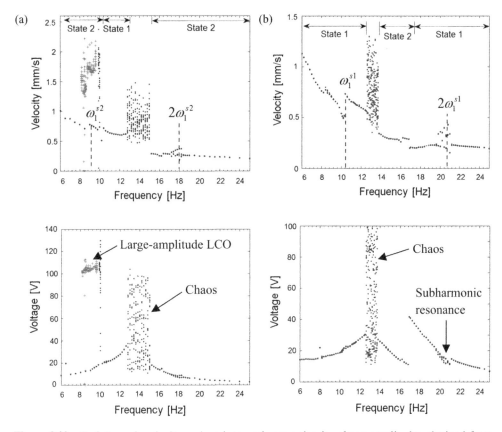

Figure 8.30 Peak-to-peak velocity and peak-to-peak open-circuit voltage amplitudes obtained from (a) forward sweep with the static initial condition at state 2 and (b) backward sweep with the static initial condition at state 1 (excitation level, $2g$; not all coexisting solutions are shown)

bistable Duffing oscillator with piezoelectric coupling is investigated. It is experimentally verified that the broadband high-energy orbits of the piezomagnetoelastic energy harvester can give an order-of-magnitude larger power compared to the conventional piezoelastic configuration over a range of frequencies. Chaotic vibration of the bistable piezoelectric energy harvester configuration is also discussed. As an alternative broadband energy harvester configuration, a bistable piezoelectric energy harvester plate is investigated experimentally. Typical voltage histories associated with various nonlinear phenomena (intermittent chaos, persistent chaos, large-amplitude limit-cycle oscillations and subharmonic resonance) in the bistable plate with piezoceramics are presented for harmonic excitations at different frequencies.

8.7 Chapter Notes

Frequency bandwidth limitations in conventional resonant energy harvesters gave rise to research on nonlinear energy harvesting toward the end of the first decade of this growing research field. As long as the excitation amplitude is large enough to have the *desired* non-linearities pronounced and/or the initial conditions are such that the *favorable* response (e.g., high-energy response in the case of coexisting solutions) is realized, the nonlinear aspects covered in this chapter can be used in order to design broadband energy harvesters that substantially outperform the conventional resonant configurations. Certain nonlinear considerations in the electrical domain of the problem can also enhance the harvested power (such as the synchronized switching on inductor technique reviewed in Section 11.3).

Prior to research on nonlinear piezoelectric energy harvesting, implementations of the hardening stiffness in the statically monostable Duffing oscillator for increasing the frequency bandwidth were discussed by Burrow *et al.* [25] and Mann and Sims [26] for electromagnetic energy harvesting. Use of the hardening stiffness in the monostable form of the Duffing oscillator was also discussed by Ramlan *et al.* [27] along with snap-through in a bistable mass–spring–damper mechanism. More recently, Daqaq [28] also considered hardening stiffness in the lumped-parameter electromagnetic energy harvester formulation and showed that the monostable Duffing oscillator does not provide any enhancement over the typical linear oscillators under white Gaussian and colored excitations (unlike the case of harmonic excitation). His theoretical observation agrees with the experimental results contemporaneously reported by Barton *et al.* [24] for electromagnetic energy harvesting with stiffness nonlinearity of hardening type (where the majority of the high-energy branch obtained under periodic forcing is not reached under random forcing). Parametric excitation [5] of a nonlinear piezoelectric energy harvester was theoretically and experimentally investigated by Daqaq *et al.* [4].

The bistable form of the Duffing oscillator was discussed by Cottone *et al.* [29] and Gammaitoni *et al.* [30] for noise excitation along with the concept of stochastic resonance [31–33] as formerly pointed out by McInnes *et al.* [34] for energy harvesting. Stanton *et al.* [35] theoretically investigated the bifurcations of a bistable configuration similar to the one tested by Cottone *et al.* [29]. They [3] also proposed a device that combines an elastic beam with different magnet configurations to create both hardening stiffness and soft hardening stiffness. Stochastic resonance in the bistable piezomagnetoelastic structure of Erturk *et al.* [11,12] was discussed theoretically by Litak *et al.* [36].

This chapter has focused on the types of nonlinearities and examples of nonlinear devices which can be exploited to enhance the performance of piezoelectric energy harvesters. Two other types of nonlinear behaviors that can also be pronounced with increasing excitation amplitude are due to inherent piezoelastic nonlinearities and nonlinearities in dissipation. The combined effect of piezoelastic and dissipation nonlinearities is such that the linear FRF predictions (Chapter 4) overestimate the electrical response for large excitation amplitudes. The effect of inherent piezoelectric nonlinearities was formerly pointed out by Crawley and Anderson [37] and Crawley and Lazarus [38] for large electric fields in structural actuation. Their observations were first implemented in energy harvesting by Triplett and Quinn [2], who presented a perturbation-based theoretical analysis using the Lindstedt–Poincaré method [1]. Stanton and Mann [7] presented a Galerkin solution by taking into account the geometric and piezoelastic nonlinearities for weak electric fields [6] (since the electric field levels in energy harvesting are not as high as in actuation). For relatively stiff cantilevers, inherent piezoelastic nonlinearities can be pronounced even when the oscillations are geometrically linear. As previously illustrated with Figure 4.35 in Section 4.5, the experiments conducted using the brass-reinforced PZT-5H bimorph of Section 4.1 show that piezoelastic nonlinearities become effective if the base excitation level exceeds a few hundreds of milli-*g* acceleration, although geometric nonlinearities are not pronounced [8]. The same work [8] also provides a mathematical technique for the identification of higher-order piezoelastic constants using the nonlinear modeling framework of Stanton and Mann [7] and presents experimental results for PZT-5H. In addition to piezoelastic nonlinearities, it is also observed [8] that nonlinear dissipative effects become pronounced with increasing excitation level even if the vibrations are geometrically linear. In the absence of other nonlinearities, the manifestation of piezoelastic nonlinearities is a softening stiffness effect, that is, the peak of the frequency response curve moves toward the left on the frequency axis with increasing base acceleration. It is interesting to note that, for flexible cantilevers (with large deflections), the softening effect of piezoelastic nonlinearities and the hardening effect of geometric nonlinearities can be pronounced together under high acceleration inputs and these effects might counteract each other. Up to an acceleration level of a few hundred milli-g (which covers the range of most sources of ambient vibration energy [39]), however, models using the standard form [40] of the linear piezoelectric constitutive equations (as in the linear electromechanical model given in Chapter 3) are safe to use.

References

1. Nayfeh, A.H. (1981) *Introduction to Perturbation Techniques*, John Wiley & Sons, Inc., New York.
2. Triplett, A. and Quinn, D.D. (2009) The effect of non-linear piezoelectric coupling on vibration-based energy harvesting. *Journal of Intelligent Material Systems and Structures*, **20**, 1959–1967.
3. Stanton, S.C., McGehee, C.C., and Mann, B.P. (2009) Reversible hysteresis for broadband magnetopiezoelastic energy harvesting. *Applied Physics Letters*, **96**, 174103.
4. Daqaq, M.F., Stabler, C., Qaroush, Y., and Seuaciuc-Osorio, T. (2009) Investigation of power harvesting via parametric excitations. *Journal of Intelligent Material Systems and Structures*, **20**, 545–557.
5. Nayfeh, A.H. and Mook, D.T. (1979) *Nonlinear Oscillations*, John Wiley & Sons, Inc., New York.
6. von Wagner, U. and Hagedorn, P. (2002) Piezo-beam systems subjected to weak electric field: experiments and modeling of nonlinearities. *Journal of Sound and Vibration*, **256**, 861–872.
7. Stanton, S.C. and Mann, B.P. (2010) Nonlinear electromechanical dynamics of piezoelectric inertial generators: modeling, analysis, and experiment. *Nonlinear Dynamics* (in review).
8. Stanton, S.C., Erturk, A., Mann, B.P., and Inman, D.J. (2010) Nonlinear piezoelectricity in electroelastic energy harvesters: modeling and experimental identification. *Journal of Applied Physics*, **108**, 074903.
9. Guckenheimer, J. and Holmes, P. (1983) *Nonlinear Oscillations, Dynamical Systems, and Bifurcations of Vector Fields*, Springer-Verlag, New York.
10. Moon, F.C. (1987) *Chaotic Vibrations*, John Wiley & Sons, Inc., New York.
11. Erturk, A., Hoffmann, J., and Inman, D.J. (2009) A piezomagnetoelastic structure for broadband vibration energy harvesting. *Applied Physics Letters*, **94**, 254102.
12. Erturk, A. and Inman, D.J. (2010) Broadband piezoelectric power generation on high-energy orbits of the bistable Duffing oscillator with piezoelectric coupling. *Journal of Sound and Vibration*, doi:10.1016/j.jsv.2010.11.018.
13. Moon, F.C. and Holmes P.J. (1979) A magnetoelastic strange attractor. *Journal of Sound and Vibration*, **65**, 275–296.
14. Poston, T. and Stewart, O. (1978) *Catastrophe Theory and Its Applications*, Pitman, London.
15. Holmes, P. (1979) A nonlinear oscillator with a strange attractor. *Philosophical Transactions of the Royal Society of London, Series A*, **292**, 419–449.
16. Ottman, G.K., Hofmann, H.F., Bhatt, A.C., and Lesieutre, G.A. (2002) Adaptive piezoelectric energy harvesting circuit for wireless remote power supply. *IEEE Transactions on Power Electronics*, **17**, 669–676.
17. Guyomar, D., Badel, A., Lefeuvre, E., and Richard, C. (2005) Toward energy harvesting using active materials and conversion improvement by nonlinear processing. *IEEE Transactions on Ultrasonics, Ferroelectrics, and Frequency Control*, **52**, 584–595.
18. Guan, M.J. and Liao, W.H. (2007) On the efficiencies of piezoelectric energy harvesting circuits towards storage device voltages. *Smart Materials and Structures*, **16**, 498–505.

19. Lefeuvre, E., Audigier, D., Richard, C., and Guyomar, D. (2007) Buck-boost converter for sensorless power optimization of piezoelectric energy harvester. *IEEE Transactions on Power Electronics*, **22**, 2018–2025.

20. Shu, Y.C., Lien, I.C., and Wu, W.J. (2007) An improved analysis of the SSHI interface in piezoelectric energy harvesting. *Smart Materials and Structures*, **16**, 2253–2264.

21. Kong, N., Ha, D.S., Erturk, A., and Inman, D.J. (2010) Resistive impedance matching circuit for piezoelectric energy harvesting. *Journal of Intelligent Material Systems and Structures*, **21**, pp. 1293–1302.

22. Arrieta, A.F., Hagedorn, P., Erturk, A., and Inman, D.J. (2010) A piezoelectric bistable plate for nonlinear broadband energy harvesting. *Applied Physics Letters*, **97**, 104102.

23. Hyer, M.W. (1981) Some observations on the cured shapes of thin unsymmetric laminates. *Journal of Composite Materials*, **15**, 175–194.

24. Barton, D.A.W., Burrow, S.G., and Clare, L.R. (2010) Energy harvesting from vibrations with a nonlinear oscillator. *ASME Journal of Vibration and Acoustics*, **132**, 021009.

25. Burrow, S.G., Clare, L.R., Carrella, A., and Barton, D. (2008) Vibration energy harvesters with non-linear compliance. *Proceedings of SPIE*, **6928**, 692807.

26. Mann, B.P. and Sims, N.D. (2009) Energy harvesting from the nonlinear oscillations of magnetic levitation. *Journal of Sound and Vibration*, **319**, 515–530.

27. Ramlan, R., Brennan, M.J., Mace, B.R., and Kovacic, I. (2009) Potential benefits of a non-linear stiffness in an energy harvesting device. *Nonlinear Dynamics*, **59**, 545–558.

28. Daqaq, M.F. (2010) Response of uni-modal Duffing-type harvesters to random forced excitations. *Journal of Sound and Vibration*, **329**, 3621–3631.

29. Cottone, F., Vocca, H., and Gammaitoni, L. (2009) Nonlinear energy harvesting. *Physical Review Letters*, **102**, 080601.

30. Gammaitoni, L., Neri, I., and Vocca, H. (2009) Nonlinear oscillators for vibration energy harvesting. *Applied Physics Letters*, **94**, 164102.

31. Benzi, R., Sutera, A., and Vulpiani, A. (1981) The mechanism of stochastic resonance. *Journal of Physics A: Mathematical and General*, **14**, L453–L457.

32. Benzi, R., Parisi, G., Sutera, A., and Vulpiani, A. (1982) Stochastic resonance in climatic change. *Tellus*, **34**, 10–16.

33. Gammaitoni, L., Hanggi, P., Jung, P., and Marchesoni, F. (1998) Stochastic resonance. *Reviews of Modern Physics*, **70**, 223–287.

34. McInnes, C.R., Gorman, D.G., and Cartmell, M.P. (2008) Enhanced vibrational energy harvesting using nonlinear stochastic resonance. *Journal of Sound and Vibration*, **318**, 655–662.

35. Stanton, S.C., McGehee, C.C., and Mann, B.P. (2010) Nonlinear dynamics for broadband energy harvesting: investigation of a bistable inertial generator. *Physica D*, **239**, 640–653.

36. Litak, G., Friswell, M.I., and Adhikari, S. (2010) Magnetopiezoelastic energy harvesting driven by random excitations. *Applied Physics Letters*, **96**, 214103.

37. Crawley, E. and Anderson, E. (1990) Detailed models for piezoelectric actuation of beams. *Journal of Intelligent Material Systems and Structures*, **1**, 4–25.

38. Crawley, E. and Lazarus, K. (1991) Induced strain actuation of isotropic and anisotropic plates. *AIAA Journal*, **29**, 944–951.

39. Roundy, S., Wright, P.K., and Rabaey, J.M. (2003) A study of low level vibrations as a power source for wireless sensor nodes. *Computer Communications*, **26**, 1131–1144.

40. Standards Committee of the IEEE Ultrasonics, Ferroelectrics, and Frequency Control Society (1987) *IEEE Standard on Piezoelectricity*, IEEE, New York.

9

Piezoelectric Energy Harvesting from Aeroelastic Vibrations

Conventional wind energy harvesting has been realized in the literature by means of windmill-type devices, which cannot easily be embedded into several engineering structures, such as aircraft wings and other flexible structural components exposed to wind. However, adding flexible piezoceramics to such structural components under aeroelastic vibrations is a feasible option. This chapter presents lumped-parameter and distributed-parameter piezoaeroelastic models for energy harvesting from airflow excitation. Lumped-parameter electromechanical modeling is given for a piezoaeroelastic section with plunge and pitch degrees of freedom by focusing on the response at the flutter boundary. Model predictions are compared against the experimental results at the flutter boundary of a modified typical section with piezoceramic patches attached to plunge spring members. The effect of piezoelectric power generation on the linear flutter speed of the aeroelastic system is discussed and the importance of having nonlinearities in the aeroelastic system is addressed. Assumed-modes formulation for piezoaeroelastic modeling of a cantilever with embedded piezoceramics is also summarized. As alternative distributed-parameter formulations, finite-element piezoaeroelastic energy harvester models using the vortex–lattice method and the doublet–lattice method are reviewed for time-domain and frequency-domain solutions, respectively, and a theoretical case study is summarized.

9.1 A Lumped-Parameter Piezoaeroelastic Energy Harvester Model for Harmonic Response

Lumped-parameter wing-section (or airfoil) models are very appealing due to their physical simplicity and the fundamental insight they provide [1–4]. Therefore, this section is based on the approach given by Erturk *et al.* [5] for lumped-parameter modeling of energy harvesting from aeroelastic vibrations of a two-degree-of-freedom system with piezoelectric coupling. The piezoaeroelastic section shown in Figure 9.1 is exposed to uniform incompressible potential flow at an airflow speed of U. The variables of the problem are the plunge displacement

---

Piezoelectric Energy Harvesting, First Edition. Alper Erturk and Daniel J. Inman.
© 2011 John Wiley & Sons, Ltd. Published 2011 by John Wiley & Sons, Ltd.

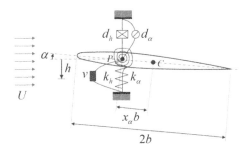

Figure 9.1   Schematic of a piezoaeroelastic section under uniform airflow

(translation) h of the reference point (P), the pitch displacement (rotation) α about the reference point, and the voltage output v across the external electrical load. The plunge degree of freedom (DOF) is piezoelectrically coupled and the electrical output generated from aeroelastic vibrations is connected to a resistive electrical load.

The extended Hamilton's principle for the piezoaeroelastic system can be expressed as

$$\int_{t_1}^{t_2} (\delta T - \delta U + \delta W_{ie} + \delta W_{nce} + \delta W_{nca} + \delta W_{sd})\, dt = 0 \qquad (9.1)$$

where δT, δU, and δW_{ie}, respectively, are the first variations of the total kinetic energy per length (T), total potential energy per length (U), and internal electrical energy per length (W_{ie}), while δW_{nce}, δW_{nca}, and δW_{sd}, respectively, are the virtual work components per length due to the non-conservative electric charge, aerodynamic loads, and structural damping.[1]

The total kinetic energy per length is

$$T = \frac{1}{2}m\left(\dot{h}^2 + 2x_\alpha b\dot{h}\dot{\alpha}\right) + \frac{1}{2}m_f\dot{h}^2 + \frac{1}{2}I_p\dot{\alpha}^2 \qquad (9.2)$$

where m is the airfoil mass per length, m_f accounts for the fixture mass per length in the experiments connecting the airfoil to the plunge springs ($m_f = 0$ for the ideal representation given in Figure 9.1), x_α is the dimensionless chord-wise offset of the reference point from the centroid (point C), b is the semi-chord length, I_p is the moment of inertia per length about the reference point, and the overdot represents differentiation with respect to time.

The total potential energy per length is

$$U = \frac{1}{2}k_h h^2 + \frac{1}{2}k_\alpha \alpha^2 - \frac{1}{2}\frac{\theta}{\ell}hv \qquad (9.3)$$

where k_h is the stiffness per length in the plunge DOF, k_α is the stiffness per length in the pitch DOF, θ is the electromechanical coupling term, and ℓ is the span length.

---

[1] These parameters are given per span length (into the page in Figure 9.1) in this two-dimensional problem.

The internal electrical energy per length is

$$W_{ie} = \frac{1}{2} \frac{C_p^{eq}}{\ell} v^2 + \frac{1}{2} \frac{\theta}{\ell} vh \tag{9.4}$$

where C_p^{eq} is the equivalent capacitance of the piezoceramic layers.

The virtual work components of the non-conservative electric charge, aerodynamic loads, and structural damping per length are

$$\delta W_{nce} = \frac{Q}{\ell} \delta v \tag{9.5}$$

$$\delta W_{nca} = -L \, \delta h + M \, \delta \alpha \tag{9.6}$$

$$\delta W_{sd} = -d_h \dot{h} \, \delta h - d_\alpha \dot{\alpha} \, \delta \alpha \tag{9.7}$$

where Q is the electric charge output, L is the aerodynamic lift per length, M is the aerodynamic pitching moment per length, and d_h and d_α, respectively, are the structural damping coefficients per length in the plunge DOF and the pitch DOF.

Following a procedure similar to that of the piezoelastic system (see Chapter 6 and Appendix H), the piezoaeroelastic Lagrange equations can be expressed as

$$\frac{d}{dt}\left(\frac{\partial T}{\partial \dot{h}}\right) - \frac{\partial T}{\partial h} + \frac{\partial U}{\partial h} - \frac{\partial W_{ie}}{\partial h} = -L - d_h \dot{h} \tag{9.8}$$

$$\frac{d}{dt}\left(\frac{\partial T}{\partial \dot{\alpha}}\right) - \frac{\partial T}{\partial \alpha} + \frac{\partial U}{\partial \alpha} - \frac{\partial W_{ie}}{\partial \alpha} = M - d_\alpha \dot{\alpha} \tag{9.9}$$

$$\frac{d}{dt}\left(\frac{\partial T}{\partial \dot{v}}\right) - \frac{\partial T}{\partial v} + \frac{\partial U}{\partial v} - \frac{\partial W_{ie}}{\partial v} = \frac{Q}{\ell} \tag{9.10}$$

Substituting the relevant terms and taking the time derivative of Equation (9.10), the foregoing Lagrange equations yield the following equations governing the piezoaeroelastic system dynamics:

$$\left(m + m_f\right)\ddot{h} + mx_\alpha b\ddot{\alpha} + d_h \dot{h} + k_h h - \frac{\theta}{\ell}v = -L \tag{9.11}$$

$$mx_\alpha b\ddot{h} + I_p \ddot{\alpha} + d_\alpha \dot{\alpha} + k_\alpha \alpha = M \tag{9.12}$$

$$C_p^{eq} \dot{v} + \frac{v}{R_l} + \theta \dot{h} = 0 \tag{9.13}$$

where R_l is the external load resistance (and $\dot{Q} = v/R_l$ is used).

Assuming harmonic response at frequency ω (i.e., $h = \bar{h}e^{j\omega t}$, $\alpha = \bar{\alpha}e^{j\omega t}$, $v = \bar{v}e^{j\omega t}$, $L = \bar{L}e^{j\omega t}$, $M = \bar{M}e^{j\omega t}$ where $j = \sqrt{-1}$) leads to the following complex eigenvalue problem for the steady-state plunge and pitch displacements:

$$
\left[
\begin{array}{cc}
\left[\beta + \dfrac{\ell_h}{\mu} - \kappa(\omega) - \sigma^2(1 + j\gamma_h)\lambda\right] & \left(x_\alpha + \dfrac{\ell_\alpha}{\mu}\right) \\[2mm]
\left(x_\alpha + \dfrac{m_h}{\mu}\right) & \left[r^2 + \dfrac{m_\alpha}{\mu} - r^2(1 + j\gamma_\alpha)\lambda\right]
\end{array}
\right]
\left[\begin{array}{c} \bar{h} \\ b \\ \bar{\alpha} \end{array}\right] = \left[\begin{array}{c} 0 \\ 0 \end{array}\right]
\tag{9.14}
$$

where the aerodynamic loads ℓ_h, ℓ_α, m_h, and m_α are [1–6]

$$
\ell_h = 1 - j\frac{2}{k}C(k)
\tag{9.15}
$$

$$
\ell_\alpha = -a - j\frac{1}{k} - \frac{2}{k^2}C(k) - j\frac{2}{k}\left(\frac{1}{2} - a\right)C(k)
\tag{9.16}
$$

$$
m_h = -a + j\frac{2}{k}\left(\frac{1}{2} + a\right)C(k)
\tag{9.17}
$$

$$
m_\alpha = \frac{1}{8} + a^2 - j\frac{1}{k}\left(\frac{1}{2} - a\right) + \frac{2}{k^2}\left(\frac{1}{2} + a\right)C(k) + j\frac{2}{k}\left(\frac{1}{4} - a^2\right)C(k)
\tag{9.18}
$$

Here, a is the dimensionless location of the reference point with respect to the mid-chord which is negative (positive) if the reference point lies toward the leading (trailing) edge of the airfoil, k is the reduced frequency given by

$$
k = \frac{b\omega}{U}
\tag{9.19}
$$

and $C(k)$ is known as the Theodorsen function [1–6]:

$$
C(k) = \frac{H_1^{(2)}(k)}{H_1^{(2)}(k) + jH_0^{(2)}(k)}
\tag{9.20}
$$

where $H_n^{(2)}(k)$ are Hankel functions of the second kind, which can be expressed as

$$
H_n^{(2)}(k) = J_n(k) - jY_n(k)
\tag{9.21}
$$

Here, $J_n(k)$ and $Y_n(k)$ are Bessel functions of the first and second kind, respectively. It is important to note that, in this linear model, the harmonic response assumption holds for the condition of neutral stability only (i.e., Equation (9.14) is valid at the classical flutter boundary only). The dimensionless terms are the complex eigenvalue (λ), the frequency ratio (σ), the dimensionless radius of gyration (r), the ratio of the airfoil to affected air mass (μ), and a mass ratio (β) that accounts for the presence of a fixture between the airfoil and the plunge springs:

$$\lambda = \left(\frac{\omega_\alpha}{\omega}\right)^2, \quad \sigma = \frac{\omega_h}{\omega_\alpha}, \quad r = \sqrt{\frac{I_p}{mb^2}}, \quad \mu = \frac{m}{\pi \rho_\infty b^2}, \quad \beta = \frac{m+m_f}{m} \tag{9.22}$$

where $\omega_h = \sqrt{k_h/m}$, $\omega_\alpha = \sqrt{k_\alpha/I_p}$, and ρ_∞ is the free-stream air mass density.[2] Furthermore, the loss factors in Equation (9.14) are assumed to obey

$$\gamma_h = \frac{d_h}{k_h}\omega, \quad \gamma_\alpha = \frac{d_\alpha}{k_\alpha}\omega \tag{9.23}$$

and they are identified at zero airflow speed. In most theoretical representations [1–4] as well as in Figure 9.1, $\beta = 1$ since only the airfoil mass contributes to the inertia that is in equilibrium with the aerodynamic lift. However, usually there is an additional fixture in the typical section experiments which makes $\beta > 1$.

The dimensionless term $\kappa(\omega)$ in Equation (9.14) is due to eliminating the voltage term using Equation (9.13) in Equation (9.11) and it depends on the eigenvalue λ since it is a function of frequency:

$$\kappa(\omega) = \frac{j\theta^2}{\omega m \ell \left(j\omega C_p^{eq} + \frac{1}{R_l}\right)} \tag{9.24}$$

Hence an iterative solution procedure is required where the frequency to be used in $\kappa(\omega)$ is obtained from the eigenvalue that becomes unstable with increasing airflow speed. The convergence of the iterative eigensolution is extremely fast if one starts with the solution of the piezoelectrically uncoupled aeroelastic problem ($\kappa(\omega) = 0$).

Once the complex eigenvector relationship between $\bar{h}$ and $\bar{\alpha}$ is obtained, $\bar{v}$ is calculated using

$$\bar{v} = \frac{-j\omega\theta\bar{h}}{j\omega C_p^{eq} + \frac{1}{R_l}} \tag{9.25}$$

For a given load resistance, the airflow speed that makes the imaginary part of the respective eigenvalue branch zero is the flutter speed ($U = U_c$) and the piezoaeroelastic eigenvector $\left[\bar{h}\ \bar{\alpha}\ \bar{v}\right]^t$ is obtained using this eigenvalue at this particular speed.

---

[2] Note that $\omega_h = \sqrt{k_h/m}$ is the undamped natural frequency in the decoupled plunge DOF for zero airflow speed if and only if $m_f = 0$ (i.e., $\beta = 1$). In the presence of a fixture mass, the undamped natural frequency in the decoupled plunge DOF for zero airflow speed is $\sqrt{k_h/(m+m_f)}$.

9.2 Experimental Validations of the Lumped-Parameter Model at the Flutter Boundary

Figure 9.2 shows the experimental setup used for investigating the piezoaeroelastic response of an airfoil section following Erturk *et al.* [5]. The system parameters are $x_\alpha = 0.260$, $r = 0.504$, $\beta = 2.597$, $\sigma = 3.33$, $\mu = 29.6$, $a = -0.5$, $b = 0.125$ m, $\ell = 0.5$ m, and $\omega_\alpha = 15.4$ rad/s. The loss factors identified for the plunge DOF and the pitch DOF at zero airflow speed are $\gamma_h = 0.007$ and $\gamma_\alpha = 0.12$. The plunge stiffness of the airfoil is due to four steel beams (in clamped–clamped end conditions) connecting the airfoil to the ground from the reference point. Two PZT-5A piezoceramics (QP10N from Midé Technology Corporation) are attached to the roots of two of these beams symmetrically and their electrodes are combined in parallel. The electromechanical coupling term is obtained based on distributed-parameter modeling[3] (Chapter 3) as $\theta = 1.55$ mN/V and the manufacturer's published equivalent capacitance of $C_p^{eq} = 120$ nF is used in the model. In the experiments, the airflow speed is slowly increased from zero until almost persistent piezoaeroelastic response is obtained for each resistive load.

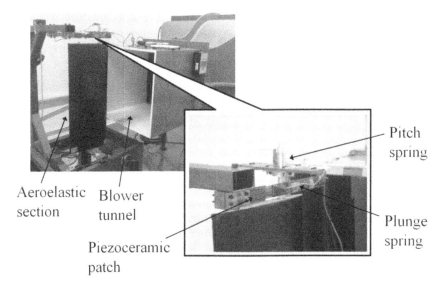

Figure 9.2 Experimental setup showing a typical aeroelastic section with piezoceramics attached to the plunge stiffness members

The short-circuit ($R_l \to 0$) and the open-circuit ($R_l \to \infty$) flutter speeds are measured as $U_c^{sc} = 8.85$ m/s and $U_c^{oc} = 8.90$ m/s, respectively. Figure 9.3 shows the piezoaeroelastic response for an electrical load resistance of $100 \, \text{k}\Omega$ with almost persistent oscillations at the flutter speed of 9.30 m/s.[4] Among the set of resistors used in the experiments, this is the electrical load that gives the maximum power output (10.7 mW). As can be expected from

---

[3] It is important to note that the plunge spring members are in clamped–clamped end conditions.

[4] The response in Figure 9.3 is decaying at a very slow rate which makes it almost persistent. A slightly larger airflow speed results in divergent oscillations as the experimental setup behaves quite linearly.

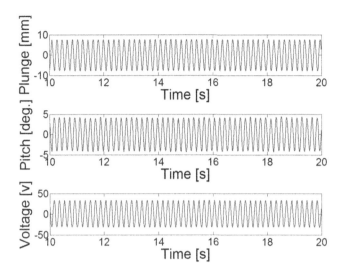

Figure 9.3 Experimental piezoaeroelastic response for $R_l = 100\,\text{k}\Omega$, $U_c = 9.30\,\text{m/s}$

the complex eigenvalue problem described previously, there is a relative phase difference between the response histories in Figure 9.3. For this electrical load, the absolute value of the normalized piezoaeroelastic eigenvector (to have the first element equal to unity) is obtained from the model as

$$\left[\,|\bar{h}|\; |\bar{\alpha}|\; |\bar{v}|\,\right]^t / |\bar{h}| = \left[\,1\;\; 0.56°/\text{mm}\;\; 4.67\,\text{V/mm}\,\right]^t$$

at the flutter speed of 9.56 m/s (where the superscript t is transpose). The experimental maximum response amplitudes in Figure 9.3 are $|\bar{h}| = 7.65\,\text{mm}$, $|\bar{\alpha}| = 4.18°$, and $|\bar{v}| = 32.7\,\text{V}$. Hence the experimental ratios

$$|\bar{\alpha}| / |\bar{h}| = 4.18/7.65 = 0.55°/\text{mm}$$

and

$$|\bar{v}| / |\bar{h}| = 32.7/7.65 = 4.27\,\text{V/mm}$$

exhibit good agreement with the model.

The flutter speed for $R_l = 100\,\Omega$ (close to short-circuit conditions) is predicted by the model as 9.06 m/s, overestimating the experimental value of 8.85 m/s by 2.4%. The model predicts the flutter speed for $R_l = 100\,\text{k}\Omega$ kΩ as 9.56 m/s, which overestimates the experimental value of 9.30 m/s by 2.8%. As shown in Figure 9.4a, the flutter speed in the model corresponds to the speed for which the respective eigenvalue branch (that becomes unstable) has zero imaginary part (the second eigenvalue branch is not shown in the figure). The flutter frequency (Figure 9.4b) is obtained by the model as 5.17 Hz for $R_l = 100\,\Omega$ (underestimating the experimental value by 1.7%) and as 5.14 Hz for $R_l = 100\,\text{k}\Omega$ (underestimating the experimental value by 2.3%).

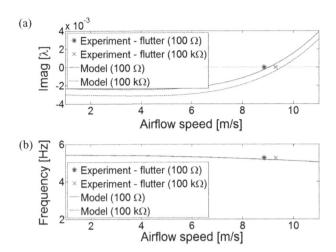

Figure 9.4 Theoretical evolutions of (a) the eigenvalue that becomes unstable and (b) the frequency obtained from this eigenvalue with changing airflow speed compared to the experimental data points at the flutter boundary (for $R_l = 100\,\Omega$ and $R_l = 100\,\text{k}\Omega$)

Figure 9.5a shows the ratio of voltage to plunge displacement amplitude, while Figure 9.5b shows the ratio of pitch to plunge displacement amplitude for the set of resistors used in the experiment along with the theoretical predictions. The curve of voltage to plunge displacement versus load resistance exhibits linear asymptotes similar to the trend in the harmonic base excitation of piezoelectric energy harvesters (Chapters 3–5), while the variation of the pitch to plunge displacement amplitude is insensitive to the changing load resistance. It should be highlighted that these theoretical and experimental data points are given for the flutter speed that corresponds to the respective load resistance (e.g., for $R_l = 10\,\text{k}\Omega$, $|\bar{h}| = 5.15\,\text{mm}$, $|\bar{\alpha}| = 2.82°$, and $|\bar{v}| = 2.42\,\text{V}$, whereas for $R_l = 1\,\text{M}\Omega$, $|\bar{h}| = 7.95\,\text{mm}$, $|\bar{\alpha}| = 4.40°$, and $|\bar{v}| = 83.1\,\text{V}$). Hence the maximum plunge and pitch amplitudes differ and it is their ratio that remains similar.

The curve for the ratio of electrical power to plunge displacement versus load resistance is shown in Figure 9.6a. The optimal load that gives the maximum power output causes an increase in the flutter speed (Figure 9.6b) due to the shunt damping effect [7] of piezoelectric power generation. The experimental increase in the flutter speed (with respect to the short-circuit flutter speed) for $R_l = 100\,\text{k}\Omega$ is 5.1% and the model predicts this increase as 5.5% in Figure 9.6b. Therefore, piezoelectric energy harvesting has the favorable effect of increasing the flutter speed of the piezoaeroelastic system.

9.3 Utilization of System Nonlinearities in Piezoaeroelastic Energy Harvesting

Often nonlinearities are present in aeroelastic systems in the forms of (1) free play, or bilinear stiffness due to loosely connected components, (2) material and geometric nonlinearity (typically yielding nonlinear stiffness effects), and (3) dry friction and other forms of nonlinear

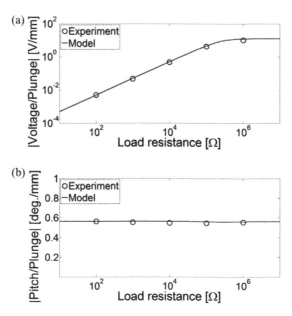

Figure 9.5   Theoretical and experimental ratios of (a) voltage output to plunge displacement and (b) pitch displacement to plunge displacement versus load resistance

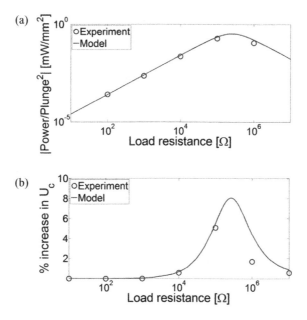

Figure 9.6   Theoretical and experimental (a) normalized power and (b) percentage increase in the flutter speed versus load resistance

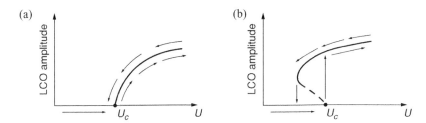

Figure 9.7 Two types of LCO response due to (a) supercritical and (b) subcritical bifurcations with changing airflow speed ($U = U_c$ is the linear flutter speed)

damping [8]. The presence of such nonlinearities may result in limit-cycle oscillations (LCOs) at airflow speeds above (Figures 9.7a and 9.7b) or below (Figure 9.7b) the linear flutter speed. The LCO mechanism in Figure 9.7b leads to both stable (solid line) and unstable (dashed line) LCOs below the linear flutter speed and is not preferred in aircraft [8,9]. However, this type of LCO might be useful in a piezoaeroelastic system designed for energy harvesting only [10,11] so that a large-amplitude response can be obtained for a wider range of airflow speeds. As suggested by Erturk *et al.* [5], a typical section similar to the one in Figure 9.1 with a nonlinear stiffness component and/or free play can be used to investigate LCOs for piezoelectric energy harvesting and theoretical tools are available to solve the resulting nonlinear equations for the limit cycles [12]. Although the focus in this chapter is placed on linear flutter, stable LCOs of acceptable amplitude in nonlinear piezoaeroelastic systems can provide an important source of persistent electrical power. A broadband piezoaeroelastic energy harvester[5] can be obtained if subcritical bifurcations can be created through mechanisms such as free play or bilinear stiffness [8,9,13].

9.4 A Distributed-Parameter Piezoaeroelastic Model for Harmonic Response: Assumed-Modes Formulation

Figure 9.8 shows a thin piezoelastic cantilever (in the form of an unswept wing) under airflow excitation along with its cross-sectional view. The cantilever (Figure 9.8a) has an airfoil section (Figure 9.8b) which is uniform along the axial direction of the structure. The electrodes of the piezoceramic layers are perpendicular to the thickness direction (direction of poling) and they can be combined either in series or in parallel (Chapter 3). The following summarizes the electromechanical assumed-modes formulation of piezoaeroelastic energy harvesting from airflow excitation of a cantilever and it is basically the piezoaeroelastic version of the aeroelastic [1–4] assumed-modes [14, 15] (Chapter 6) formulation.

In the absence of mechanical (structural) damping, the extended Hamilton's principle for the piezoaeroelastic system can be expressed as

$$\int_{t_1}^{t_2} (\delta T - \delta U + \delta W_{ie} + \delta W_{nce} + \delta W_{nca}) \, dt = 0 \qquad (9.26)$$

---

[5] The term *broadband* here refers to the broad range of airflow speeds that yields LCOs (analogous to the frequency-wise broadband energy harvesters – see Chapter 8 – that respond to harmonic base excitation in a wider frequency range compared to resonant energy harvesters).

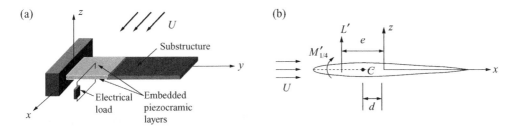

Figure 9.8 (a) An unswept cantilever with embedded piezoceramics and (b) its cross-sectional view with an airfoil profile

where δT, δU and δW_{ie} are the first variations of the total kinetic energy, total potential energy, and internal electrical energy, respectively, while δW_{nce} and δW_{nca} are the virtual work components due to the non-conservative electric charge and aerodynamic loads, respectively.

The total kinetic energy (T) and the total potential energy (U) are

$$T = \frac{1}{2}\left(\int_{V_s} \rho_s \frac{\partial \mathbf{u}^t}{\partial t} \frac{\partial \mathbf{u}}{\partial t} dV_s + \int_{V_p} \rho_p \frac{\partial \mathbf{u}^t}{\partial t} \frac{\partial \mathbf{u}}{\partial t} dV_p \right) \tag{9.27}$$

$$U = \frac{1}{2}\left(\int_{V_s} \mathbf{S}^t \mathbf{T} dV_s + \int_{V_p} \mathbf{S}^t \mathbf{T} dV_p \right) \tag{9.28}$$

where ρ_s and ρ_p are the mass densities of the substructure and piezoceramic layers, respectively, $\mathbf{u}$ is the vector of displacement components, $\mathbf{S}$ is the vector of strain components, and $\mathbf{T}$ is the vector of stress components. The integrations are performed over the volume of the structure and the subscripts s and p stand for the substructure and the piezoceramic, respectively, while the superscript t stands for the transpose.

Equations (9.27) and (9.28) can be reduced to

$$T = \frac{1}{2} \int_0^\ell \left[m\left(\frac{\partial w}{\partial t}\right)^2 + 2md\frac{\partial w}{\partial t}\frac{\partial \alpha}{\partial t} + mb^2r^2\left(\frac{\partial \alpha}{\partial t}\right)^2 \right] dy \tag{9.29}$$

$$U = \frac{1}{2} \int_0^\ell \left[YI\left(\frac{\partial^2 w}{\partial y^2}\right)^2 + GJ\left(\frac{\partial \alpha}{\partial y}\right)^2 \right] dy - \frac{1}{2}\int_0^{\ell_p} \vartheta v \frac{\partial^2 w}{\partial y^2} dy \tag{9.30}$$

where $w = w(y,t)$ is the transverse displacement, $\alpha = \alpha(y,t)$ is the torsional rotation, m is the mass per length, d is the offset of the centroid from the elastic axis, br is the radius of gyration about the elastic axis, YI is the bending stiffness and GJ is the effective torsional stiffness of the cross-section at position y, ϑ is the electromechanical coupling term (similar to Equations (3.8) and (3.9)), ℓ is the total length of the cantilever, and ℓ_p is the length of the piezoceramic layers.

The internal electrical energy is

$$W_{ie} = \frac{1}{2} \int\limits_{V_p} \mathbf{E}^t \mathbf{D} dV_p = \frac{1}{2} \int\limits_0^{\ell_p} \vartheta v \frac{\partial^2 w}{\partial y^2} dy + \frac{1}{2} C_p^{eq} v^2 \tag{9.31}$$

where $\mathbf{E}$ is the vector of electric field components, $\mathbf{D}$ is the vector of electric displacement components (having the only non-zero components in the thickness (poling) direction as discussed in Chapters 3 and 6), and C_p^{eq} is the equivalent capacitance of the piezoceramic layers.

The virtual work expressions are

$$\delta W_{nce} = Q \, \delta v \tag{9.32}$$

$$\delta W_{nca} = \int\limits_0^{\ell} \left[L' \, \delta w + \left(M'_{1/4} + eL' \right) \delta \alpha \right] dy \tag{9.33}$$

where Q is the net electric charge flowing to the load, v is the voltage across the load, L' is the aerodynamic lift per length, and $M'_{1/4}$ is the aerodynamic pitching moment per length. Following Hodges and Pierce [4], the distributed-parameter formulation can be related to the lumped-parameter formulation since $d = -b x_\alpha$, $e = (1/2 + a)b$, and $w = -h$. Then the Theodorsen model [6] leads to [4]

$$L' = 2\pi \rho_\infty U b C(k) \left[U\alpha - \frac{\partial w}{\partial t} + b \left(\frac{1}{2} - a \right) \frac{\partial \alpha}{\partial t} \right] + \pi \rho_\infty b^2 \left(U \frac{\partial \alpha}{\partial t} - \frac{\partial^2 w}{\partial t^2} - ba \frac{\partial^2 \alpha}{\partial t^2} \right) \tag{9.34}$$

$$M'_{1/4} = -\pi \rho_\infty b^3 \left[U \frac{\partial \alpha}{\partial t} - \frac{1}{2} \frac{\partial^2 w}{\partial t^2} + b \left(\frac{1}{8} - \frac{a}{2} \right) \frac{\partial^2 \alpha}{\partial t^2} \right] \tag{9.35}$$

The structural response components are

$$w(y, t) = \sum_{r=1}^{N_w} \phi_r(y) \eta_r(t) \tag{9.36}$$

$$\alpha(y, t) = \sum_{r=1}^{N_\alpha} \psi_r(y) \chi_r(t) \tag{9.37}$$

where $\eta_r(t)$ and $\chi_r(t)$ are the generalized modal coordinates associated with the bending and torsion admissible functions $\phi_r(y)$ and $\psi_r(y)$, respectively. The generalized coordinates of the piezoaeroelastic system are therefore η_r, χ_r, and v. The admissible functions satisfy the kinematic boundary conditions at the fixed end:[6]

$$\phi_r(0) = \left.\frac{d\phi_r(y)}{dy}\right|_{y=0} = 0, \quad \psi_r(0) = 0 \tag{9.38}$$

Finally, the mechanically undamped piezoaeroelastic Lagrange equations are obtained from

$$\frac{d}{dt}\left(\frac{\partial T}{\partial \dot{\eta}_r}\right) - \frac{\partial T}{\partial \eta_r} + \frac{\partial U}{\partial \eta_r} - \frac{\partial W_{ie}}{\partial \eta_r} = F_w \tag{9.39}$$

$$\frac{d}{dt}\left(\frac{\partial T}{\partial \dot{\chi}_r}\right) - \frac{\partial T}{\partial \chi_r} + \frac{\partial U}{\partial \chi_r} - \frac{\partial W_{ie}}{\partial \chi_r} = F_\alpha \tag{9.40}$$

$$\frac{d}{dt}\left(\frac{\partial T}{\partial \dot{v}}\right) - \frac{\partial T}{\partial v} + \frac{\partial U}{\partial v} - \frac{\partial W_{ie}}{\partial v} = Q \tag{9.41}$$

where the generalized forces F_w and F_α, respectively, are related to the generalized coordinates η_r and χ_r while the generalized force of the electrical equation is nothing but the electric charge Q (that is associated with the generalized coordinate v, the voltage across the load). After obtaining the governing piezoaeroelastic equations, proportional damping can be introduced as shown in Chapter 6. Depending on the number of assumed vibration modes (one bending and one torsional mode in the simplest case: $N_w = N_\alpha = 1$), Equations (9.39) and (9.40) lead to equations analogous to Equations (9.11) and (9.12), respectively, while it is the time derivative of Equation (9.41) that results in an electrical circuit equation of the form given by Equation (9.13) since $\dot{Q} = v/R_l$.

The nature of the aeroelastic problem in distributed-parameter systems often requires consideration of multiple bending and torsion modes so their coupling with increasing airflow speed is predicted correctly. Moreover, the assumed-modes formulation is an approximate analytical solution and, depending on the admissible functions used in the solution procedure, it is usually necessary to employ several modes to ensure convergence of the natural frequencies to the exact frequencies of interest (see Sections 6.8 and 6.9) [14, 15]. Unlike in the lumped-parameter problem (Figure 9.1) where the piezoceramic layers undergo only bending vibrations even at the condition of flutter, the piezoceramic layers in the distributed-parameter problem (Figure 9.8a) undergo coupled bending and torsion vibrations at the condition of flutter. Hence it becomes necessary to use segmented electrodes [16, 17] to avoid electrical cancellations due to strain distribution.

9.5 Time-Domain and Frequency-Domain Piezoaeroelastic Formulations with Finite-Element Modeling

The following two subsections combine a finite-element piezoelastic model [18] with unsteady aerodynamic models to obtain linear piezoaeroelastic energy harvester models. The

---

[6] If the independent variable x in Equations (6.232) and (6.235) is replaced by y, these expressions can be used as $\phi_r(y)$ and $\psi_r(y)$, respectively, as they satisfy Equations (9.38).

first approach follows De Marqui *et al.* [17] and is a time-domain formulation based on the vortex–lattice method (VLM) of aeroelasticity. The second approach is a frequency-domain piezoaeroelastic formulation by the same group [19] and uses the doublet–lattice method (DLM). Both methods summarized in the following are given for the finite-element model of a rectangular, thin cantilevered plate (like an unswept wing) with embedded piezoceramics under incompressible potential flow (similar to Figure 9.8). Therefore the following sections require that the finite-element model of the cantilever plate with piezoceramics is obtained [18] so the governing electromechanical equations can be expressed as

$$\mathbf{M}\ddot{\mathbf{\Psi}} + \mathbf{C}\dot{\mathbf{\Psi}} + \mathbf{K}\mathbf{\Psi} - \tilde{\Theta}v = \mathbf{F} \tag{9.42}$$

$$C_p\dot{v} + \frac{v}{R_l} + \tilde{\Theta}^t\dot{\mathbf{\Psi}} = 0 \tag{9.43}$$

where $\mathbf{M}$ is the global mass matrix, $\mathbf{K}$ is the global stiffness matrix, $\mathbf{C}$ is the global damping matrix (assumed here to be proportional to the mass and stiffness matrices, $\mathbf{C} = \tau\mathbf{M} + \upsilon\mathbf{K}$, where τ and υ are the constants of proportionality), $\tilde{\Theta}$ is the effective electromechanical coupling vector, $\mathbf{\Psi}$ is the vector of mechanical coordinates (nodal mechanical variables), C_p is the equivalent capacitance of the piezoceramic layers, R_l is the load resistance, and v is the voltage across the load [18]. The term $\mathbf{F}$ on the right hand side of Equation (9.42) represents the unsteady aerodynamic loads and is obtained from the respective aerodynamic model.

9.5.1   Time-Domain Formulation Based on the VLM

In the VLM approach [20], the cantilevered plate is represented as a thin lifting surface divided into a number of elements (panels). A planar vortex ring is associated with each rectangular panel of the body itself and its wake. The vortex singularity is a solution of the Laplace equation [20] and the aerodynamic loads acting on the cantilever can be obtained by combining these singularities with the incompressible potential flow around the body.

Figure 9.9 shows a typical vortex–lattice mesh used in the formulation. The leading segment of each vortex ring is placed at the quarter chord point of each panel and a control point is placed at the three-quarter chord of each panel where the boundary condition is satisfied. The surface of the rectangular plate is divided into m panels ($m = R \times S$, where R and S are the number of panels along the chord and the span, respectively) and consequently m vortex rings and control points. Then the condition at the fluid–structure boundary is

$$\mathbf{a}_{KL}\Gamma_{m,1} = \left[\mathbf{v}_{m_{m,1}} + \mathbf{v}_{w_{m,1}}\right] \cdot \mathbf{n}_{m,1} \tag{9.44}$$

where $\mathbf{a}_{KL}$ is the matrix of influence coefficients that relates the circulation at the vortex ring K to the inner product of the perturbed velocity at point L. The velocity field $\mathbf{v}_m$ depends on the free-stream velocity and the velocities of the control points due to structural deformations and $\mathbf{v}_w$ is the velocity field induced by the wake. Therefore the unknowns in this linear set of equations are the circulations Γ_m of each vortex ring. In the time-domain solution process, new vortex rings are formed and shed from the trailing edge to the wake at each time step and

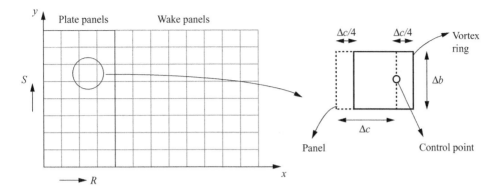

Figure 9.9 Vortex–lattice mesh for a cantilevered plate

the Kutta condition [20] is satisfied, imposing the condition that the circulation values of the most recently shed vortex rings are the same as those at the trailing edge at the previous time step. The unsteady Bernoulli equation [20] is then employed to calculate the aerodynamic load on each panel.

Since proportional damping is assumed (normal-mode system [21]), the equations of motion obtained from the finite-element formulation can be represented in the modal domain as

$$\bar{\mathbf{M}}\ddot{\boldsymbol{\eta}} + \bar{\mathbf{C}}\dot{\boldsymbol{\eta}} + \bar{\mathbf{K}}\boldsymbol{\eta} - \boldsymbol{\Phi}^t\tilde{\boldsymbol{\Theta}}v = \boldsymbol{\Phi}^t\mathbf{F} \tag{9.45}$$

$$C_p\dot{v} + \frac{v}{R_l} + \tilde{\boldsymbol{\Theta}}^t\boldsymbol{\Phi}\dot{\boldsymbol{\eta}} = 0 \tag{9.46}$$

where $\boldsymbol{\eta}$ is the vector of modal coordinates, $\boldsymbol{\Phi}$ is the modal matrix (mass normalized [14] so that the modal mass matrix $\bar{\mathbf{M}}$ is the identity matrix), $\bar{\mathbf{C}}$ is the diagonal modal damping matrix, $\bar{\mathbf{K}}$ is the diagonal modal stiffness matrix, and $\mathbf{F}$ is the vector of aerodynamic loads.

As described by De Marqui *et al.* [17], the electromechanical problem and the aerodynamic problem are originally solved for distinct meshes (nodes of the finite-element mesh and control points of the VLM mesh). In order to obtain the aerodynamic loads, it is required to know the structural response (and consequently the electrical response), which depends on the aerodynamic loads. Hence, an iterative solution procedure that accounts for the interaction between the aerodynamic and the electromechanical domains is employed to solve the governing piezoaeroelastic equations. The structural finite-element nodes and the aerodynamic control points can be related as

$$\boldsymbol{\Psi}_a = \mathbf{G}\boldsymbol{\Psi} \tag{9.47}$$

where $\boldsymbol{\Psi}_a$ is the vector of mechanical coordinates of control points in the aerodynamic mesh and $\mathbf{G}$ is a transformation matrix. The same transformation matrix can also be used to write the structural mode shapes in terms of aerodynamic coordinates

$$\mathbf{\Phi}_a = \mathbf{G}\mathbf{\Phi} \tag{9.48}$$

where $\mathbf{\Phi}_a$ is the modal matrix in aerodynamic coordinates.

Since the virtual work done by the aerodynamic forces is the same for the representations in both domains, one can write

$$\delta\mathbf{\Psi}_a^t \mathbf{F}_a = \delta\mathbf{\Psi}^t \mathbf{F} \tag{9.49}$$

where $\mathbf{F}_a$ is the vector of aerodynamic loads at the control points and $\mathbf{F}$ is the vector of aerodynamic loads on the structural mesh (finite-element nodes). Using Equations (9.48) and (9.49), the equations of motion in modal coordinates can be written as

$$\bar{\mathbf{M}}\ddot{\eta} + \bar{\mathbf{C}}\dot{\eta} + \bar{\mathbf{K}}\eta - \mathbf{\Phi}^t \tilde{\mathbf{\Theta}}v = \mathbf{\Phi}_a^t \mathbf{F}_a \tag{9.50}$$

$$C_p \dot{v} + \frac{v}{R_l} + \tilde{\mathbf{\Theta}}^t \mathbf{\Phi}\dot{\eta} = 0 \tag{9.51}$$

where the aerodynamic loads are transformed to the nodes of the structural mesh. In addition, the structural displacements obtained at the structural nodes (in the finite-element mesh) at each time step are obtained at the corners of vortex rings (in the aerodynamic mesh) for calculation of the aerodynamic loads [17].

Finally, the equations of motion are written as a system of $2N + 1$ first-order ordinary differential equations, where N is the number of vibration modes taken into account in the solution:

$$\dot{\mathbf{u}}_1 = \mathbf{u}_2 \tag{9.52}$$

$$\dot{\mathbf{u}}_2 = \mathbf{\Phi}_a^t \mathbf{F}_a - \bar{\mathbf{C}}\mathbf{u}_2 - \bar{\mathbf{K}}\mathbf{u}_1 - \mathbf{\Phi}^t \tilde{\mathbf{\Theta}}u_3 \tag{9.53}$$

$$\dot{u}_3 = \frac{1}{C_p}\left(-\tilde{\mathbf{\Theta}}^t \mathbf{\Phi}\mathbf{u}_2 - \frac{u_3}{R_l}\right) \tag{9.54}$$

where $\mathbf{u}_1 = \eta$, $\mathbf{u}_2 = \dot{\eta}$, and $u_3 = v$. The $2N + 1$ ordinary differential equations with the aerodynamic loads applied at the finite-element nodes (after the transformation) are then solved using a predictor–corrector scheme that accounts for the interaction between the aerodynamic and electromechanical domains [17].

9.5.2   Frequency-Domain Formulation Based on the DLM

The DLM approach [22] uses the unsteady Euler equations of the surrounding fluid with the linearized, oscillatory, inviscid, subsonic lifting surface theory in order to relate the normal component of the velocity field on the body to the aerodynamic loads. The doublet singularity is a solution of the aerodynamic potential equation and is used to represent the unsteady aeroelastic behavior of the system.

The velocity field normal to the surface of a plate is given by [22]

$$\bar{w}(x, y, z, t) = \frac{-1}{4U\pi\rho_\infty} \iint_S \Delta p(x, y, z, t) K(x - \mu, y - \sigma, z) d\lambda \, dy \tag{9.55}$$

where $\Delta p(x, y, z, t)$ is the differential pressure, U is the free-stream velocity, ρ_∞ is the density of the air, μ and σ are the dummy variables of integration over the area S of the plate, z is the transverse direction, and K is the kernel function that provides the relation between the differential pressure and the velocity field normal to the surface of the plate:

$$K(x - \mu, y - \sigma, z) = \exp\left(\frac{-j\omega(x - \mu)}{U}\right) \frac{\partial^2}{\partial z^2} \left\{ \int_{-\infty}^{x-\mu} \frac{1}{\bar{R}} \exp\left[\frac{j\omega}{U\beta^2}(\varsigma - M\bar{R})\right] d\varsigma \right\} \tag{9.56}$$

Here, $\beta^2 = 1 - M^2$ and $\bar{R} = \sqrt{(x - \mu)^2 + (x - \sigma)^2 + z^2}$, ω is the frequency of harmonic oscillation, M is the Mach number, and ς is a dummy variable of integration.

The DLM approach provides a numerical approximation for the solution of the kernel function. The cantilevered plate is represented by a thin lifting surface and is divided into a number of panels (boxes) with doublet singularities of constant strength in the chord-wise and parabolic strength in the span-wise direction. A line of doublets is assumed at the quarter chord line of each panel, which is equivalent to a pressure jump across the surface. A control point is defined in the half span of each box at the three-quarter chord line (the point where the boundary condition is verified). The strength values of the oscillating potential placed at the quarter chord lines are the unknowns of the problem. The prescribed downwash introduced by the lifting lines is checked at each control point and the solution of the resulting matrix equation is

$$\frac{\bar{\mathbf{w}}}{U} = \mathbf{A}\Delta\mathbf{C_p} \tag{9.57}$$

where $\mathbf{A}$ is the influence matrix (related to the kernel function) between the normal velocity field $\bar{\mathbf{w}}$ (in matrix form) and the non-dimensional pressure distribution ΔC_p. Equation (9.57) gives the strength of the lifting line at each panel and consequently the pressure distribution across the surface. Therefore, integration over the surface gives the resulting aerodynamic loads. These loads can be included in the piezoaeroelastic equations as an aerodynamic matrix of influence coefficients. Just like the VLM-based formulation, the aerodynamic loads and the structural motion are obtained from distinct numerical methods with distinct meshes. Hence the resulting displacements of the structural mesh are also interpolated at the corners of the aerodynamic mesh.

Since the unsteady aerodynamic solution is assumed to be harmonic ($\eta = \mathbf{H}e^{j\omega t}$ and $v = Ve^{j\omega t}$), the piezoaeroelastic equations in modal coordinates can be expressed as [19]

$$\left(-\omega^2 \bar{\mathbf{M}} + j\omega \bar{\mathbf{C}} + \bar{\mathbf{K}} - q\mathbf{Q}\right)\mathbf{H} - \mathbf{\Phi}'\tilde{\mathbf{\Theta}}V = 0 \tag{9.58}$$

$$\left(j\omega C_p + \frac{1}{R_l}\right)V + j\omega\,\tilde{\mathbf{\Theta}}^t\mathbf{\Phi}\mathbf{H} = 0 \tag{9.59}$$

where q is the dynamic pressure and $\mathbf{Q}$ is the aerodynamic influence matrix.

The conventional p–k scheme can be used based on the harmonic motion assumption [23]. In this method, the evolution of the frequencies and damping is iteratively investigated for different airflow speeds (or reduced frequencies) to solve the resulting eigenvalue problem. Since the governing equations have electromechanical coupling, the modified form of the piezoaeroelastic eigenvalue problem is

$$p\begin{bmatrix}\mathbf{h}_1\\ \mathbf{h}_2\\ h_3\end{bmatrix} = \begin{bmatrix} \mathbf{0} & \mathbf{I} & \mathbf{0}\\ -\bar{\mathbf{M}}^{-1}\left(\bar{\mathbf{K}} - q\mathbf{Q}^R\right) & -\bar{\mathbf{M}}^{-1}\left(\bar{\mathbf{C}} - q\mathbf{Q}^I\right) & \bar{\mathbf{M}}^{-1}\mathbf{\Phi}'\tilde{\mathbf{\Theta}}\\ \mathbf{0} & -\tilde{\mathbf{\Theta}}^t\mathbf{\Phi}/C_p & -1/R_l C_p\end{bmatrix}\begin{bmatrix}\mathbf{h}_1\\ \mathbf{h}_2\\ h_3\end{bmatrix} \tag{9.60}$$

where $\mathbf{h}_1 = \mathbf{H}, \mathbf{h}_2 = p\mathbf{H}$, and $h_3 = V$, the superscripts R and I stand for the real and imaginary parts of the aerodynamic influence matrix, and p is the eigenvalue which gives the frequency (related to the imaginary part) and damping (related to the real part). The p–k solution provides the evolution of frequency and damping of the modes with increasing airflow speed accounting for the presence of piezoelectric coupling and the external resistive load.

In addition to the p–k scheme that gives the neutral stability limit, the piezoaeroelastic behavior can be investigated in terms of piezoaeroelastic FRFs for the case of harmonic response. Such FRFs can be defined if the cantilever is exposed to harmonic base motion in addition to the airflow excitation. The forcing term in Equation (9.45) is modified as

$$\mathbf{F} = \mathbf{F}_a + \mathbf{F}_b \tag{9.61}$$

where $\mathbf{F}_a$ is the unsteady aerodynamic loads determined using the DLM solution and $\mathbf{F}_b$ is due to harmonic base excitation. If the base is vibrating in the transverse direction (z-direction), the effective force on the structure is due to the inertia of the structure acting on the structure in the opposite direction [24]. Therefore, the forcing term $\mathbf{F}_b$ is represented as

$$\mathbf{F}_b = -\mathbf{m}^* a_b \tag{9.62}$$

where $\mathbf{m}^*$ is the vector of effective mass per unit area of the plate obtained from the finite-element solution and a_b is the base acceleration. The piezoaeroelastic FRFs are defined by the matrix equation

$$\Lambda = -\begin{bmatrix} -\omega^2\bar{\mathbf{M}} + j\omega\bar{\mathbf{C}} + \bar{\mathbf{K}} - q\mathbf{Q} & -\mathbf{\Phi}'\tilde{\mathbf{\Theta}}\\ j\omega\tilde{\mathbf{\Theta}}^t\mathbf{\Phi} & j\omega C_p + 1/R_l\end{bmatrix}^{-1}\mathbf{m}^* \tag{9.63}$$

Here, Λ is an $(N + 1) \times 1$ vector of FRFs containing N complex modal displacements ($\mathbf{H}$) per base acceleration and the voltage output (V) across the resistive load per base acceleration for a desired airflow speed. Having the voltage FRF and the load resistance value, one can obtain the FRF for the power output to base acceleration.

9.6   Theoretical Case Study for Airflow Excitation of a Cantilevered Plate

Following De Marqui *et al.* [17,19], a theoretical case study is given in this section using the VLM-based and DLM-based piezoaeroelastic formulations for the same structural configuration. The overall dimensions of the rectangular cantilevered plate considered in this case study are 1200 mm $\times$ 240 mm $\times$ 3 mm and the plate has two identical layers of PZT-5A embedded into the top and bottom of the aluminum structure as in Figure 9.8a. The piezoceramic layers are poled in the opposite thickness directions and their electrical outputs are combined in series (Figure 3.1a). These embedded piezoceramic layers cover 30 % of the length (at the root) and each one has a thickness of 0.5 mm (therefore the aluminum thickness at the root is 2 mm). The identical piezoceramic layers have the same width as the aluminum substructure. Typical properties of PZT-5A are as given in Appendix E while the aluminum substructure has an elastic modulus of 70 GPa and mass density of 2750 kg/m^3. The damping proportionality constants of the structure used in the finite-element formulation are $\tau = 0.1635$ rad/s and $\upsilon = 4.1711 \times 10^{-4}$ s/rad and the mass density of free-stream air is 1.225 kg/m^3. The spanwise elastic axis of the cantilever and the center of gravity are coincident at the mid-chord.

9.6.1   Simulations Based on the VLM Formulation

In the VLM simulations [17], the piezoaeroelastic response of the cantilevered plate is presented as time histories of the electrical power output and the tip displacement of the mid-chord line in the transverse direction. The input assumed in the simulations is a variation of 3° in the flow direction for five time steps representing a sharp edge gust, while the initial conditions are assumed to be zero.

The mode sequence and the undamped natural frequencies for the cantilevered plate obtained for short-circuit conditions (very low load resistance) are shown in Figure 9.10: the first bending (1.68 Hz), the second bending (10.46 Hz), the first torsion (16.66 Hz), the third bending (27.74 Hz), and the second torsion (48.65 Hz).

Three regions of piezoaeroelastic response are considered in the simulations: (1) the low airflow speed region with small structural deformations and aerodynamic loads; (2) the region of maximum aerodynamic damping; and (3) the flutter boundary. Investigating the total damping evolution with increasing airflow speed, it is observed that the maximum aerodynamic damping is observed for airflow speeds around 30 m/s [17]. As the airflow speed is further increased, the aerodynamic damping decreases and eventually vanishes at the flutter speed (40 m/s). It is well known that mechanical losses (damping) are undesired in vibration-based energy harvesting so that large structural deformations and electrical power outputs can be obtained. Therefore, it can be anticipated that the flutter boundary (with large aerodynamic loads and low damping) is the most effective region for energy harvesting.

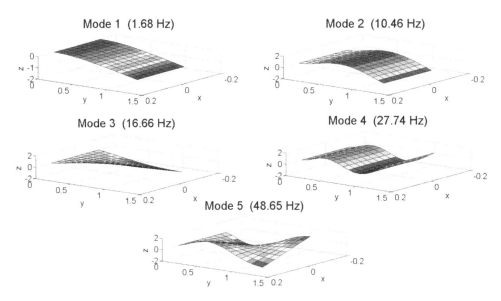

Figure 9.10 The mode sequence and the undamped natural frequencies of the cantilever close to short-circuit conditions ($R_l = 100\,\Omega$)

Figure 9.11a shows the piezoelectric power output extracted for an airflow speed of 10 m/s. At this airflow speed, the total damping in the system (the resultant of structural and aerodynamic damping) is dominated by the structural damping and the aerodynamic loads are not large compared to higher speeds, yielding a peak power output less than 0.1 µW. Among the set of resistors used in the simulations, the maximum power output at this airflow speed is observed for 100 kΩ. The displacement response at the mid-chord of the tip is presented in Figure 9.11b. De Marqui *et al.* [17] report the oscillation frequencies of the smallest (100 Ω) and the largest (1 MΩ) electrical loads in Figure 9.11b as 1.83 Hz and 2.14 Hz, respectively, which are the short-circuit and open-circuit resonance frequencies. The shunt damping effect of piezoelectric power generation is also observed in Figure 9.11b (the maximum piezoelectric damping is for the 100 kΩ load among the five resistors used in the simulations).

Since the maximum aerodynamic damping is around 30 m/s, any oscillation due to airflow excitation is rapidly damped out at this airflow speed as shown in Figures 9.12a and 9.12b, for the electrical power and the vibration response, respectively. Although the oscillations decay faster than the first case (10 m/s), the peak power is larger in Figure 9.12a compared to that of Figure 9.11a since the aerodynamic forces are larger around 30 m/s. However, due to the large amount of total damping, this airflow speed is not preferable for energy harvesting.

The piezoaeroelastic response of the system at the short-circuit flutter speed (40 m/s) is shown in Figure 9.13. Since the flutter boundary has the least amount of damping but the largest aerodynamic loads, this is the best condition for power generation (as far as the linear system is concerned). As shown in Figure 9.13a, among the set of resistors, 10 kΩ provides the largest power output. The time history of the largest power output case exhibits a decaying behavior due to the shunt damping effect of power generation (Figure 9.13b). Therefore, for a resistive load of 10 kΩ, 40 m/s is no longer the critical speed of neutral stability, implying

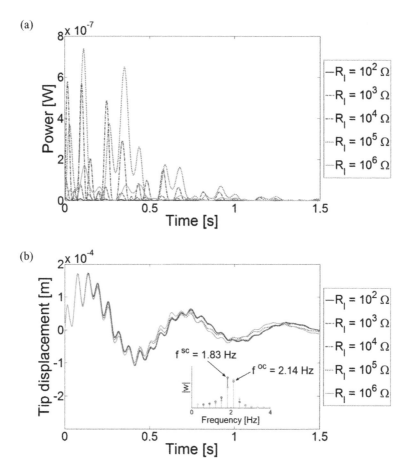

Figure 9.11   (a) Electrical power output and (b) tip displacement for five different values of load resistance at an airflow speed of 10 m/s

a slight increase in the linear flutter speed [17] (similar to the typical section case given in Figure 9.6).

9.6.2   Simulations Based on the DLM Formulation

Based on the p–k solution scheme with electromechanical coupling used in conjunction with the DLM formulation [19], the damping and frequency evolutions with increasing airflow speed are shown in Figures 9.14a and 9.14b, respectively, for an external load resistance close to short-circuit conditions. In agreement with the time-domain VLM solution, the linear flutter speed of the system in short-circuit conditions is 40 m/s (Figure 9.14a). The frequency evolution and coalescence of the second bending and the first torsion modes with increasing airflow speed are given in Figure 9.14b (the first bending and the second torsion modes are excluded for clarity).

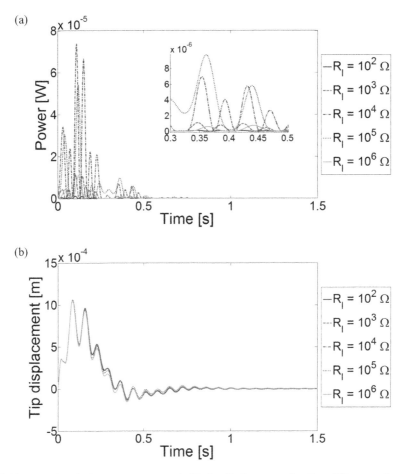

Figure 9.12 (a) Electrical power output and (b) tip displacement for five different values of load resistance at an airflow speed of 30 m/s

For the case of harmonic base excitation along with airflow excitation, Figures 9.15a and 9.15b, respectively, show the relative tip displacement FRFs (the ratio of the transverse displacement at the tip of the leading edge to the base displacement) and the electrical power output FRFs (per base acceleration) for several airflow speeds below the linear flutter speed. The peaks of the first and second bending modes are observed for zero airflow speed. Since the structure is symmetric with respect to the mid-chord, torsional modes are not excited by base motion in the absence of airflow. The modal frequency of the first torsional mode is 16.66 Hz (Figure 9.10) and no peak is observed at this frequency in Figures 9.15a and 9.15b when $U = 0$ m/s. This behavior is modified with increasing airflow speed as the bending and torsion modes are coupled. A peak is observed around 16 Hz for the airflow speed of 20 m/s in Figure 9.15a, which is not observed in Figure 9.15b. At this airflow speed, the resulting response is a coupled bending–torsion mode dominated by torsional motion. Hence the electrical output from torsional vibrations is canceled when continuous electrodes cover the piezoceramic

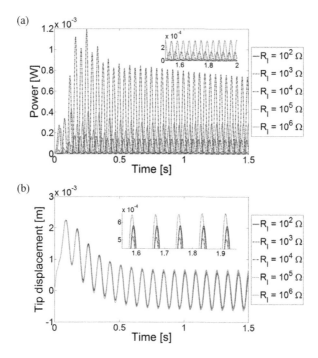

Figure 9.13 (a) Electrical power output and (b) tip displacement histories for five different values of load resistance at the short-circuit flutter speed (40 m/s)

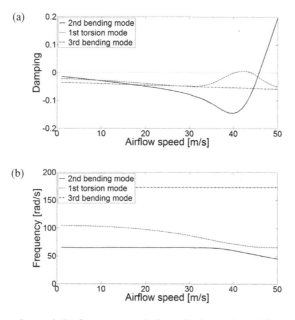

Figure 9.14 (a) Damping and (b) frequency evolution with increasing airflow speed close to short-circuit conditions ($R_l = 100\,\Omega$)

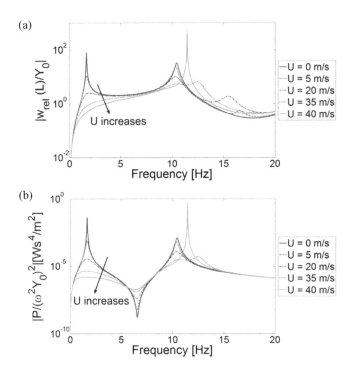

Figure 9.15 (a) Relative tip displacement (transmissibility) FRFs and (b) electrical power FRFs for several airflow speeds close to short-circuit conditions ($R_l = 100\,\Omega$)

layers [19]. For an airflow speed of 35 m/s, this peak of the coupled bending–torsion mode shifts to 13 Hz. However, the response at this airflow speed is dominated by bending motion, resulting in a peak in the power FRF as well (that is, no complete cancellation takes place). The aerodynamic damping vanishes at the flutter speed (40 m/s) and the modes are coupled at the flutter frequency (11.47 Hz), yielding the maximum tip displacement and power output. Note that the power could be optimized if segmented electrodes were used to avoid the cancellation of electrical outputs from the torsional component of the coupled bending–torsion motions of flutter (along with shorter piezoceramics due to the strain node of the second bending mode [16] – see Appendix D). The effect of increasing airflow speed on the overall damping is also observed in Figure 9.15. As discussed in the VLM simulations, aerodynamic damping increases from 5 m/s to 35 m/s dramatically due to the unsteady aerodynamic effects. The maximum aerodynamic damping is observed at an airflow speed of 35 m/s (among the speeds given in Figure 9.15), which is not preferred for energy harvesting.

Figures 9.16a and 9.16b, respectively, display FRFs of the relative tip displacement and the electrical power at the short-circuit flutter speed for two resistive loads. The first load resistance is close to short-circuit conditions (100 Ω) while the second one is the optimal load resistance (15.8 kΩ) for the maximum power output. Similar to the case of simple harmonic excitation (Chapter 3), the resonance frequency depends on the load resistance and the electrical power generation results in a shunt damping effect. Although there is a significant

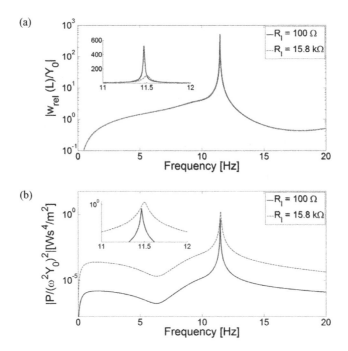

Figure 9.16   (a) Relative tip displacement FRFs and (b) electrical power FRFs with close-up views around the flutter frequency for a load resistance close to short-circuit conditions and for the optimal load resistance that gives the maximum power output

vibration attenuation for the 15.8 kΩ load in the vibration amplitude as shown in the close-up view of Figure 9.16a, the increase in the flutter speed is less than 1 m/s [19].

9.7   Summary

Lumped-parameter and distributed-parameter piezoaeroelastic models are presented for energy harvesting from airflow excitation. Lumped-parameter electromechanical modeling is given for a piezoaeroelastic section with plunge and pitch degrees of freedom. The focus is placed on the harmonic response at the flutter boundary (neutral stability condition). Model predictions are compared against the experimental results for a modified typical section with piezoceramic patches attached to plunge spring members. The effect of piezoelectric power generation on the linear flutter speed of the aeroelastic system is discussed. The importance of having nonlinearities in the aeroelastic system is addressed. Assumed-modes formulation for piezoaeroelastic modeling of a cantilever with embedded piezoceramics is also summarized as a distributed-parameter modeling approach. In addition, finite-element piezoaeroelastic energy harvester models using the vortex–lattice method and the doublet–lattice method are reviewed for time-domain and frequency-domain solutions, respectively, and a theoretical case study is discussed for the same structural configuration using these two solutions.

9.8 Chapter Notes

Other than possible integration into aerospace structures and flexible structural components, the concept of piezoaeroelastic energy harvesting discussed in this chapter is considered as a scalable alternative to bulky and relatively complicated windmill [25] design for low-power applications. In order to extract continuous electrical power from a piezoaeroelastic system, it is necessary to reach the minimum airflow speed of persistent oscillations. Hence, the designer should come up with an aeroelastic energy harvester that can undergo persistent oscillations at airflow speeds available in the application of interest. Just like the fact that reducing the dimensions of resonant energy harvesters operating under harmonic excitation increases the resonance frequencies, scaling down piezoaeroelastic energy harvesters tends to make them stiffer, shifting the flutter speed to higher values, and constituting the main design problem. As suggested by Erturk *et al.* [5] and reviewed in Section 9.3, the designer can introduce nonlinearities into the system (such as free play) for reducing the onset of LCOs through subcritical bifurcations. Recently, De Melo Anicézio *et al.* [26] modified the linear setup studied by Erturk *et al.* [5] (Section 9.2) by introducing a free play to the torsional DOF and obtained bilinear stiffness behavior. Lumped-vortex formulation is used in modeling [26] and it is shown theoretically and experimentally that the configuration with free play can reduce the minimum speed of persistent power compared to the linear configuration.

Prior to the work covered in this chapter, an early experimental effort of generating electricity from thin curved airfoils with macro-fiber composite (MFC) piezoceramics was presented by Erturk *et al.* [27]. For an alternative typical section configuration and aeroelastic modeling approach that uses the finite-state theory of Peters *et al.* [28], the reader is referred to papers by Bryant and Garcia [10, 11]. An extensive analysis of the energy harvesting potential for a foil–damper system is presented by Peng and Zhu [29] using a Navier–Stokes model without focusing on a specific transduction mechanism. A simple scalable configuration for energy harvesting from airflow excitation is proposed by St. Clair *et al.* [30] based on the self-excited vibrations of a piezoelectric beam embedded within a cavity under airflow. Tang *et al.* [31] present a rigorous analysis of the energy transfer from the fluid to the structure for self-excited vibrations due to axial flow over a cantilever. Kwon [32] investigates a simple and scalable T-shaped cantilever design with MFC piezoceramics and extracts persistent power under axial flow at relatively low airflow speeds. Robbins *et al.* [33], Pobering *et al.* [34], and Akaydin *et al.* [35, 36] discuss vortex-induced oscillations of piezoelectric cantilevers located behind bluff bodies through experiments and numerical simulations. The idea of utilizing the von Kármán vortex street [37] for piezoelectric energy harvesting dates back to the early work of Allen and Smits [38], who located a PVDF membrane behind a bluff body under water flow. Elvin and Elvin [39] theoretically investigate the flutter response of a cantilevered pipe with piezoceramics for power generation from liquid flow and its effect on the flutter instability.

References

1. Bisplinghoff, R.L. and Ashley, H. (1962) *Principles of Aeroelasticity*, John Wiley & Sons, Inc., New York.
2. Fung, Y.C. (1969) *Introduction to the Theory of Aeroelasticity*, Dover, New York.
3. Dowell, E.H., Curtiss, H.C. Jr., Scalan, R.H., and Sisto, F. (1978) *A Modern Course in Aeroelasticity*, Sijthoff and Noordhoff, The Hagues.
4. Hodges, D.H. and Pierce, G.A. (2002) *Introduction to Structural Dynamics and Aeroelasticity*, Cambridge University Press, New York.

5. Erturk, A., Vieira, W.G.R, De Marqui, C. Jr., and Inman, D.J. (2010) On the energy harvesting potential of piezoaeroelastic systems. *Applied Physics Letters*, **96**, 184103.

6. Theodorsen, T. (1935) General theory of aerodynamic instability and mechanism of flutter. Langley Memorial Aeronautical Laboratory, NACA-TR-496.

7. Agneni, A., Mastroddi, F., and Polli, G.M. (2003) Shunted piezoelectric patches in elastic and aeroelastic vibrations. *Computers and Structures*, **81**, 91–105.

8. Dowell, E.H. and Tang, D. (2002) Nonlinear aeroelasticity and unsteady aerodynamics. *AIAA Journal*, **40**, 1697–1707.

9. Dowell, E.H., Edwards, J., and Strganac, T. (2003) Nonlinear aeroelasticity. *AIAA Journal of Aircraft*, **40**, 857–874.

10. Bryant, M. and Garcia, E. (2009) Development of an aeroelastic vibration power harvester. *Proceedings of SPIE*, **7288**, 728812.

11. Bryant, M. and Garcia, E. (2009) Energy harvesting: a key to wireless sensor nodes. *Proceedings of SPIE*, **7493**, 74931W.

12. Chen, Y.M. and Liu, J.K. (2010) Homotopy analysis method for limit cycle oscillations of an airfoil with cubic nonlinearities. *Journal of Vibration and Control*, **16**, 163–179.

13. Conner, M.C., Tang, D.M., Dowell, E.H., and Virgin, L.N. (1997) Nonlinear behavior of a typical airfoil section with control surface freeplay: a numerical and experimental study. *Journal of Fluids and Structures*, **11**, 89–109.

14. Meirovitch, L. (2001) *Fundamentals of Vibrations*, McGraw-Hill, New York.

15. Erturk, A. and Inman, D.J. (2010) Assumed-modes formulation of piezoelectric energy harvesters: Euler-Bernoulli, Rayleigh and Timoshenko models with axial deformations. Proceedings of the ASME ESDA 10th Biennial Conference on Engineering Systems, Design and Analysis, Istanbul, Turkey, July 12–14, 2010.

16. Erturk, A., Tarazaga, P., Farmer, J.R., and Inman, D.J. (2009) Effect of strain nodes and electrode configuration on piezoelectric energy harvesting from cantilevered beams. *ASME Journal of Vibration and Acoustics*, **131**, 011010.

17. De Marqui, C. Jr., Erturk, A., and Inman, D.J. (2010) Piezoaeroelastic modeling and analysis of a generator wing with continuous and segmented electrodes. *Journal of Intelligent Material Systems and Structures*, **21**, 983–993.

18. De Marqui, C. Jr., Erturk, A., and Inman, D.J. (2009) An electromechanical finite element model for piezoelectric energy harvester plates. *Journal of Sound and Vibration*, **327**, 9–25.

19. De Marqui, C. Jr., Vieira, W.G.R., Erturk, A., and Inman, D.J. (2011) Modeling and analysis of piezoelectric energy harvesting from aeroelastic vibrations using the doublet-lattice method. *ASME Journal of Vibration and Acoustics*, **133**, 011003.

20. Katz, J. and Plotkin, A. (2001) *Low Speed Aerodynamics*, Cambridge University Press, Cambridge.

21. Caughey, T.K. and O'Kelly, M.E.J. (1965) Classical normal modes in damped linear dynamic systems. *ASME Journal of Applied Mechanics*, **32**, 583–588.

22. Albano, E. and Rodden, W. P. (1969) A doublet-lattice method for calculating lift distributions on oscillating surfaces in subsonic flow. *AIAA Journal*, **7**, 279–285.

23. Hassig, H. J. (1971) An approximate true damping solution of the flutter equation by determinant iteration. *Journal of Aircraft*, **8**, 885–889.

24. Erturk, A. and Inman, D.J. (2008) Issues in mathematical modeling of piezoelectric energy harvesters. *Smart Materials and Structures*, **17**, 065016.

25. Priya, S., Chen, C.T., Fye, D., and Zahnd, J. (2005) Piezoelectric windmill: a novel solution to remote sensing. *Japanese Journal of Applied Physics*, **44**, L104–L107.

26. De Melo Anicézio, M., De Marqui, Jr., C., Erturk, A., and Inman, D.J. (2011) Nonlinear modeling and analysis of a piezoaeroelastic energy harvester, Proceedings of the 14th International Symposium on Dynamic Problems of Mechanics, Sao Sebastiao, SP, Brazil, March 13–18 2011.

27. Erturk, A., Bilgen, O., Fontenille, M., and Inman, D.J. (2008) Piezoelectric energy harvesting from macro-fiber composites with an application to morphing wing aircraft. Proceedings of the 19th International Conference on Adaptive Structures and Technologies, Monte Verità, Ascona, Switzerland, October 6–9, 2008.

28. Peters, D.A., Karunamoorthy, S., and Cao, W.M. (1995) Finite state induced flow models, Part I: two dimensional thin airfoil. *Journal of Aircraft*, **32**, 313–322.

29. Peng, Z. and Zhu, Q. (2009) Energy harvesting through flow-induced oscillations of a foil. *Physics of Fluids*, **21**, 123602.

30. St. Clair, D., Bibo, A., Sennakesavababu, V.R., Daqaq, M.F., and Li, G. (2010) A scalable concept for micropower generation using flow-induced self-excited oscillations. *Applied Physics Letters*, **96**, 144103.

31. Tang, L., Paidoussis, M., and Jiang, J. (2009) Cantilevered flexible plates in axial flow: energy transfer and the concept of flutter-mill. *Journal of Sound and Vibration*, **326**, 263–276.

32. Kwon, S.D. (2010) A T-shaped piezoelectric cantilever for fluid energy harvesting. *Applied Physics Letters*, **97**, 164102.

33. Robbins, W.P., Morris, D., Marusic, I., and Novak, T.O. (2006) Wind-generated electrical energy using flexible piezoelectric materials. Proceedings of ASME IMECE 2006, Chicago, IL, October 5–10, 2006.

34. Pobering, S., Ebermeyer, S., and Schwesinger, N. (2009) Generation of electrical energy using short piezoelectric cantilevers in flowing media. *Proceedings of SPIE*, **7288**, 728807.

35. Akaydin, H.D., Elvin, N., and Andreopoulos, Y. (2010) Wake of a cylinder: a paradigm for energy harvesting with piezoelectric materials. *Experiments in Fluids*, **49**, 291–304.

36. Akaydin, H.D., Elvin, N., and Andreopoulos, Y. (2010) Energy harvesting from highly unsteady fluid flows using piezoelectric materials. *Journal of Intelligent Material Systems and Structures*, **21**, pp. 1263–1278.

37. von Kármán, T. (1963) *Aerodynamics*, McGraw-Hill, New York.

38. Allen, J.J. and Smits, A.J. (2001) Energy harvesting eel. *Journal of Fluids and Structures*, **15**, 629–640.

39. Elvin, N.G. and Elvin, A.A. (2009) The flutter response of a piezoelectrically damped cantilever pipe. *Journal of Intelligent Material Systems and Structures*, **20**, 2017–2026.

10

Effects of Material Constants and Mechanical Damping on Power Generation

This chapter investigates the effects of material constants and mechanical damping on piezoelectric energy harvesting. The first part of the discussion compares the resonant power generation performances of soft ceramics and soft single crystals. In these comparisons, the focus is placed on the most popular piezoceramics (PZT-5A and PZT-5H) and the single crystals (PMN-PT and PMN-PZT).[1] The goal is to understand the effects of piezoelectric, elastic, and dielectric constants as well as mechanical damping on piezoelectric power generation and to clarify whether or not the substantially large piezoelectric strain constants (particularly the d_{31} constant) of single crystals result in a substantially large power generation performance compared to commonly employed piezoceramics. The second part of the discussion compares soft and hard ceramics (PZT-5H and PZT-8) and soft and hard crystals (PMN-PZT and PMN-PZT-Mn)[2] for resonant and off-resonant energy harvesting. Both of these discussions comparing ceramics against singe crystals as well as their soft and hard counterparts are associated with extensive theoretical case studies by using the experimentally validated analytical model for the simulations. Finally, an experimental comparison between brass-reinforced PZT-5A and PZT-5H bimorphs is given in order to verify some of the conclusions drawn in this chapter.

10.1 Effective Parameters of Various Soft Ceramics and Single Crystals

10.1.1 Properties of Various Soft Ceramics and Single Crystals

Typical properties of PZT-5A and PZT-5H piezoceramics can be found in several sources on the Web [1] or in the literature [2] (see Appendix E). At room temperature, PZT-5A

---

[1] PMN stands for *lead magnesium niobate*.

[2] Mn stands for *manganese*.

Piezoelectric Energy Harvesting, First Edition. Alper Erturk and Daniel J. Inman.
© 2011 John Wiley & Sons, Ltd. Published 2011 by John Wiley & Sons, Ltd.

Table 10.1 Properties of interest for various ceramics and single crystals

	d_{31} (pm/V)	s_{11}^E (pm²/N)	$\varepsilon_{33}^T/\varepsilon_0$	ρ (kg/m³)
PZT-5A	−171	16.4	1700	7750
PZT-5H	−274	16.5	3400	7500
PMN-PT (30% PT)	−921	52	7800	8040
PMN-PT (33% PT)	−1330	69	8200	8060
PMN-PZT	−2252	127	5000	7900
Average	**−989.6**	**56.2**	**5220**	**7850**

has $d_{31} = -171$ pm/V (piezoelectric strain constant), $s_{11}^E = 16.4$ pm²/N (elastic compliance at constant electric field), and $\varepsilon_{33}^T/\varepsilon_0 = 1700$ (dielectric constant), whereas PZT-5H has $d_{31} = -274$ pm/V, $s_{11}^E = 16.5$ pm²/N, and $\varepsilon_{33}^T/\varepsilon_0 = 3400$ (where $\varepsilon_0 = 8.854$ pF/m is the permittivity of free space [3]).

The single-crystal technology is relatively new compared to the commonly employed PZT-5A and PZT-5H piezoceramics. Limited literature is available for the material data of single crystals. Cao *et al.* [4] reported the relevant constants for a PMN-PT with 30% PT as $d_{31} = -921$ pm/V, $s_{11}^E = 52$ pm²/N, and $\varepsilon_{33}^T/\varepsilon_0 = 7800$, whereas they reported $d_{31} = -1330$ pm/V, $s_{11}^E = 69$ pm²/N, and $\varepsilon_{33}^T/\varepsilon_0 = 8200$ for a PMN-PT with 33% PT (poled along [001]). Properties of a relatively new single crystal PMN-PZT tested by Erturk *et al.* [5] are reported by the manufacturer [6] in the data sheet as $d_{31} = -2252$ pm/V, $s_{11}^E = 127$ pm²/N, and $\varepsilon_{33}^T/\varepsilon_0 = 5000$.[3]

For use in the thin-beam type of analytical model (given in Chapter 3), the aforementioned properties of these five different piezoelectric materials (PZT-5A, PZT-5H, PMN-PT with 30% PT, PMN-PT with 33% PT, and PMN-PZT) are as listed in Table 10.1 along with the average value of each property. The mass densities of these piezoelectric materials are also given in this table.

As can be seen in Figure 10.1, the piezoelectric strain constant increases by more than an order of magnitude from PZT-5A to PMN-PZT. It is observed that the large piezoelectric strain constant comes with large elastic compliance as plotted in Figure 10.2. A difference of approximately an order of magnitude is seen between the elastic compliance values of PZT-5A to PMN-PZT. The dielectric constants are larger for single crystals; however, the difference is not as dramatic as in the piezoelectric strain constants or the elastic compliance values. The mass densities of these piezoelectric materials are very similar to each other as given in Table 10.1. It is worth adding that the mechanical quality factors ($Q_m = 1/2\zeta$ where ζ is the mechanical damping ratio for the vibration mode of interest) of these relatively soft ceramics and single crystals are typically in the range $Q_m < 100$ [7, 8].[4]

---

[3] Composition of this PMN-PZT is the company's proprietary information.

[4] The mechanical quality factors of hard ceramics and hard single crystals are significantly different compared to their soft counterparts. This is discussed in Sections 10.3 and 10.4.

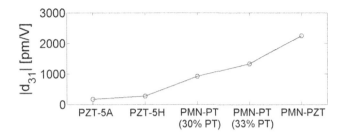

Figure 10.1   Variation of the piezoelectric strain constant for various ceramics and crystals

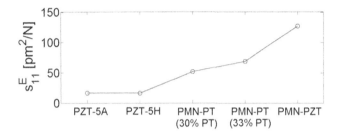

Figure 10.2   Variation of the elastic compliance for various ceramics and crystals

10.1.2   Plane-Stress Piezoelectric, Elastic, and Permittivity Constants for a Thin Beam

For the plane-stress conditions of a thin beam, the elastic modulus at constant electric field is obtained from $\bar{c}_{11}^E = 1/s_{11}^E$, the effective piezoelectric stress constant is obtained from $\bar{e}_{31} = d_{31}/s_{11}^E$, and the effective permittivity value at constant electric field is given by $\bar{\varepsilon}_{33}^S = \varepsilon_{33}^T - d_{31}^2/s_{11}^E$ (Appendix A.2). These parameters are calculated using the data in Table 10.1 and are listed in Table 10.2. It is important to note that $\bar{e}_{31}$ is the piezoelectric constant that appears in the reduced (plane-stress) piezoelectric constitutive relations of the analytical model (Chapter 3), hence in the modal electromechanical coupling terms (Table 3.1). The elastic modulus at constant electric field is simply the reciprocal of the elastic compliance shown in Figure 10.2 and is plotted in Figure 10.3 for these piezoelectric materials. Clearly, the single crystals with large d_{31} constants have low elastic moduli. Since it is the product of d_{31} and $\bar{c}_{11}^E$ (which exhibit opposite trends in Figures 10.1 and 10.3), the plane-stress piezoelectric constant $\bar{e}_{31}$ for the piezoelectric materials considered here has values in the same order of magnitude (Figure 10.4). For instance, even though the d_{31} constant of PMN-PZT is more than 10 times that of PZT-5A, its $\bar{e}_{31}$ constant is only 1.7 times that of the latter. From the mathematical point of view, d_{31} never appears alone in the formulation given in Chapter 3. It is the $\bar{e}_{31}$ constant that appears in the coupling terms (i.e., the multiplication of d_{31} and $\bar{c}_{11}^E$ for a thin beam). For this reason, *very high d_{31} constants of single crystals may not necessarily imply very high power output* as their large elastic compliance negatively affects the resulting electromechanical coupling. The PMN-PT with 33% PT has the largest $\bar{e}_{31}$ among the samples considered here.

Table 10.2   Beam-type plane-stress properties of various ceramics and single crystals

	$\bar{c}_{11}^{E}$ (GPa)	$\bar{e}_{31}$ (C/m^2)	$\bar{\varepsilon}_{33}^{S}$ (nF/m)
PZT-5A	61.0	−10.4	13.3
PZT-5H	60.6	−16.6	25.6
PMN-PT (30% PT)	19.2	−17.7	52.7
PMN-PT (33% PT)	14.5	−19.3	47.0
PMN-PZT	7.9	−17.7	4.34
Average	**32.6**	**−16.3**	**28.6**

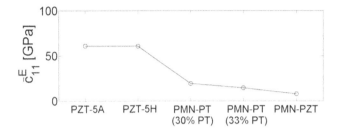

Figure 10.3   Variation of the plane-stress elastic modulus for various ceramics and crystals

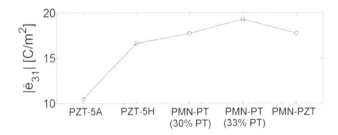

Figure 10.4   Variation of the plane-stress piezoelectric stress constant for various ceramics and crystals

Variation of the plane-stress permittivity at constant strain is also plotted for different piezoelectric materials in Figure 10.5. The dielectric constant ($\varepsilon_{33}^{T}/\varepsilon_{0}$) increases from PZT-5A until PMN-PT with 33% PT in Table 10.2. However, the permittivity component at constant strain ($\bar{\varepsilon}_{33}^{S}$) decreases after PMN-PT with 30% PT. It is interesting to note that the plane-stress permittivity of PMN-PZT at constant strain is lower than that of all the other materials considered here (which is due to its large d_{31} value).

10.2   Theoretical Case Study for Performance Comparison of Soft Ceramics and Single Crystals

10.2.1   Properties of the Bimorph Cantilevers

In this section, power generation performances of bimorph cantilevers with identical dimensions and substructure materials using PZT-5A, PZT-5H, PMN-PT (with 30% PT and with

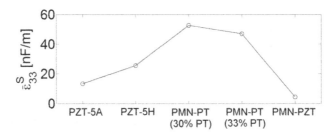

Figure 10.5 Variation of the plane-stress permittivity constant for various ceramics and crystals

33% PT), and PMN-PZT are compared. The piezoelectric layers of each bimorph cantilever are assumed to be connected in parallel (Figure 3.1b) and no tip mass is used. The substructure material in all cases is aluminum with identical dimensions and properties ($Y_s = 70$ GPa, $\rho = 2700$ kg/m^3). For all cases considered here, the dimensions of the bimorph cantilever are as given in Table 10.3. Recall that all these ceramics and single crystals are rather soft and their mechanical quality factors are usually in the range $Q_m < 100$ [7, 8]. Therefore, the overall modal mechanical damping ratio is assumed to be identical ($\zeta_1 = 0.01$) for all bimorphs (except for Section 10.2.7) to study the effects of other parameters involved.

The analytical model developed in Chapter 3 and experimentally validated in Chapter 4 is used here. As the electrical load resistance is increased from zero to infinity, the resonance frequencies of a bimorph shift from the short-circuit resonance frequencies to the open-circuit resonance frequencies. The short-circuit resonance frequency ω_r^{sc} (for $R_l \rightarrow 0$) of the rth mode in the voltage FRF is the undamped natural frequency ω_r as shown in Chapter 5. As $R_l \rightarrow \infty$, the resonance frequency moves to the open-circuit resonance frequency (ω_r^{oc}). The short-circuit and open-circuit resonance frequencies of the fundamental vibration mode ($r = 1$) of the five bimorphs with identical dimensions are listed in Table 10.4 (where

Table 10.3 Geometric properties of the bimorph cantilevers

	Piezoelectric	Substructure
Length (L) (mm)	40	40
Width (b) (mm)	6	6
Thickness ($h_{\bar{p}}, h_{\bar{s}}$) (mm)	0.2 (each)	0.1

Table 10.4 Fundamental short-circuit and open-circuit resonance frequencies of the bimorph cantilevers

	f_1^{sc} (Hz)	f_1^{oc} (Hz)
PZT-5A cantilever	151.9	157.0
PZT-5H cantilever	153.8	160.5
PMN-PT (30% PT) cantilever	84.7	90.7
PMN-PT (33% PT) cantilever	73.8	82.2
PMN-PZT cantilever	55.8	83.6

$f = \omega/2\pi$). Recall from Chapter 5 that the resonance frequency shift from the short-circuit to the open-circuit conditions is a measure of electromechanical coupling (Equation (5.31)). Among the piezoelectric materials considered here, with the second largest $\bar{e}_{31}$ constant and the smallest $\bar{\varepsilon}_{33}^S$ (as well as the smallest ω_r due to low stiffness), PMN-PZT bimorph exhibits the largest relative resonance frequency shift (as expected from Equation (5.14)).

10.2.2 Performance Comparison of the Original Configurations

Figure 10.6a displays the power versus load resistance curves of these five bimorphs for excitation at the fundamental short-circuit resonance frequency of each bimorph. The outputs are normalized with respect to the base acceleration (in terms of the gravitation acceleration, $g = 9.81$ m/s^2). The PMN-PZT bimorph gives the largest power output (1.73 mW/g^2) and it is followed by the PMN-PT bimorph with 33% PT (1.33mW/g^2) and the PMN-PT bimorph with 30% PT (1.15 mW/g^2). The PZT-5A bimorph generates 0.61 mW/g^2 whereas the PZT-5H bimorph generates 0.59 mW/g^2. The order of the power outputs might seem to agree with the order of the d_{31} constants (Figure 10.1). However, in spite of its larger d_{31} constant, the PZT-5H bimorph gives slightly lower power output compared to the PZT-5A bimorph. Moreover, the largest power output is less than only three times the smallest power output (unlike the difference of an order of magnitude between the d_{31} constants).

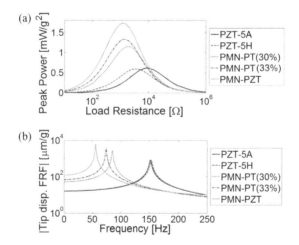

Figure 10.6 (a) Power vs. load resistance curves for excitation at the fundamental short-circuit resonance frequency of each bimorph and (b) vibration FRFs of the bimorphs for $R_l \rightarrow 0$ ($\zeta_1 = 1\%$ for all bimorphs)

Due to the large variance of the elastic compliance constants (Figure 10.2), the natural frequencies of the bimorphs differ considerably, except for the PZT-5A and PZT-5H bimorphs, which have similar elastic compliances. Figure 10.6b shows the tip vibration response of the bimorphs in short-circuit conditions. Clearly, as it has the lowest stiffness (and therefore the lowest natural frequency), the PMN-PZT bimorph exhibits the largest dynamic flexibility. The tip deflection of the PMN-PZT bimorph at resonance is about 7.5 times that of the PZT-5H bimorph for the same mechanical damping ratio (in the absence of any piezoelectrically induced damping, since $R_l \rightarrow 0$ in Figure 10.6b). Therefore, since the power curves in Figure 10.6a are obtained for the resonant excitation of each bimorph, it is not possible to claim that the order of the maximum power outputs in Figure 10.6a is due to the order of the d_{31} constants. Considering the average values of the parameters listed in Table 10.1, artificial case studies are discussed next to visualize the effect of each parameter.

10.2.3 Effect of the Piezoelectric Strain Constant

In order to understand the role of the d_{31} constant in piezoelectric power generation, $d_{31} = -989.6$ pm/V is assumed for all bimorphs (which is the average of the d_{31} constants of these five piezoelectric materials as shown in Table 10.1). The geometry of the bimorph and the modal mechanical damping ratio are unchanged. Figure 10.7 shows the simulation results for this *artificial* case. The optimum loads are highly affected (as the optimum electrical load depends on the electromechanical coupling in Equation (5.42)) but the maximum power outputs are affected very slightly and the order of the power outputs is the same for the bimorphs. This observation supports the idea that the large power outputs in Figure 10.6a can be due to large dynamic flexibilities (originating from the large elastic compliance values) of the respective bimorphs at their individual resonances rather than the substantial difference in the d_{31} constants.

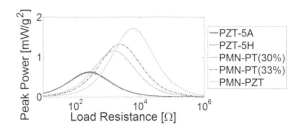

Figure 10.7 Power vs. load resistance curves for excitation at the fundamental short-circuit resonance frequency of each bimorph ($\zeta_1 = 1\%$ and $d_{31} = -989.6$ pm/V for all bimorphs)

10.2.4 Effect of the Elastic Compliance

It is useful practice to assume that all these bimorphs have the average compliance ($s_{11}^E = 56.2$ pm^2/N) and the average mass density ($\rho = 7850$ kg/m^3) values given in Table 10.1. Then the fundamental short-circuit resonance frequencies of these bimorphs become identical in this second artificial case. Thus, the dynamic flexibilities of these bimorphs for resonant excitation are expected to be very similar in this case. Along with the remaining parameters, d_{31} values are kept at their original values and the power curves in Figure 10.8a are obtained for resonant excitation. Figure 10.8b verifies that the dynamic flexibilities of these samples are indeed identical for $R_l \rightarrow 0$. The maximum power output is obtained with the PMN-PZT bimorph as 1.16 mW/g^2 and the minimum power output is obtained with the PZT-5A bimorph as 0.94 mW/g^2. In this particular case where the natural frequencies and therefore the dynamic flexibilities are forced to be identical, the power outputs of the bimorphs are very similar (although the optimum electrical loads differ considerably).

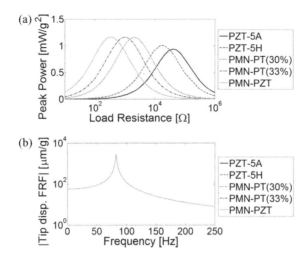

Figure 10.8 (a) Power vs. load resistance curves for excitation at the fundamental short-circuit res-onance frequency of each bimorph and (b) vibration FRFs of the bimorphs for $R_l \to 0$ ($\zeta_1 = 1\%$, $s^E_{11} = 56.2$ pm^2/N, and $\rho = 7850$ kg/m^3 for all bimorphs)

The discussion so far shows that the difference of an order of magnitude in the electrical power outputs is *not* like that between the d_{31} constants. The electrical power outputs differ in the same order of magnitude (just like the $\bar{e}_{31}$ constants) and the dynamic flexibility of the cantilever plays an important role.

10.2.5 Effect of the Permittivity Constant

The last artificial case is to assume that the constant stress dielectric constants are identical and equal to the average value given in Table 10.1 ($\varepsilon^T_{33}/\varepsilon_0 = 5220$) to study how the difference in the relative permittivity affects the results. As can be seen in Figure 10.9, the qualitative order (in terms of the amplitude-wise results) is not changed significantly compared to the original graph given by Figure 10.6a except for the PZT-5A and PZT-5H bimorphs. In Figure 10.9, the PZT-5H bimorph gives slightly larger power output. Therefore the slightly larger power output of the PZT-5A bimorph (compared to the PZT-5H bimorph) in the original case (Figure 10.6a) is partially due to its smaller relative permittivity, in addition to its slightly larger mass and dynamic flexibility.

10.2.6 Effect of the Overhang Length

For the original piezoelectric, elastic, and dielectric properties of these bimorphs (given in Tables 10.1 and 10.2), it is possible to obtain the same short-circuit natural frequencies for different overhang lengths. Suppose that one is interested in designing these bimorphs for an excitation frequency of 60 Hz and the length dimension can be varied for this purpose (all the other parameters are the original parameters in Tables 10.1–10.3). In order to obtain a short-circuit resonance frequency of 60 Hz for the identical cross-sectional geometry described in Table 10.3, the lengths of the bimorphs must be as given in Figure 10.10a. Note that the tip

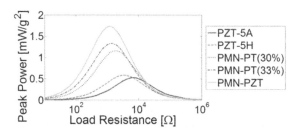

Figure 10.9   Power vs. load resistance curves for excitation at the fundamental short-circuit resonance frequency of each bimorph ($\zeta_1 = 1\%$ and $\varepsilon_{33}^T/\varepsilon_0 = 5220$ for all bimorphs)

vibration response amplitudes of the bimorphs to the same excitation input are almost identical (Figure 10.10b). The original picture (Figure 10.6a) for the identical lengths (of 40 mm) is now reversed in Figure 10.10a. In this particular design problem of tuning the resonance frequency of a bimorph cantilever to an excitation frequency (with design flexibility in the length dimension), the PZT-5A and PZT-5H bimorphs generate larger power. It is important to note that, since the base acceleration is the constant input, larger overhang length means larger excitation of the structure due to increased mass (see Equation (3.32)). This is the particular reason for the substantial increase in the power outputs of the PZT-5A and PZT-5H bimorphs.

Similar to the discussion given in Section 10.2.4 (identical elastic compliances and mass densities), this demonstration also agrees with the fact that the larger power outputs of the single crystals in the original case of Figure 10.6a are mainly due to their larger dynamic flexibilities depicted in Figure 10.6b. When the overhang lengths of the bimorphs with ceramics are increased to achieve the same dynamic flexibility, they generate larger power compared to

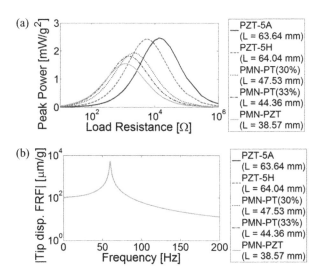

Figure 10.10   (a) Power vs. load resistance curves for excitation at the fundamental short-circuit resonance frequency of each bimorph and (b) vibration FRFs of the bimorphs for $R_l \rightarrow 0$ ($\zeta_1 = 1\%$ and the lengths are chosen to satisfy $f_1^{sc} = 60$ Hz for all bimorphs)

those with single crystals due to the increased mass (hence increased mechanical forcing due to base excitation).

10.2.7 Effect of the Mechanical Damping

The final discussion in this section demonstrates the sensitivity of the results to mechanical damping (which is not very easy to predict and control in practice). So far the simulations have assumed identical mechanical damping since all of the five piezoelectric materials are soft and the substructure material is identical in all cases. For the original parameters of this case study (which yield Figure 10.6a), if one assumes the damping ratios shown in Figure 10.11 (instead of $\zeta_1 = 1\%$ for all bimorphs), identical maximum power outputs are obtained. Note that these damping ratios are typical values one can identify in practice due to clamping conditions, bonding layer, and so on. Although the individual mechanical loss factors (related to the reciprocal of the mechanical quality factors) of the piezoelectric and the substructure layers can provide some insight, adhesive layers and clamped interfaces are important sources of mechanical damping which can dominate the damping due to the loss factors of the individual layers. The particular reason for the sensitivity of the results to mechanical damping is because the electrical power outputs already have the same order of magnitude. For instance, the favorable power output of the flexible PMN-PZT bimorph in Figure 10.6a can easily become less than that of the PZT-5H bimorph if the bonding layer and/or the clamped boundary of the PMN-PZT bimorph causes larger mechanical damping.

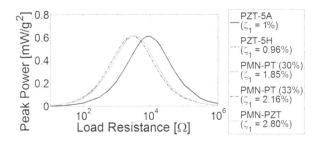

Figure 10.11 Power vs. load resistance curves for excitation at the fundamental short-circuit resonance frequency of each bimorph showing the sensitivity of the power output to mechanical damping ratio

10.3 Effective Parameters of Typical Soft and Hard Ceramics and Single Crystals

10.3.1 Properties of the Soft Ceramic PZT-5H and the Hard Ceramic PZT-8

The properties of the soft ceramic PZT-5H and the hard ceramic PZT-8 are listed in Table 10.5 [7].[5] As can be seen from the table, the mechanical quality factor (Q_m) of the hard

---

[5] The properties of PZT-5H and PZT-8 in this table are from a specific manufacturer [7]. Note that the elastic compliance of PZT-5H is slightly different from that given in the typical PZT-5H parameters list of Appendix E (the manufacturer's data is used for the comparisons given here).

Table 10.5 Properties of the soft ceramic PZT-5H and the hard ceramic PZT-8

	PZT-5H	PZT-8
d_{31} (pm/V)	-274	-97
s_{11}^E (pm^2/N)	16.4	11.5
$\varepsilon_{33}^T/\varepsilon_0$	3400	1000
$\bar{c}_{11}^E$ (GPa)	61.0	87.0
$\bar{e}_{31}$ (C/m^2)	-16.7	-8.4
$\bar{\varepsilon}_{33}^S$ (nF/m)	25.5	8.0
Q_m	65	1000

ceramic PZT-8 is more than 15 times that of the soft ceramic PZT-5H. On the other hand, the piezoelectric strain constant d_{31} of PZT-5H is about three times that of PZT-8. Due to the difference between the elastic compliance values, the effective (plane-stress) piezoelectric stress constant of PZT-5H is only about twice that of PZT-8 ($\bar{e}_{31} = d_{31}/s_{11}^E$). The constant stress dielectric constant of PZT-5H is about three times that of PZT-8 and its reflection to the plane-stress permittivity at constant strain ($\bar{\varepsilon}_{33}^S = \varepsilon_{33}^T - d_{31}^2/s_{11}^E$) is in the same order of magnitude.

10.3.2   Properties of the Soft Single-Crystal PMN-PZT and the Hard Single-Crystal PMN-PZT-Mn

The properties of the soft single-crystal PMN-PZT and the hard single-crystal PMN-PZT-Mn are listed in Table 10.6 [8].[6] The mechanical quality factor of PMN-PZT-Mn is more than 10 times that of PMN-PZT while the d_{31} constant of PMN-PZT is only 40% larger than that of PMN-PZT-Mn. Due to the difference between the elastic compliances, the plane-stress piezoelectric stress constants are very similar. The constant stress dielectric constant of the hard counterpart is larger, as in the case of the ceramics given in Table 10.5. The resulting plane-stress permittivity of PMN-PZT at constant strain is about 44% larger than that of PMN-PZT-Mn. Therefore, as in the previous case (PZT-5H versus PZT-8), the dramatic difference between the properties is for the mechanical quality factor (which is very important for resonant energy harvesting as it controls the resonance region).

10.4   Theoretical Case Study for Performance Comparison of Soft and Hard Ceramics and Single Crystals

10.4.1   Properties of the Bimorph Cantilevers

This theoretical case study focuses on bimorph cantilevers made of PZT-5H, PZT-8, PMN-PZT, and PMN-PZT-Mn types of piezoelectric materials. The geometric properties of the bimorph cantilevers are given in Table 10.7 and the relevant material properties can be found in Tables

---

[6] The properties of the soft PMN-PZT and its Mn-doped hard version PMN-PZT-Mn are from a paper by Zhang *et al.* [8]. It should be clear from Tables 10.1 and 10.6 that that the PMN-PZT investigated by Zhang *et al.* [8] has a different composition compared to the one quoted from the data sheet of a manufacturer in Section 10.1.1.

Table 10.6 Properties of the soft crystal PMN-PZT and the hard crystal PMN-PZT-Mn

	PMN-PZT	PMN-PZT-Mn
d_{31} (pm/V)	-718	-513
s_{11}^E (pm^2/N)	62.0	42.6
$\varepsilon_{33}^T/\varepsilon_0$	4850	3410
$\bar{c}_{11}^E$ (GPa)	16.1	23.5
$\bar{e}_{31}$ (C/m^2)	-11.6	-12.0
$\bar{\varepsilon}_{33}^S$ (nF/m)	34.6	24.0
Q_m	100	1050

10.5 and 10.6. Series connection (Figure 3.1a) is used for each bimorph cantilever. It is assumed that the entire mechanical loss of each piezoelectric cantilever is due to its mechanical quality factor and the bimorph cantilevers studied in this section have no substructure layer to be close to this assumption. Therefore, the dissipative effects of external damping and clamping conditions are assumed to be identical (and negligible), and the bonding between the layers for all cantilevers is assumed to be perfect (no sliding or dissipation), so that the active materials can be compared under the same conditions by considering their internal mechanical loss only. The damping ratio of the fundamental vibration mode for each cantilever is given in Table 10.8 (based on the quality factors listed in Tables 10.5 and 10.6).

Using the parameters given in Tables 10.5–10.8, the fundamental short-circuit and open-circuit resonance frequencies of the bimorph cantilevers are obtained as listed in Table 10.9. Recall that the fundamental short-circuit resonance frequency is the first resonance frequency for $R_l \to 0$ (short-circuit conditions) while the fundamental open-circuit resonance frequency is that for $R_l \to \infty$ (open-circuit conditions) in the voltage FRF.

Table 10.7 Geometric properties of the bimorph cantilevers (no substructure layer)

	Piezoelectric	Substructure
Length (L) (mm)	25	—
Width (b) (mm)	3	—
Thickness ($h_{\tilde{p}}$, $h_{\tilde{s}}$) (mm)	0.1	—

Table 10.8 Mechanical damping ratio for the fundamental vibration mode of each bimorph cantilever

Cantilever type	$\zeta_1 \cong 1/2Q_m$
PZT-5H	0.77%
PZT-8	0.050%
PMN-PZT	0.50%
PMN-PZT-Mn	0.048%

Table 10.9 Fundamental short-circuit and open-circuit resonance frequencies of the bimorph cantilevers

	f_1^{sc} (Hz)	f_1^{oc} (Hz)
PZT-5H	147.9	153.6
PZT-8	174.9	178.8
PMN-PZT	74.3	78.1
PMN-PZT-Mn	89.7	94.5

In the following comparisons, two excitation frequencies are considered for each cantilever to demonstrate resonant and off-resonant excitation. The resonant excitation case corresponds to excitation at the fundamental short-circuit resonance frequency ($\omega = \omega_1^{sc}$). The off-resonant excitation case corresponds to excitation at a quarter of the fundamental short-circuit resonance frequency ($\omega = \omega_1^{sc}/4$) of the respective cantilever. Therefore, the off-resonant excitation case is selected arbitrarily (but below the fundamental resonance frequency) and is sufficiently far away from the region of resonance. Note that, since the geometries of the bimorph configurations are forced to be identical, these two frequencies are different for each cantilever. However, they are all in the frequency range of 0–200 Hz.

10.4.2 Comparison of Soft and Hard Ceramics: PZT-5H vs. PZT-8

For $R_l \rightarrow 0$, the tip displacement - to - base acceleration FRFs of the PZT-5H and PZT-8 cantilevers are shown in Figure 10.12a. Since the cantilevers are close to short-circuit conditions, the frequencies of the fundamental peaks correspond to the fundamental short-circuit resonance frequencies. Therefore, the *circles* and the *squares* on each FRF are the points of resonant and off-resonant excitation, respectively. The open-circuit voltage FRFs are given in Figure 10.12b. The frequencies of peak response in Figure 10.12b are the fundamental open-circuit resonance frequencies of each cantilever. In both figures, the relatively sharp peaks

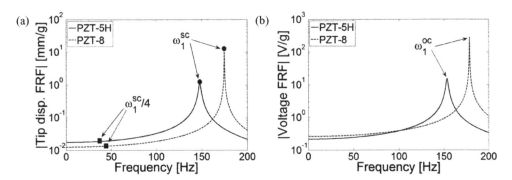

Figure 10.12 (a) Tip displacement - to - base acceleration FRFs in short-circuit conditions displaying the resonant (circle) and off-resonant (square) excitation frequencies of interest; (b) Open-circuit voltage output - to - base acceleration FRFs of the (soft) PZT-5H and the (hard) PZT-8 bimorph cantilevers

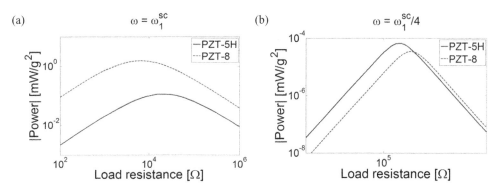

Figure 10.13 Comparison of the (soft) PZT-5H and the (hard) PZT-8 bimorph cantilevers for power generation under (a) resonant excitation and (b) low-frequency off-resonant excitation

of the hard PZT-8 cantilever are due to its low mechanical damping (i.e., high mechanical quality factor).

For resonant excitation ($\omega = \omega_1^{sc}$) of each cantilever, the variation of the power amplitude with load resistance is shown in Figure 10.13a. The peak power of the PZT-8 cantilever is about 13.4 times that of the PZT-5H cantilever for resonant excitation (due to the low mechanical damping in the former). On the other hand, for the low-frequency off-resonant excitation ($\omega = \omega_1^{sc}/4$) of each cantilever, the peak power of the PZT-5H cantilever is about two times that of the PZT-8 cantilever as shown in Figure 10.13b. Since the mechanical damping is not expected to be effective for this off-resonant excitation case, the superior performance of the cantilever with soft piezoceramic is attributed to its large effective piezoelectric constant and elastic compliance (Table 10.5).

10.4.3 Comparison of Soft and Hard Single Crystals: PMN-PZT vs. PMN-PZT-Mn

The tip displacement - to - base acceleration FRFs of the PMN-PZT and PMN-PZT-Mn cantilevers are shown in Figure 10.14a for $R_l \rightarrow 0$. The frequencies of the fundamental peaks in these FRFs correspond to the fundamental short-circuit resonance frequencies. The resonant and the off-resonant excitation frequencies are shown in Figure 10.14a. The open-circuit voltage FRFs are given in Figure 10.14b (where the fundamental open-circuit resonance frequencies are also shown). The relatively sharp peaks in the mechanical and the electrical FRFs of the PMN-PZT-Mn cantilever are expected due to its low mechanical damping.

For resonant excitation ($\omega = \omega_1^{sc}$) of each cantilever, the variation of the power amplitude with load resistance is shown in Figure 10.15a. The peak power of the PMN-PZT-Mn cantilever is about 8.6 times that of the PMN-PZT cantilever for resonant excitation as a result of the difference between the mechanical quality factors. For the low-frequency off-resonant excitation ($\omega = \omega_1^{sc}/4$) of each cantilever, however, the peak power of the PMN-PZT cantilever is only 14% larger than that of the PMN-PZT-Mn cantilever (Figure 10.15b). Therefore, in this case, the soft cantilever (PMN-PZT) does not exhibit a considerably superior off-resonant performance. According to Table 10.5, the effective piezoelectric constants of these soft and

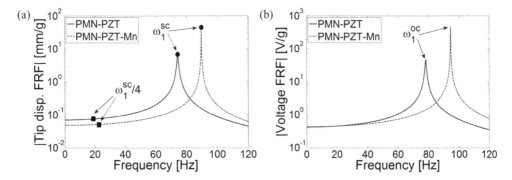

Figure 10.14   (a) Tip displacement - to - base acceleration FRFs in short-circuit conditions displaying the resonant (circle) and off-resonant (square) excitation frequencies of interest; (b) Open-circuit voltage output - to - base acceleration FRFs of the (soft) PMN-PZT and the (hard) PMN-PZT-Mn bimorph cantilevers

hard single crystals are very similar, yielding the similar low-frequency performance results shown in Figure 10.15b.

10.4.4   Overall Comparison of Ceramics (PZT-5H, PZT-8) and Single Crystals (PMN-PZT, PMN-PZT-Mn)

For resonant excitation ($\omega = \omega_1^{sc}$) of each ceramic and single-crystal cantilever, the power versus load resistance curves are shown in Figure 10.16a. Clearly, the cantilevers made of the hard single-crystal PMN-PZT-Mn and the hard ceramic PZT-8 perform much better than their soft counterparts for power generation under resonant excitation. The reason is the importance of mechanical damping (or quality factor) for resonant energy harvesting (since the vicinity of a natural frequency is strongly controlled by mechanical damping). The hard single crystals

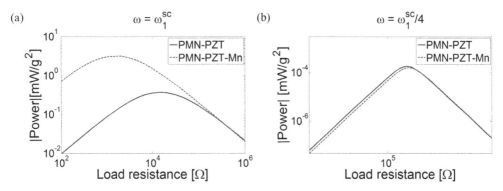

Figure 10.15   Comparison of the (soft) PMN-PZT and the (hard) PMN-PZT-Mn bimorph cantilevers for power generation under (a) resonant excitation and (b) low-frequency off-resonant excitation

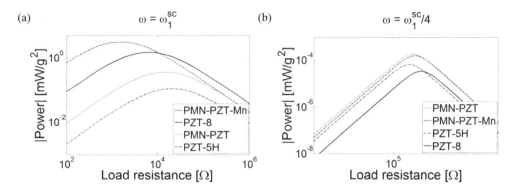

Figure 10.16 Comparison of the PZT-5H, PZT-8, PMN-PZT, and PMN-PZT-Mn cantilevers for power generation under (a) resonant excitation and (b) low-frequency off-resonant excitation

and hard ceramics have one order of magnitude less mechanical damping, and they can produce about one order of magnitude larger power than their soft counterparts for excitations at their respective resonance frequencies. Since the PMN-PZT-Mn cantilever exhibits larger dynamic flexibility due to its larger elastic compliance compared to PZT-8 (see their fundamental resonance frequencies in Table 10.9 as a measure of dynamic flexibility), the former generates the largest power output. The fact that the soft single-crystal PMN-PZT generates more power than the soft ceramic PZT-5H is also related to the larger dynamic flexibility of the former.[7] The conclusion is that *hard single crystals and hard ceramics should be preferred for resonant energy harvesting*. Cantilevers made of hard ceramics can reach the power level of those made of hard single crystals if they are tuned (e.g., by increasing their length or by using a tip mass) to have the same natural frequency so their dynamic flexibilities are similar (analogous to the discussion given in Section 10.2.6).

Figure 10.16b shows the power versus load resistance curves for the low-frequency off-resonant excitation ($\omega = \omega_1^{sc}/4$) of each cantilever. Single crystals PMN-PZT and PMN-PZT-Mn exhibit the best performance for excitations at their off-resonant frequencies (below the fundamental resonance frequency of each cantilever). The elastically stiff cantilever made of the hard ceramic PZT-8 (although a very effective device for resonant energy harvesting) gives the least power output at its off-resonance frequency. PZT-8 has the lowest elastic compliance and effective piezoelectric constant among these samples, which makes it a poor alternative for low-frequency off-resonant energy harvesting. The better performance of the hard single-crystal PMN-PZT-Mn compared to the soft ceramic PZT-5H is again due to the larger elastic compliance of the former. Overall, the single crystals PMN-PZT and PMN-PZT-Mn generate more power for off-resonant energy harvesting at low frequencies. However, it is important to note that the energy that can be harvested at these off-resonant frequencies is several orders of magnitude less than the energy that can be harvested at the fundamental resonance frequency (compare the amplitudes in Figures 10.16a and 10.16b).[8]

---

[7] These effects are discussed in Section 10.2 by focusing on soft ceramics and crystals.

[8] The reader is referred to Section 7.7.1 for an example case of base excitation of a bimorph cantilever at its off-resonant low frequencies.

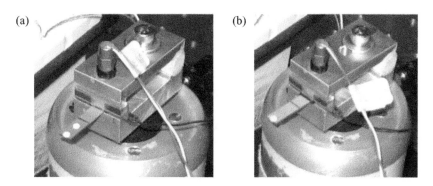

Figure 10.17   (a) PZT-5A and (b) PZT-5H bimorph cantilevers under base excitation

10.5   Experimental Demonstration for PZT-5A and PZT-5H Cantilevers

10.5.1   Experimental Setup

Two brass-reinforced bimorphs with PZT-5A and PZT-5H layers are tested under base excitation as shown in Figure 10.17 (the same model as those tested in Chapter 4). Both PZT-5A and PZT-5H are relatively soft ceramics (with mechanical quality factors less than 100) compared to hard ceramics such as PZT-4 or PZT-8 (with mechanical quality factors in the range of 600–1000). However, PZT-5A is slightly harder than PZT-5H (and would yield less mechanical damping if the entire mechanical damping were ideally due to losses in the piezoceramics, ignoring the clamping conditions, the presence of a brass layer, bonding layers, etc.).

The setup used for base excitation is very similar to the one described in Section 4.1.1 and the focus is placed here directly on the voltage frequency response measurements per base acceleration input. The excitation is provided using an electromagnetic shaker (LDS Test and Measurement Ltd.) and the base acceleration is measured with a small accelerometer (PCB Piezotronics, Inc.). The PZT-5A and PZT-5H bimorphs (T226-A4-203X and T226-H4-203X models) are manufactured by Piezo Systems, Inc. Typical properties of PZT-5A and the PZT-5H piezoceramics are listed in Tables 10.1 and 10.2. The layers of each bimorph are oppositely poled and the brass substructure layer provides electrical conductivity (for the series connection of the piezoceramic layers). When clamping these cantilevers, the overhang lengths are adjusted to have similar short-circuit resonance frequencies. As listed in Table 10.10, the PZT-5A cantilever is clamped with an overhang length of approximately 24.20 mm whereas the PZT-5H is clamped with an overhang length of approximately 24.39 mm.

Table 10.10   Geometric properties of the PZT-5A and PZT-5H bimorph cantilevers

	PZT-5A cantilever	PZT-5H cantilever
Length (L) (mm)	24.20	24.39
Width (b) (mm)	6.4	6.4
Thickness ($h_{\hat{p}}$, $h_{\hat{s}}$) (mm)	0.265 (each)	0.140

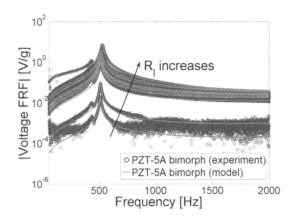

Figure 10.18   Voltage FRFs of the PZT-5A cantilever for a set of resistive loads

10.5.2   Identification of Mechanical Damping and Model Predictions

The fundamental short-circuit resonance frequencies of the PZT-5A and PZT-5H cantilevers are measured for a 100 Ω load resistance as 511.3 Hz and 508.1 Hz. For this lowest value of load resistance, the mechanical damping ratios of the fundamental vibration mode are identified for the PZT-5A and the PZT-5H cantilevers as $\zeta_1 = 0.0091$ and $\zeta_1 = 0.0141$, respectively.[9] The possible sources of the difference in these identified damping ratios might be differences between the mechanical quality factors of the composite structures as well as slightly different clamping conditions in the experiments. As can be seen from Figure 10.17, both cantilevers have non-conductive black tapes[10] at their roots, which might act as a source of uneven mechanical damping in these two separate configurations. Therefore these identified damping ratios are not purposely tuned and they are not easy to control. Nevertheless, the identified values are in agreement with the fact that PZT-5A exhibits harder behavior since the device using PZT-5A has less mechanical damping. The mechanical quality factors of the PZT-5A and PZT-5H bimorph devices can be obtained as 54.9 and 35.5, respectively. Therefore the overall mechanical quality factor of the PZT-5A bimorph device is 55% larger than that of the PZT-5H bimorph device.

The voltage FRFs of the PZT-5A cantilever obtained for a set of resistive loads (ranging from 100 Ω to 248 kΩ) are shown in Figure 10.18. The fundamental short-circuit resonance frequency of the PZT-5A cantilever is predicted by the model as 511.5 Hz. As the load resistance is increased from 100 Ω to 248 kΩ, the experimental value of the resonance frequency in Figure 10.18 moves from 511.3 Hz to 528.1 Hz and the analytical model predicts this frequency as 527.9 Hz. Note that, for the bandwidth of 0–2000 Hz, the frequency resolution of the data acquisition system used here automatically adjusts itself to give a frequency increment of 0.625 Hz. Therefore, both the short-circuit and the open-circuit resonance frequency

---

[9] See Chapters 4 and 5 for discussions regarding the accurate identification of mechanical damping ratio in the electromechanical system (in the presence of an external load).

[10] These non-conductive tapes are preferred in the experiments not only to avoid shorting of the electrodes through the conductive clamp, but also to minimize the possible stress concentration due to direct contact of the aluminum clamp and the brittle piezoceramic layers.

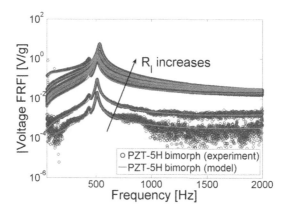

Figure 10.19 Voltage FRFs of the PZT-5H cantilever for a set of resistive loads

readings from the FRFs include experimental error as in the cases discussed in Chapter 4. A clamp-related imperfection can be seen in the FRFs (as in the experiments of Section 4.1.1) but it is sufficiently away from the fundamental resonance frequency.

For the same set of resistors, the model predictions for the PZT-5H cantilever are plotted in Figure 10.19 along with the experimental voltage FRFs. The model prediction of the fundamental short-circuit resonance frequency is 508.4 Hz. In Figure 10.19, as the load resistance is increased from 100 Ω to 248 kΩ, the experimental value of the resonance frequency moves from 508.1 Hz to 530.6 Hz and the model predicts this frequency as 530.3 Hz. These results are listed in Table 10.11 and the model predictions agree very well with the experimental results.

10.5.3   Performance Comparison of the PZT-5A and PZT-5H Cantilevers

The focus is placed on the fundamental short-circuit resonance frequencies of these cantilevers in order to compare their electrical performance results. For excitations at these frequencies, the variations of the voltage and current outputs with load resistance are shown in Figures 10.20 and 10.21, respectively. Only for very low values of load resistance does the PZT-5H cantilever generate slightly larger voltage and current. For a wide range of load resistance, the PZT-5A cantilever generates larger voltage and current.

Using the voltage data for excitations at the fundamental short-circuit resonance frequencies of the cantilevers, variations of the power output with load resistance are plotted in Figure 10.22. The maximum experimental power output of the PZT-5A cantilever is obtained as 0.202 mW/g^2 (for a load resistance of 17.8 kΩ – the optimum value among the resistors used). The maximum experimental power output of the PZT-5H cantilever is 0.140 mW/g^2 (for a

Table 10.11 Fundamental short-circuit resonance frequencies of the PZT-5A and PZT-5H bimorph cantilevers

	PZT-5A cantilever	PZT-5H cantilever
Experiment (Hz)	511.3	508.1
Model (Hz)	511.5	508.4

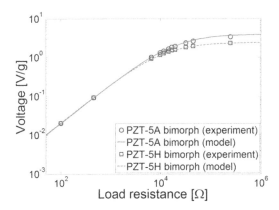

Figure 10.20 Voltage vs. load resistance curves of the PZT-5A and PZT-5H cantilevers for excitation at the fundamental short-circuit resonance frequency

load resistance of $11.7\,\text{k}\Omega$ – the optimum value among the resistors used). Since the overhang volumes and masses of these cantilevers are slightly different, the power density and the specific power values are also reported in Table 10.12.

As can be seen from Table 10.12, the maximum power density of the PZT-5A cantilever is about 45% larger than that of the PZT-5H cantilever. If the maximum specific power outputs are compared, it is found that the PZT-5A cantilever generates 42% larger specific power. Therefore, choosing the PZT-5H cantilever due to its larger d_{31} constant (larger by 60% compared to that of PZT-5A) could result in surprising results in terms of the power output as the mechanical damping can change the entire picture. Since the mechanical quality factor of the PZT-5A device is 55% larger than that of the PZT-5H device, its power output for resonant excitation is larger roughly by the same amount. The power output for resonant excitation is strongly controlled by mechanical damping. Compared to these two devices using rather

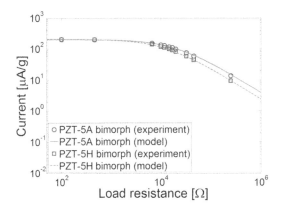

Figure 10.21 Current vs. load resistance curves of the PZT-5A and PZT-5H cantilevers for excitation at the fundamental short-circuit resonance frequency

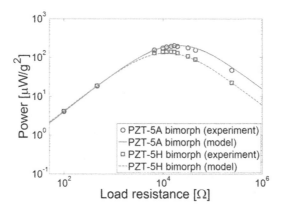

Figure 10.22   Power vs. load resistance curves of the PZT-5A and PZT-5H cantilevers for excitation at the fundamental short-circuit resonance frequency

soft ceramics, one can expect an order-of-magnitude larger power from a similar bimorph cantilever using the hard ceramic PZT-8.

In agreement with the theoretical simulations given in this chapter, the foregoing simple experiment shows the importance of mechanical damping in *resonant* piezoelectric energy harvesting (and resonant excitation is preferred for the maximum power output). Mechanical damping is probably the most difficult parameter to control in the real system (as in several other vibration engineering problems) and it can change the entire picture regardless of the piezoelectric material being used.

Table 10.12   Maximum power outputs and identified mechanical damping ratios

	PZT-5A cantilever	PZT-5H cantilever
Max. power (mW/g^2)	0.202	0.140
Max. power density (mW/(g^2 cm^3))	1.95	1.34
Max. specific power (mW/(g^2 g))	0.243	0.171
Mechanical damping (%)	0.91	1.41
Mechanical quality factor	54.9	35.5

10.6   Summary

The effects of material constants and mechanical damping on piezoelectric power generation are discussed in this chapter. Among the several material constants defining piezoceramics and single crystals, the focus is placed on the parameters required for thin-beam formulations given in Chapter 3. The discussion covered in this chapter can be considered in two parts. The first part investigates whether the well-known d_{31} strain constant (which can be very large in single crystals) alone is a sufficient parameter to choose an active material for piezoelectric energy harvesting. Performance comparisons are presented for bimorphs using the soft ceramics PZT-5A and PZT-5H, and the soft crystals PMN-PT (with 30% PT), PMN-PT (with 33% PT), and

PMN-PZT. The effects of piezoelectric, elastic, and dielectric constants as well as mechanical damping are discussed. It is shown that the difference of an order of magnitude in the d_{31} value of PMN-PZT is not the case for its power output. Importantly, it is observed that the amount of mechanical damping can change the entire picture dramatically regardless of the piezoelectric material used. The second part of the discussion therefore compares soft and hard ceramics (PZT-5H and PZT-8) and soft and hard crystals (PMN-PZT and PMN-PZT-Mn) for resonant and off-resonant energy harvesting. Soft ceramics and crystals typically have mechanical quality factors less than 100 whereas their hard counterparts can have mechanical quality factors as high as 1000. It is observed that the hard ceramics and crystals generate an order-of-magnitude larger power compared to their soft counterparts. That is, the resonant power output is directly proportional to the mechanical quality factor (assuming that no other sources of mechanical dissipation are present). Soft ceramics and crystals can give larger power than their hard counterparts only for off-resonant excitation, which is usually not preferred in energy harvesting. An experimental case study is given for resonant excitation of brass-reinforced PZT-5A and PZT-5H cantilevers. Due to its 55% larger mechanical quality factor (or 35% less mechanical damping), the PZT-5A bimorph is observed to give 45% larger power density (although its d_{31} value is 38% less), showing the importance of mechanical damping in resonant energy harvesting and verifying the theoretical conclusions.

10.7   Chapter Notes

Since piezoelectric devices have several piezoelectric, elastic, and dielectric constants, the discussion and the results given here can help the designer in choosing the best type of piezoelectric for a given application. It should be noted that the entire discussion is given for thin-beam type of configurations under base excitation (in agreement with the physical problem of energy harvesting) so that the focus could be placed on a few parameters and the mechanical quality factor.

For resonant excitation (which is often the case considered in the literature for the maximum power output), the mechanical quality factor becomes a very critical parameter. *Therefore, hard ceramics should be preferred to soft ceramics, and hard crystals should be preferred to soft crystals under resonant excitation.* Moreover, hard ceramics (such as PZT-8) can generate more power than soft single crystals (such as PMN-PZT) under resonant excitation. Soft ceramics and single crystals might generate more power than their hard counterparts under off-resonant excitation in the low-frequency region. However, the power output at such frequencies is typically orders of magnitude lower than the power output obtained at the fundamental resonance frequency. Therefore, resonant excitation is of primary interest and the effective parameter is the mechanical quality factor weighted form of the electromechanical coupling given by Equation (5.14): $Q_m \bar{e}_{31}^2 s_{11}^E / \bar{\varepsilon}_{33}^S$ (where the permittivity term is due to the capacitance while the elastic compliance is due to the square of the natural frequency in Equation (5.14)). It can easily be shown using the data given in Tables 10.5 and 10.6 that the order of this parameter for PZT-5H, PZT-8, PMN-PZT, and PMN-PZT-Mn agrees with the order of their maximum power outputs (Figure 10.16a) under resonant excitation.

From another point of view, for resonant excitation of two different energy harvesters with strong electromechanical coupling, the power generation superiority of one over the other depends on mechanical damping. For instance, larger mechanical damping in a

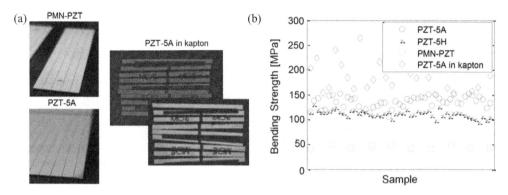

Figure 10.23   (a) Some of the ceramics and single crystals tested by Anton *et al.* [9] and (b) the measured bending strength values for a set of samples

single-crystal-based energy harvester due to poor bonding or clamping conditions may result in much less power output compared to a tightly clamped ceramic-based energy harvester with light mechanical damping.

Comparing ceramics and single crystals, the fragile nature of the latter can be an issue in practical applications with relatively large excitation amplitudes or in relatively harsh environments. Figure 10.23 shows a comparison for the bending strengths of PZT-5A, PZT-5H, and PMN-PZT reported by Anton *et al.* [9]. Three-point bending tests show that the bending strength of the single-crystal PMN-PZT is less than the bending strengths of PZT-5A and PZT-5H ceramics by a factor of about three.

Although hard ceramics and crystals perform better than their soft counterparts under resonant excitation, other materials aspects of the problem should also be investigated. *Aging* and *performance degradation* can take place due to large changes in mechanical stresses and temperature [10] (consider, for instance, a piezoelectric patch or stack located on a road and exposed to large stress variations due to vehicle loads). In addition, as far as single crystals are concerned, an interesting problem is to exploit their *anisotropy* for piezoelectric energy harvesting (by matching their optimal direction(s) of parameters for power generation with the direction(s) of principal strain(s) in the application). Another important materials science research topic is to develop *lead-free piezoelectric materials* [11, 12] with sufficient electromechanical coupling to be used in energy harvesting, since the lead content in piezoelectric materials remains an environmental issue.

References

1. Engineering Fundamentals, Inc. (www.efunda.com).
2. Heinonen, E., Juuti, J., and Leppavuori, S. (2005) Characterization and modelling of 3D piezoelectric ceramic structures with ATILA software. *Journal of European Ceramic Society*, **25**, 2467–2470.
3. Standards Committee of the IEEE Ultrasonics, Ferroelectrics, and Frequency Control Society (1987) *IEEE Standard on Piezoelectricity*, IEEE, New York.
4. Cao, H., Schmidt, V.H., Zhang, R., Cao, W., and Luo, H. (2004) Elastic, piezoelectric, and dielectric properties of $0.58Pb(Mg_{1/3}Nb_{2/3})O_3$-$0.42PbTiO_3$ single crystal. *Journal of Applied Physics*, **96**, 549–554.

5. Erturk, A., Bilgen, O., and Inman, D.J. (2008) Power generation and shunt damping performance of a single crystal lead magnesium niobate – lead zirconate titanate unimorph: analysis and experiment. *Applied Physics Letters*, **93**, 224102.

6. Ceracomp Co., Ltd. (www.ceracomp.com).

7. Morgan Technical Ceramics (www.morgantechnicalceramics.com).

8. Zhang, R., Lee, S.M., Kim, D.H., Lee, H.Y., and Shrout, T.R. (2008) Characterization of Mn-modified $Pb(Mg_{1/3}Nb_{2/3})O_3$-$PbZrO_3$-$PbTiO_3$ single crystals for high power broad bandwidth transducers. *Applied Physics Letters*, **93**, 122908.

9. Anton, S.R., Erturk, A., and Inman, D.J. (2010) Strength analysis of piezoceramic materials for structural considerations in energy harvesting for UAVs. *Proceedings of SPIE*, **7643**, 76430E.

10. Zhang, Q.M. and Zhao, J. (1999) Electromechanical properties of lead zirconate titanate piezoceramics under the influence of mechanical stresses. *IEEE Transactions on Ultrasonics, Ferroelectrics, and Frequency Control*, **46**, 1518–1526.

11. Saito, Y., Takao, H. Tani, I., Nonoyama, T., Takatori, K., Homma, T., Nagaya, T., and Nakamura, M. (2004) Lead-free piezoceramics. *Nature*, **432**, 84–87.

12. Shrout, T.R. and Zhang, S.J. (2007) Lead-free piezoelectric ceramics – alternatives for PZT?. *Journal of Electroceramics*, **19**, 111–124.

11

A Brief Review of the Literature of Piezoelectric Energy Harvesting Circuits

Although the focus is placed on the AC output in the previous chapters to investigate various forms of mechanical excitation and the fundamental trends of the electromechanical response in detail, charging a storage component such as a battery or a capacitor requires a stable DC signal to be obtained. For this purpose, it is necessary to use an AC–DC converter that consists of a rectifier bridge and a smoothing capacitor. The rectified voltage level usually depends on the vibration amplitude. Therefore, one cannot achieve the optimal rectified voltage for all vibration levels if a simple AC–DC converter is used. Often a DC–DC converter is connected after the AC–DC converter to regulate its DC output for the maximum power transfer to the battery through impedance matching. This chapter briefly reviews some of the major papers from the literature of piezoelectric energy harvesting circuits. First, lumped-parameter modeling of a piezoelectric energy harvester with a standard (one-stage) energy harvesting interface (AC–DC converter) is summarized to express the rectified voltage and the average power in terms of the vibration input. Then, the two-stage approach of combining a DC–DC converter with the AC–DC converter is discussed with examples from the literature. Finally, the technique of synchronous switch harvesting on inductor (SSHI) is reviewed as an effective way of increasing the power output particularly for weakly coupled energy harvesting systems using a switch and an inductor. Lumped-parameter modeling of an energy harvester using the SSHI interface is summarized along with simulation results. This brief review ends with the chapter notes, referring the reader to other papers on piezoelectric energy harvesting circuits.

11.1 AC–DC Rectification and Analysis of the Rectified Output

This section discusses an analysis of the standard interface used for converting the AC output of the harvester to a stable DC output. Figure 11.1a shows a cantilevered piezoelectric energy harvester connected to a standard AC–DC converter (one-stage energy harvesting interface)

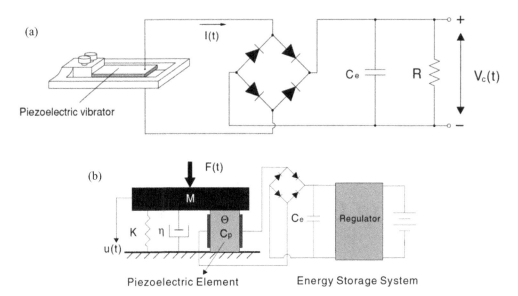

Figure 11.1 (a) Standard AC–DC energy harvesting circuit (from Shu *et al.* [4], reproduced by permission of IOP © 2007) and (b) an equivalent circuit representation (from Shu and Lien [1], reproduced by permission of IOP © 2006)

prior to the resistive load R. The full-wave rectifier is followed by a smoothing capacitor (C_e) to obtain a constant rectified voltage (V_c). Shu and Lien [1] represent the piezoelectric energy harvester by a lumped-parameter system as depicted in Figure 11.1b and provide derivations for estimating the harvested power considering the rectified DC voltage. The fundamental steps of their derivation are summarized here while the details and further discussions can be found in the original work [1]. This derivation improves the early circuit models given by Ottman *et al.* [2] (who assumed that the current generated by the piezoelectric energy harvester does not depend on the load resistance)[1] and Guyomar *et al.* [3] (who assumed that the external periodic excitation is in phase with the velocity response of the harvester). Using a lumped-parameter electromechanical model with a single mechanical degree of freedom, Shu and Lien [1] take into account the effect of piezoelectric power generation on the harvester so that the current output of the harvester is a dependent variable. Moreover, they [1] consider the phase angle between the external forcing and the vibration response.

The electromechanical equations for the lumped-parameter representation given in Figure 11.1 are [1]

$$M\ddot{u}(t) + \eta\dot{u}(t) + Ku(t) + \Theta V_p(t) = F(t) \tag{11.1}$$

$$-\Theta\dot{u}(t) + C_p\dot{V}_p(t) = -I(t) \tag{11.2}$$

---

[1] See Section 5.7.6.

where M is the effective mass, K is the effective stiffness, η is the effective damping coefficient, Θ is the effective piezoelectric coupling coefficient, C_p is the effective capacitance, $u(t)$ is the transverse displacement at the tip of the cantilever, $V_p(t)$ is the voltage across the piezoceramic, and $F(t)$ is the excitation force.[2]

The current $I(t)$ flowing into the AC–DC harvesting circuit is given by

$$
I(t) =
\begin{cases}
C_e \dot{V}_c(t) + \dfrac{V_c(t)}{R} & \text{if} \quad V_p = V_c \\[2mm]
-C_e \dot{V}_c(t) - \dfrac{V_c(t)}{R} & \text{if} \quad V_p = -V_c \\[2mm]
0 & \text{if} \quad |V_p| < V_c
\end{cases}
\tag{11.3}
$$

where R is the external resistive load. It is well known that the rectified voltage is independent of the smoothing capacitor if the time constant RC_e is much larger than the period of oscillation of the harvester. To describe the meaning of the current expression given by Equation (11.3), Shu and Lien [1] summarize the working principle of an ideal rectifying bridge as follows. The bridge is open-circuited if $|V_p| < V_c$, hence there is no current flow. When $|V_p|$ reaches V_c, the bridge conducts and the piezoelectric voltage is blocked at the rectified voltage: $|V_p| = V_c$. Then, the conductance in the diodes is blocked again when $|V_p|$ starts decreasing. Therefore $V_p(t)$ either changes proportionally with $u(t)$ when the bridge blocks, or is equal to V_c when the bridge conducts. Hence, for a sinusoidal excitation of the form

$$
F(t) = F_0 \sin \omega t
\tag{11.4}
$$

where F_0 is the constant excitation amplitude and ω is very close to the resonance frequency, the steady-state displacement and the piezoelectric voltage are assumed to have the following forms:

$$
u(t) = u_0 \sin(\omega t - \theta), \quad V_p(t) = g(\omega t - \theta)
\tag{11.5}
$$

Here, u_0 is the constant displacement amplitude, $g(t)$ is a periodic function with 2π period ($|g(t)| \le V_c$, see Figure 11.2b) and θ is the phase relative to the forcing function. Therefore, the resulting waveforms of the displacement response and the piezoelectric voltage response for the standard interface shown in Figure 11.1a are as depicted in Figures 11.2a and 11.2b, respectively. Note that the period of vibration response is the same as the period of the forcing function: $T = 2\pi/\omega$.

---

[2] Note that these equations are simplified equations from the mechanical point of view. The implicit assumption is that the forcing function should have an excitation frequency at or very close to the resonance frequency. The reader can match the terms of these equations with those of the single-mode expressions given in Chapter 3 for improved accuracy.

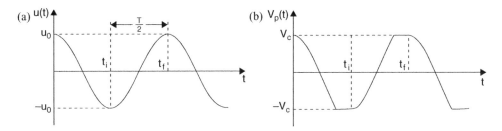

Figure 11.2 (a) Displacement and (b) piezoelectric voltage waveforms for the standard energy harvesting interface (from Shu *et al.* [4], reproduced by permission of IOP © 2007)

The time history of the current is then integrated for half of the period of the motion $(T/2)$ in Figure 11.2a, as the displacement goes from the minimum (at $t = t_i$) to the maximum (at $t = t_f$), and gives

$$\int_{t_i}^{t_f} I(t)\,dt = \frac{T}{2}\frac{V_c}{R} \tag{11.6}$$

Therefore, integrating Equation (11.2) over the same interval leads to

$$V_c = \frac{2\Theta R\omega}{2C_p R\omega + \pi}u_0 \tag{11.7}$$

Hence the average power output is

$$P = \frac{V_c^2}{R} = \frac{4R\Theta^2\omega^2}{\left(2C_p R\omega + \pi\right)^2}u_0^2 \tag{11.8}$$

which requires the vibration amplitude u_0 to be known. The approximation of Ottman *et al.* [2] at this stage is to neglect the $\Theta V_p(t)$ term in Equation (11.1) while Guyomar *et al.* [3] assume the velocity response to be in phase with the force input so that $\dot{u}(t) = u_0\omega \sin \omega t$. Shu and Lien [1] keep the governing equations and the response in their general form and proceed to obtain an energy equation combining Equations (11.1) and (11.2). For this, Equation (11.1) is multiplied by $\dot{u}(t)$ while Equation (11.2) is multiplied by $V_p(t)$, and the resulting expressions are integrated over the semi-period $t_i \le t \le t_f$. Eventually, the displacement amplitude is obtained in terms of the forcing amplitude and the system parameters as [1]

$$u_0 = \frac{F_0}{\left\{\omega^2\left[\eta + \dfrac{2\Theta^2 R}{\left(C_p R\omega + \pi/2\right)^2}\right]^2 + \left(K - \omega^2 M + \dfrac{\Theta^2 R\omega}{C_p R\omega + \pi/2}\right)^2\right\}^{1/2}} \tag{11.9}$$

which is normalized to give

$$\bar{u}_0 = \frac{u_0}{F_0/K} = \frac{1}{\left\{ \Omega^2 \left[2\zeta + \frac{2k_e^2 r}{(r\Omega + \pi/2)^2} \right]^2 + \left(1 - \Omega^2 + \frac{k_e^2 r\Omega}{r\Omega + \pi/2} \right)^2 \right\}^{1/2}} \tag{11.10}$$

yielding

$$\bar{V}_c = \frac{V_c}{F_0/\Theta} = \left(\frac{r\Omega}{r\Omega + \pi/2} \right) \frac{k_e^2}{\left\{ \Omega^2 \left[2\zeta + \frac{2k_e^2 r}{(r\Omega + \pi/2)^2} \right]^2 + \left(1 - \Omega^2 + \frac{k_e^2 r\Omega}{r\Omega + \pi/2} \right)^2 \right\}^{1/2}} \tag{11.11}$$

$$\bar{P} = \frac{P}{F_0^2/(M\omega_n)} = \frac{1}{(r\Omega + \pi/2)^2} \frac{k_e^2 \Omega^2 r}{\Omega^2 \left[2\zeta + \frac{2k_e^2 r}{(r\Omega + \pi/2)^2} \right]^2 + \left(1 - \Omega^2 + \frac{k_e^2 r\Omega}{r\Omega + \pi/2} \right)^2} \tag{11.12}$$

The dimensionless terms are

$$k_e^2 = \frac{\Theta^2}{KC_p}, \quad \zeta = \frac{\eta}{2\sqrt{KM}}, \quad r = C_p R\omega_n, \quad \Omega = \frac{\omega}{\omega_n} \tag{11.13}$$

where $\omega_n = \sqrt{K/M}$ is the short-circuit natural frequency (therefore K is evaluated at constant electric field), k_e^2 is the electromechanical coupling factor, r is the dimensionless resistance, Ω is the dimensionless frequency, and ζ is the mechanical damping ratio.[3]

Shu and Lien [1] then provide expressions for the optimal conditions and compare the results to the uncoupled solution [2] as well as in-phase [3] and modified in-phase [5] solutions. Their [1] graphical results for a theoretical configuration with large electromechanical coupling ($k_e^2/\zeta \gg 1$) are summarized in Figure 11.3. Figures 11.3a–11.3c show the dimensionless displacement, rectified voltage, and average power obtained from Equations (11.10)–(11.12) for $k_e^2/\zeta = 43.3$ (this is very large, yielding an open-circuit resonance frequency of $\Omega_{oc} = 1.52$ [1]). Predictions of the in-phase approximation [3] are shown in Figures 11.3d–11.3f but they fail for this strongly coupled configuration. In particular, the in-phase model predicts two optimal loads for every excitation frequency (Figure 11.3f), which is not the case in the physical system.

The basic trends in the vibration, rectified voltage, and average power output given by Figures 11.3a–11.3c are indeed very similar to those of the AC simulations given in Chapter 3 as well as the experimental results given in Chapter 4. The short-circuit and the open-circuit resonance frequencies reported [1] for the theoretical configuration of the simulations

---

[3] It is useful to compare these terms to the dimensionless modal parameters given by Equations (5.13)–(5.15) in Section 5.1.3. In particular, k_e^2 corresponds to γ_r in the single-mode derivations based on distributed-parameter modeling.

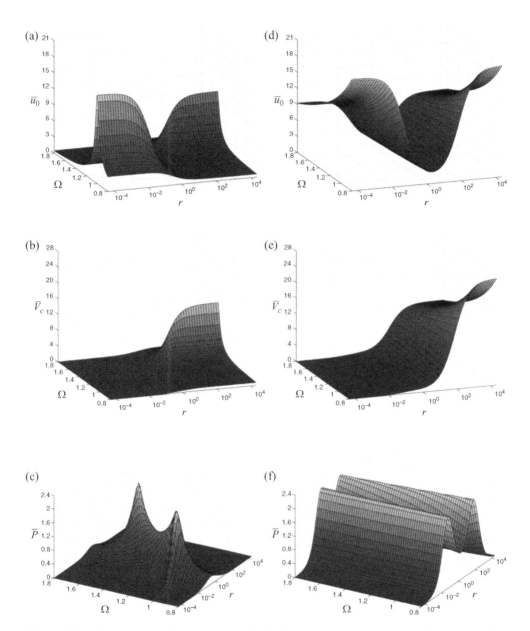

Figure 11.3 Dimensionless displacement, rectified voltage and average power vs. dimensionless frequency and resistance (for $k_e^2/\zeta = 43.3$) obtained from (a)–(c) the complete solution reviewed here and from (d)–(f) the in-phase approximation [3] (from Shu and Lien [1], reproduced by permission of IOP © 2006)

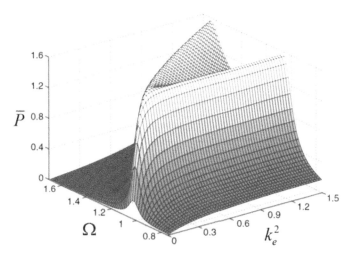

Figure 11.4 Optimal dimensionless power vs. dimensionless frequency and electromechanical coupling (for $\zeta = 0.04$) showing its saturation for very high values of k_e^2 (from Shu *et al.* [4], reproduced by permission of IOP © 2007)

given in Figure 11.3 are $\Omega_{sc} = 1$ and $\Omega_{oc} = 1.52$, respectively. Among other examples in Chapter 4, one can compare the experimental and analytical vibration response for excitations at the fundamental short-circuit and open-circuit resonance frequencies given in Figure 4.14 with Figure 11.3a given here. Likewise, the voltage output vs. load resistance trends at these two frequencies presented in Figure 4.11 are similar to the rectified voltage trends in Figure 11.3b for $\Omega_{sc} = 1$ and $\Omega_{oc} = 1.52$. The AC power output for excitations at the short-circuit and the open-circuit resonance frequencies given by Figure 4.13 has trends similar to the case of Figure 11.3c with two peaks of substantially different matched resistance. It should be noted that the experimental case of Section 4.1 has a modal coupling factor of $\gamma_1 = 0.094$ (Chapter 5) and modal mechanical damping ratio of $\zeta_1 = 0.00874$, yielding $\gamma_1/\zeta_1 = 10.8$ (which is equivalent to k_e^2/ζ [1] as a figure of merit). Although not as strongly coupled as the theoretical case study of Shu and Lien [1] discussed with Figure 11.3, the experimental device of Section 4.1 has fairly strong electromechanical coupling, yielding double peaks in the power curve of Figure 4.13.

In another paper, Shu *et al.* [4] simulate the variation of the average power output with increased electromechanical coupling for the standard energy harvesting interface (Figure 11.1). It is observed from Figure 11.4 that the average power becomes saturated for very large values of electromechanical coupling. They [4] also point out the importance of mechanical damping and report that the average power output for excitations at the short-circuit and open-circuit resonance frequencies is inversely proportional to the mechanical damping ratio ($\bar{P} \propto 1/\zeta$) in agreement with the discussion of Chapter 10 for the AC problem (see Section 10.6).

11.2 Two-Stage Energy Harvesting Circuits: DC–DC Conversion for Impedance Matching

Since the mechanical excitation level and therefore the rectified voltage level are not constant in most practical applications, Ottman *et al.* [2] proposed using a DC–DC converter right after

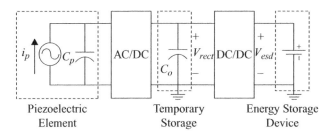

Figure 11.5 Generalized two-stage piezoelectric energy harvesting circuit (from Guan and Liao [7], reproduced by permission of IOP © 2007)

the AC–DC converter of the standard one-stage interface (discussed in Section 11.1) so that the DC output of the rectifier can be regulated to maximize the power transfer to the storage device. They introduced an adaptive DC–DC converter for this purpose and observed a 400% increase of energy flow to the battery compared to the simple AC–DC conversion interface. In another work, they [6] introduced an optimized step-down DC–DC converter operating in discontinuous conduction mode (DCM). Step-down conversion is considered assuming that the piezoelectric material can generate a higher voltage than the required voltage level of the battery. Ottman *et al.* [6] optimized the duty cycle for the step-down converter and observed three times more energy flow to the battery (and the controller was self-powered).

Guan and Liao [7] summarize the optimized two-stage energy harvesting process introduced by Ottman *et al.* [6] through Figure 11.5 and provide an efficiency estimate for the step-down DC–DC converter in terms of the voltage levels of the storage device and other components. In the generalized two-stage representation given by Figure 11.5, the piezoelectric power output is first rectified to the temporary storage capacitor C_0, whose voltage level is kept at the optimal rectifier voltage. The input power is then transferred to the storage device (which has a voltage level of V_{esd}) through the DC–DC converter for the maximum power transfer.

Guan and Liao [7] then focus on the DC–DC converter (Figure 11.6) to estimate its efficiency (which was assumed to operate with 100% efficiency in Ottman *et al.* [6]). The resulting converter efficiency is obtained in their work [7] as

$$\eta_c = \frac{V_{rect} + V_D - V_{ces}}{V_{rect}} \frac{V_{esd}}{V_{esd} + V_D} \tag{11.14}$$

where V_{rect} is the rectifier voltage, V_{esd} is the voltage of the energy storage device, V_D is the forward bias of the diode D_1 in Figure 11.6, and V_{ces} is the voltage drop of the internal switch T_r. Hence the efficiency of the converter depends on several voltage parameters (some of them are simply read from data sheets, such as the bias voltage of the diode) including the voltage level of the storage device. In their paper, Guan and Liao [7] provide an improved estimate of the optimal duty cycle as well.

After validating their expression of the converter efficiency against SPICE[4] simulations (Figure 11.7a), Guan and Liao [7] show theoretically and experimentally that the standard one-stage interface can have a higher efficiency than the two-stage interface depending on the voltage of the storage device (Figure 11.7b). They conclude that, if an optimal voltage level

---

[4] SPICE stands for Simulation Program with Integrated Circuit Emphasis.

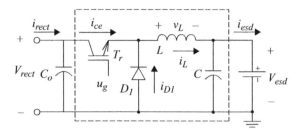

Figure 11.6 DC–DC converter in the two-stage energy harvesting scheme (from Guan and Liao [7], reproduced by permission of IOP © 2007)

is chosen for the energy storage device, the standard one-stage interface (Figure 11.1) can be more efficient than the two-stage interface (Figure 11.5) as the amount of energy collected in the DC–DC converter might cancel the benefits of using the two-stage interface. However, very often the voltage level of the storage device is a fixed parameter due to the application requirements.

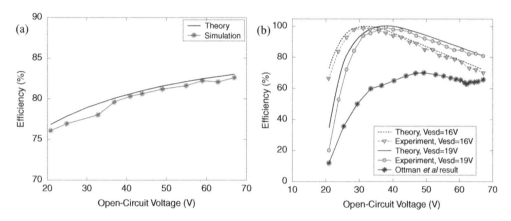

Figure 11.7 (a) Efficiency of the DC–DC converter obtained from Equation (11.14) and SPICE simulations and (b) efficiency comparison of the one-stage and two-stage interfaces (from Guan and Liao [7], reproduced by permission of IOP © 2007)

Kong *et al.* [8] present the resistive impedance matching circuit shown in Figure 11.8, where the buck–boost converter running in DCM is connected right after the rectifier to emulate a resistive load. Therefore, unlike the approaches given by Ottman *et al.* [2,6] and Lefeuvre *et al.* [9], the big smoothing capacitor right after the rectifier bridge is eliminated in this configuration. The power switch is driven by a low-power oscillator and the duty cycle as well as the frequency of the oscillator can be tuned to match the source impedance over a wide range. Kong *et al.* [8] state that the source impedance is more resistive (than reactive) provided that the resonance frequency of the harvester is tuned to the mechanical excitation frequency. Resistive impedance matching is then an acceptable and practical compromise to extract the maximum power (compared to conjugate impedance matching). Unlike the previous impedance matching approaches [2,6], the piezoelectric energy harvester is not represented as a current source with constant amplitude. In order to account for the damping in the energy harvester as a result of power generation, the piezoelectric energy harvester is

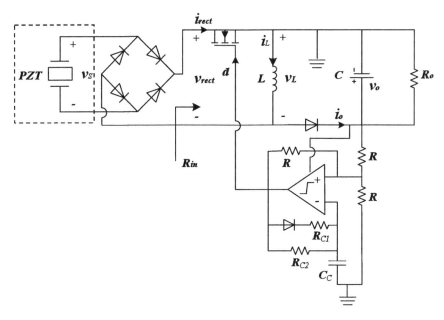

Figure 11.8 Resistive impedance matching circuit with a buck–boost converter operating in DCM by Kong *et al.* [8]

represented by a second-order circuit with its structural and piezoelectric coupling parameters as shown in Figure 11.9. The lumped-parameters M_{11}, K_{11}, and D_{11} are the effective mass, stiffness, and damping of the energy harvester, m^*a is the effective forcing (due to base acceleration a), and the transformer turns ratio n is related to piezoelectric coupling. This representation follows Elvin and Elvin [10] (who employed such lumped-parameter circuits for SPICE simulations), and the parameters of the piezoelectric energy harvester can be taken from the Rayleigh–Ritz solution [10], the analytical solution [11] (single-mode equations in Chapter 3), the assumed-modes solution [12] given in Chapter 6, or the finite-element modeling [13, 14].

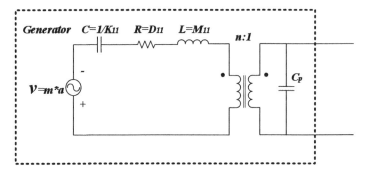

Figure 11.9 Equivalent circuit representation of the piezoelectric energy harvester as a second-order circuit with a transformer [8]

Assuming lossless power switch, diodes and LC filters, the effective input resistance of the DCM buck-boost converter is obtained from [8]

$$R_{in} = \frac{v_{rect}}{\frac{1}{T_S}\int_0^{D_1 T_S} i_L dt} = \frac{v_{rect}}{\frac{1}{T_S}\int_0^{D_1 T_S} \frac{v_{rect}}{L} t dt} = \frac{v_{rect}}{\frac{1}{T_S}\frac{v_{rect}}{L}\frac{(D_1 T_S)^2}{2}} = \frac{2L}{D_1^2 T_S} \qquad (11.15)$$

where D_1 is the duty cycle, T_S is the switching period and L is the inductance (see Figures 11.8 and 11.10). The assumption of Equation (11.15) is given by

$$(R_{dson} + R_{dcr})\sqrt{\frac{2T_S}{R_{in,opt} L}} \ll 1 \qquad (11.16)$$

where R_{dson} is the resistance of the MOSFET during the on-time and R_{dcr} is the parasitic resistance of the inductor [8]. The duty cycle and the switching frequency are tuned by using the parameters of the RC network shown in Figure 11.8. Kong et al. [8] validate the foregoing expressions experimentally. The voltage and current waveforms of the impedance matching circuit shown in Figure 11.8 are displayed in Figure 11.10 for one period of harmonic base input applied to the energy harvester.

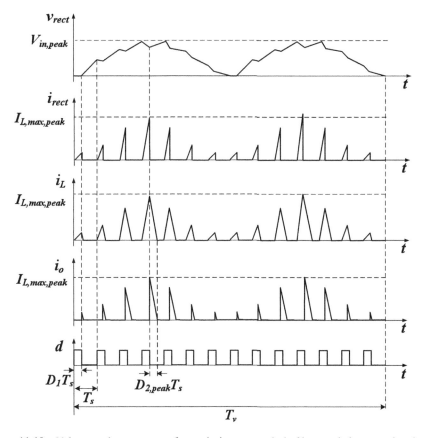

Figure 11.10   Voltage and current waveforms during one period of harmonic base acceleration [8]

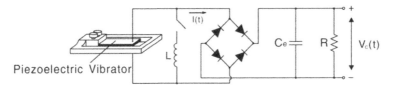

Figure 11.11   Schematic of an SSHI energy harvesting circuit (from Shu *et al.* [4], reproduced by permission of IOP ⓒ 2007)

11.3   Synchronized Switching on Inductor for Piezoelectric Energy Harvesting

Guyomar *et al.* [3] extended the *synchronized switch damping* technique of Richard *et al.* [15] to energy harvesting and introduced an effective technique called *synchronized switch harvesting on inductor* (SSHI). They reported that the harvested power can be increased (compared to the standard AC–DC conversion scheme covered in Section 11.1) as much as 250 to 900% depending on the electromechanical coupling in the system [3,16,17]. An analysis given with the in-phase approximation [3] was followed by a complete lumped-parameter analysis in Shu *et al.* [4], and the fundamental derivation steps from the latter are shown with a graphical simulation in the following.

Compared to the standard interface shown in Figure 11.1, the SSHI interface depicted by Figure 11.11 adds a switch and an inductor (connected in series) which are in parallel with the piezoelectric element and the rectifier bridge. The switch is triggered according to the displacement response so that the piezoelectric voltage is synchronized with the maximum values of displacement [3]. As in the standard interface case discussed with Figure 11.2, $T = 2\pi/\omega$ is the period of the vibration response and the instants of the minimum and the maximum displacement amplitudes are denoted by t_i and t_f, respectively (Figure 11.12a). When the switch is turned on at $t = t_i$, $|V_p| < V_c$, hence the rectifier bridge is open-circuited and an oscillating current due to the inductance L and the piezoelectric capacitance C_p is obtained, which results in an *inversion process* for the piezoelectric voltage. Half of this oscillation period is given by [3,4]

$$\Delta t = \pi \sqrt{LC_p} \tag{11.17}$$

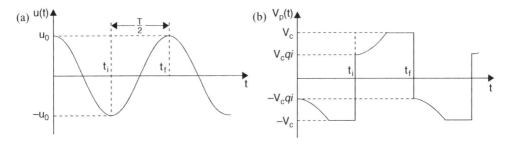

Figure 11.12   (a) Displacement and (b) piezoelectric voltage waveforms for the SSHI interface (from Shu *et al.* [4], reproduced by permission of IOP ⓒ 2007)

where the inversion time is much smaller than the period of vibration, that is, $\Delta t = t_i^+ - t_i \ll T$. In this small period, the switch is kept closed and the piezoelectric voltage is inverted. Therefore,

$$V_p(t_i^+) = -V_p(t_i)e^{-\pi/(2Q_I)} = V_c q_I \tag{11.18}$$

Here,

$$q_I = e^{-\pi/(2Q_I)} \tag{11.19}$$

where Q_I is the inversion quality factor that accounts for the electrical losses in the inversion process [3,4]. The waveform of the piezoelectric voltage is shown in Figure 11.12b.

The electric current is integrated in the semi-period of vibration $(t_i^+ \le t \le t_i)$ to give

$$\int_{t_i^+}^{t_f} I(t)\,dt = \frac{T}{2}\frac{V_c}{R} \tag{11.20}$$

Therefore, integrating Equation (11.2) for the same interval leads to [4]

$$V_c = \frac{2\Theta R\omega}{(1 - q_I)C_p R\omega + \pi}u_0 \tag{11.21}$$

and the average power delivered to the load becomes

$$P = \frac{V_c^2}{R} = \frac{4R\Theta^2\omega^2}{\left[(1 - q_I)C_p R\omega + \pi\right]^2}u_0^2 \tag{11.22}$$

Following steps similar to the standard AC–DC interface case, Shu et al. [4] obtain the displacement amplitude as

$$u_0 = \frac{F_0}{\left\{\left[\omega^2\left(\eta + \frac{2\left[1 + \frac{C_p R\omega}{2\pi}\left(1 - q_I^2\right)\right]\Theta^2 R}{\left(\frac{1 - q_I}{2}C_p R\omega + \frac{\pi}{2}\right)^2}\right)\right]^2 + \left(K - \omega^2 M + \frac{\frac{1 - q_I}{2}\Theta^2 R\omega}{\frac{1 - q_I}{2}C_p R\omega + \frac{\pi}{2}}\right)^2\right\}^{1/2}} \tag{11.23}$$

which is normalized to give

$$\bar{u}_0 = \frac{u_0}{F_0/K} = \cfrac{1}{\left\{ \Omega^2 \left[2\zeta + \cfrac{2\left[1 + \cfrac{r\Omega}{2\pi}\left(1 - q_I^2\right)\right]k_e^2 r}{\left(\cfrac{1 - q_I}{2}r\Omega + \cfrac{\pi}{2}\right)^2} \right]^2 + \left(1 - \Omega^2 + \cfrac{\cfrac{1 - q_I}{2}k_e^2 r\Omega}{\cfrac{1 - q_I}{2}r\Omega + \cfrac{\pi}{2}}\right)^2 \right\}^{1/2}}$$

(11.24)

yielding

$$\bar{V}_c = \frac{V_c}{F_0/\Theta} = \cfrac{k_e^2\left(\cfrac{r\Omega}{\cfrac{1 - q_I}{2}r\Omega + \cfrac{\pi}{2}}\right)}{\left\{ \Omega^2 \left[2\zeta + \cfrac{2\left[1 + \cfrac{r\Omega}{2\pi}\left(1 - q_I^2\right)\right]k_e^2 r}{\left(\cfrac{1 - q_I}{2}r\Omega + \cfrac{\pi}{2}\right)^2} \right]^2 + \left(1 - \Omega^2 + \cfrac{\cfrac{1 - q_I}{2}k_e^2 r\Omega}{\cfrac{1 - q_I}{2}r\Omega + \cfrac{\pi}{2}}\right)^2 \right\}^{1/2}}$$

(11.25)

$$\bar{P} = \frac{P}{F_0^2/(M\omega_n)} = \cfrac{\cfrac{k_e^2\Omega^2 r}{\left(\cfrac{1 - q_I}{2}r\Omega + \cfrac{\pi}{2}\right)^2}}{\Omega^2 \left[2\zeta + \cfrac{2\left[1 + \cfrac{r\Omega}{2\pi}\left(1 - q_I^2\right)\right]k_e^2 r}{\left(\cfrac{1 - q_I}{2}r\Omega + \cfrac{\pi}{2}\right)^2} \right]^2 + \left(1 - \Omega^2 + \cfrac{\cfrac{1 - q_I}{2}k_e^2 r\Omega}{\cfrac{1 - q_I}{2}r\Omega + \cfrac{\pi}{2}}\right)^2}$$

(11.26)

where the dimensionless parameters are as given by Equations (11.13). Shu et al. [4] then compare their results to the in-phase approximation by Guyomar et al. [3] and show that the latter cannot capture the frequency dependence of the outputs. This is further verified by SPICE simulations in the same work [4]. Comparing the optimal conditions of the SSHI case, Shu et al. [4] conclude that *the energy harvesting system using the SSHI interface is similar to that of a strongly coupled system with the standard interface operated at the short-circuit resonance frequency.* This statement is explained in the following discussion.

Since the original piezoelectric capacitance is modified due to the inversion process, the effective capacitance in the SSHI derivation is

$$\bar{C}_p = \frac{1 - q_I}{2}C_p$$

(11.27)

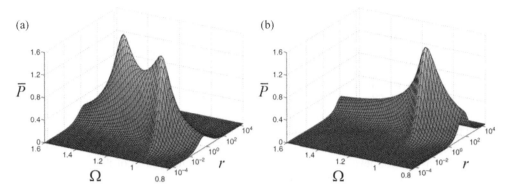

Figure 11.13 Dimensionless average power vs. dimensionless frequency and resistance: (a) standard AC–DC interface with strong coupling ($k_e^2/\zeta = 25$); and (b) ideal SSHI interface with weak coupling ($k_e^2/\zeta = 0.25$, $Q_I \to \infty$) (from Shu et al. [4], reproduced by permission of IOP © 2007)

yielding the effective electromechanical coupling factor of

$$\bar{k}_e^2 = \frac{2\Theta^2}{K\,(1-q_I)\,C_p} \tag{11.28}$$

Shu et al. [4] immediately note that, for the ideal inversion case ($Q_I \to \infty$, hence $q_I \to 1$), $\bar{k}_e^2 \to \infty$. Therefore, the open-circuit resonance frequency tends to infinity; that is, there is no second peak of the power curve for finite values of r and Ω (regardless of the amount of electromechanical coupling). Shu et al. [4] demonstrate this result along with a comparison for the standard interface. Figure 11.13a is the standard interface covered in Section 11.1 for a strongly coupled case ($k_e^2/\zeta = 25$), with two peaks for the power output for different excitation frequencies and matched loads (as in Figure 11.3c). The configuration simulated in Figure 11.13b uses the SSHI interface (with ideal inversion, i.e., $Q_I \to \infty$) for a weakly coupled harvester ($k_e^2/\zeta = 0.25$). If one uses the standard interface for this weakly coupled system, the average power output is very low. However, the average harvested power is *boosted* in the SSHI case and the maximum harvested power obtained for the weakly coupled system connected to the SSHI interface (Figure 11.13b) is the same as the maximum power output of the strongly coupled system connected to the standard interface (Figure 11.13a) [4]. One can see this in Figure 11.13 for the short-circuit resonance ($\Omega = 1$) excitation of both configurations (note that the matched load of the weakly coupled system with the SSHI interface is much larger). The main conclusion of Shu et al. [4] is that, using the SSHI interface, the harvested power of a weakly coupled system can be increased to the level of the strongly coupled system at the cost of using a much larger load. In the same work [4], the reader can also find a discussion related to the non-ideal voltage inversion process for finite values of Q_I, where it is observed that the performance degradation due to non-ideal voltage inversion can be significant if the system is weakly coupled.

11.4 Summary

This chapter briefly reviews some of the major papers from the literature of piezoelectric energy harvesting circuits. First, lumped-parameter modeling of a piezoelectric energy harvester with a standard energy harvesting interface (AC–DC converter) is summarized. The rectified voltage and the average power are given in terms of the vibration input. Simulations are then reviewed for this standard AC–DC conversion case. Often a DC–DC converter is used to regulate the DC output of the rectifier for the maximum power transfer to the storage device. Therefore, the two-stage approach that combines an AC–DC converter and a DC–DC converter is discussed with examples from the literature. The importance of the DC–DC conversion efficiency and its power consumption level as well as the effect of the voltage level of the storage device are pointed out based on the published results. Finally, the synchronous switch harvesting on inductor technique is reviewed as an effective way of increasing the power output if the system is weakly coupled. This technique can increase tremendously the power output of a weakly coupled system to the level of a strongly coupled system that uses the standard AC–DC conversion interface. Lumped-parameter modeling of an energy harvester using the SSHI interface is summarized considering the voltage inversion process and simulation results are reviewed.

11.5 Chapter Notes

Designing energy harvesting circuits for charging storage components efficiently is an important task. It is crucial to develop an energy harvesting circuit with minimal energy losses caused by the circuit components, such as conduction losses in diodes. As can be concluded from the examples covered in this chapter, there is no unique energy harvesting circuit that will be the best option of all energy harvesters and for all excitation levels. However, the designer can choose the most appropriate one depending on the system parameters and the level of excitation. For instance, the SSHI interface is a good option for a weakly coupled energy harvester (Figure 11.13b) while the standard interface can be preferred for a strongly coupled one (Figure 11.13a). As far as two-stage energy harvesting interfaces are concerned, the designer should check to see if the vibration level is sufficiently high to use only a reasonable portion of the harvested energy for powering the controller that optimizes the duty cycle. If the vibration level is not strong enough to power the controller for impedance matching, a simple linear regulator might be a better option.

Regarding the relevant literature, several other papers can be found on the simulations and experiments of the SSHI interface (series and parallel configurations) by Badel et al. [18], Lefeuvre et al. [19], and Lallart and Guyomar [20]. Makihara et al. [21] suggested an SSHI interface that reduces the number of diodes by a factor of two so that the voltage drop in the harvesting process can be reduced. Lefeuvre et al. [9] proposed to use a sensor-less buck–boost converter running in DCM in order to track the optimal working points of the energy harvester. Wu et al. [22] developed models for different synchronized discharging techniques used in piezoelectric energy harvesting. Guan and Liao [23] investigated the characteristics of several storage devices and observed that the leakage resistance of storage devices is a dominant factor affecting the charge and discharge efficiency in piezoelectric energy harvesting systems. Lallart et al. [24] proposed a double synchronized switch harvesting (DSSH) technique by integrating an intermediate switching stage to the SSHI circuit in order to optimize the power output independent of the load resistance. More recently, Shen *et al.* [25] introduced an enhanced synchronized switch harvesting (ESSH) architecture for optimizing the energy conversion with a substantial reduction in the overall dissipation of the control circuitry. Lien *et al.* [26] revisited the SSHI circuits (parallel and series) and analyzed the problem in detail by accounting for the diode loss. More papers from the literature of piezoelectric energy harvesting circuits can be found in their references.

Simulation software (such as SPICE) is commonly used by researchers in order to simulate circuit dynamics [10,13,14] or to validate circuit models [1,4]. Such software is also employed for building powerful models that can cover the entire electromechanical problem. Elvin and Elvin [10] represented the piezoelectric energy harvester as a second-order circuit and accounted for the piezoelectric coupling as a transformer. The lumped parameters and the transformer ratio were obtained from the Rayleigh–Ritz method (which is similar to the assumed-modes method covered in Chapter 6). After validating their results against the analytical solution for the AC problem, Elvin and Elvin [10] used the lumped parameters in SPICE for simulating a capacitor charging problem. Elvin and Elvin [13] and Yang and Tang [14] suggested combining the system parameters obtained from finite-element analysis (FEA) with SPICE for the purpose of accurately representing the mechanics of the harvester using FEA while being able to simulate nonlinear circuits in SPICE. The system parameters taken from electromechanical analytical, assumed-modes, Rayleigh-Ritz or finite-element solutions can be conveniently used for simulating the entire electromechanical problem using circuit-simulating software for the case of storage applications with nonlinear circuits.

References

1. Shu, Y.C. and Lien, I.C. (2006) Analysis of power outputs for piezoelectric energy harvesting systems. *Smart Materials and Structures*, **15**, 1499–1502.
2. Ottman, G.K., Hofmann, H.F., Bhatt, A.C., and Lesieutre, G.A. (2002) Adaptive piezoelectric energy harvesting circuit for wireless remote power supply. *IEEE Transactions on Power Electronics*, **17**, 669–676.
3. Guyomar, D., Badel, A., Lefeuvre, E., and Richard, C. (2005) Toward energy harvesting using active materials and conversion improvement by nonlinear processing. *IEEE Transactions of Ultrasonics, Ferroelectrics, and Frequency Control*, **52**, 584–595.
4. Shu, Y.C., Lien, I.C., and Wu, W.J. (2007) An improved analysis of the SSHI interface in piezoelectric energy harvesting. *Smart Materials and Structures*, **16**, 2253–2264.
5. Lefeuvre, E., Badel, A., Richard, C., and Guyomar, D. (2005) Piezoelectric energy harvesting device optimization by synchronous charge extraction. *Journal of Intelligent Material Systems and Structures*, **16**, 865–876.
6. Ottman, G.K., Hofmann, H.F., and Lesieutre, G.A. (2003) Optimized piezoelectric energy harvesting circuit using step-down converter in discontinuous conduction mode. *IEEE Transactions on Power Electronics*, **18**, 696–703.
7. Guan, M.J. and Liao, W.H. (2007) On the efficiencies of piezoelectric energy harvesting circuits towards storage device voltages. *Smart Materials and Structures*, **16**, 498–505.
8. Kong, N., Ha, D.S., Erturk, A., and Inman, D.J. (2010) Resistive impedance matching circuit for piezoelectric energy harvesting. *Journal of Intelligent Material Systems and Structures*, **21**. doi: 10.1177/1045389X09357971.
9. Lefeuvre, E., Audigier, D., Richard, C., and Guyomar, D. (2007) Buck-boost converter for sensorless power optimization of piezoelectric energy harvester. *IEEE Transactions on Power Electronics*, **22**, 2018–2025.
10. Elvin, N.G. and Elvin, A.A. (2009) A general equivalent circuit model for piezoelectric generators. *Journal of Intelligent Material Systems and Structures*, **20**, 3–9.
11. Erturk, A. and Inman, D.J. (2009) An experimentally validated bimorph cantilever model for piezoelectric energy harvesting from base excitations. *Smart Materials and Structures*, **18**, 025009.
12. Erturk, A. and Inman, D.J. (2010) Assumed-modes formulation of piezoelectric energy harvesters: Euler-Bernoulli, Rayleigh and Timoshenko models with axial deformations. Proceedings of the ASME 2010 ESDA 10th Biennial Conference on Engineering Systems, Design and Analysis, Istanbul, Turkey, July 12–14, 2010.

13. Elvin, N.G. and Elvin, A.A. (2009) A coupled finite element – circuit simulation model for analyzing piezoelectric energy generators. *Journal of Intelligent Material Systems and Structures*, **20**, 587–595.
14. Yang, Y. and Tang, L. (2009) Equivalent circuit modeling of piezoelectric energy harvesters. *Journal of Intelligent Material Systems and Structures*, **20**, 2223–2235.
15. Richard, C., Guyomar, D., Audigier, D., and Ching, G. (1998) Semi-passive damping using continuous switching of a piezoelectric device. *Proceedings of the SPIE*, **3672**, 104–111.
16. Badel, A., Benayad, A., Lefeuvre, E., Leburn, L., Richard, C., and Guyomar, D. (2006) Single crystals and nonlinear process for outstanding vibration-powered electrical generators. *IEEE Transactions of Ultrasonics, Ferroelectrics, and Frequency Control*, **53**, 673–684.
17. Badel, A., Guyomar, D., Lefeuvre, E., and Richard, C. (2005) Efficiency enhancement of a piezoelectric energy harvesting device in pulsed operation by synchronous charge inversion. *Journal of Intelligent Material Systems and Structures*, **16**, 889–901.
18. Badel, A., Guyomar, D., Lefeuvre, E., and Richard, C. (2006) Piezoelectric energy harvesting using a synchronous switch technique. *Journal of Intelligent Material Systems and Structures*, **17**, 831–839.
19. Lefeuvre, E., Badel, A., Richard, C., Petit, L., and Guyomar, D. (2006) A comparison between several vibration-powered piezoelectric generators for standalone systems. *Sensors and Actuators A*, **126**, 405–416.
20. Lallart, M. and Guyomar, D. (2008) An optimized self-powered switching circuit for non-linear energy harvesting with low voltage output. *Smart Materials and Structures*, **17**, 035030.
21. Makihara, K., Onoda, J., and Miyakawa, T. (2006) Low energy dissipation electric circuit for energy harvesting. *Smart Materials and Structures*, **15**, 1493–1498.
22. Wu, W.J., Wickenheiser, A.M., Reissman, T., and Garcia, E. (2009) Modeling and experimental verification of synchronized discharging techniques for boosting power harvesting from piezoelectric transducers. *Smart Materials and Structures*, **18**, 055012.
23. Guan, M.J. and Liao, W.H. (2008) Characteristics of energy storage devices in piezoelectric energy harvesting systems. *Journal of Intelligent Material Systems and Structures*, **19**, 671–679.
24. Lallart, M., Grabuio, L., Petit, L., Richard, C., and Guyomar, D. (2008) Double synchronized switch harvesting (DSSH): a new energy harvesting scheme for efficient energy extraction. *IEEE Transactions on Ultrasonics, Ferroelectrics, and Frequency Control*, **5**, 2119–2130.
25. Shen, H., Qiu, J., Ji, H., Zhu, K., and Balsi, M. (2010) Enhanced synchronized switch harvesting: a new energy harvesting scheme for efficient energy extraction. *Smart Materials and Structures*, **19**, 115017.
26. Lien, I.C., Shu, Y.C., Wu, W.J., Shiu, S.M., and Lin, H.C. (2010) Revisit of series-SSHI with comparisons to other interfacing circuits in piezoelectric energy harvesting, **19**, 125009.

Appendix A

Piezoelectric Constitutive Equations

A.1 Three-Dimensional Form of the Linear Piezoelectric Constitutive Equations

In general, poled piezoceramics (such as PZT-5A and PZT-5H) are transversely isotropic materials. To be in agreement with the IEEE Standard on Piezoelectricity [1], the plane of isotropy is defined here as the 12-plane (or the xy-plane). The piezoelectric material therefore exhibits symmetry about the 3-axis (or the z-axis), which is the poling axis of the material. The field variables are the stress components (T_{ij}), strain components (S_{ij}), electric field components (E_k), and the electric displacement components (D_k).

The standard form of the piezoelectric constitutive equations can be given in four different forms by taking either two of the four field variables as the independent variables. Consider the tensorial representation of the strain–electric displacement form [1] where the independent variables are the stress components and the electric field components (and the remaining terms are as defined in Section 1.4):

$$S_{ij} = s^E_{ijkl} T_{kl} + d_{kij} E_k \tag{A.1}$$

$$D_i = d_{ikl} T_{kl} + \varepsilon^T_{ik} E_k \tag{A.2}$$

which is the preferred form of the piezoelectric constitutive equations for bounded media (to eliminate some of the stress components depending on the geometry and some of the electric field components depending on the placement of the electrodes). Equations (A.1) and (A.2) can be given in matrix form as

$$\begin{bmatrix} \mathbf{S} \\ \mathbf{D} \end{bmatrix} = \begin{bmatrix} \mathbf{s}^E & \mathbf{d}^t \\ \mathbf{d} & \varepsilon^T \end{bmatrix} \begin{bmatrix} \mathbf{T} \\ \mathbf{E} \end{bmatrix} \tag{A.3}$$

where the superscripts E and T denote that the respective constants are evaluated at constant electric field and constant stress, respectively, and the superscript t stands for the transpose.

Piezoelectric Energy Harvesting, First Edition. Alper Erturk and Daniel J. Inman.
© 2011 John Wiley & Sons, Ltd. Published 2011 by John Wiley & Sons, Ltd.

The expanded form of Equation (A.3) is

$$
\begin{bmatrix} S_1 \\ S_2 \\ S_3 \\ S_4 \\ S_5 \\ S_6 \\ D_1 \\ D_2 \\ D_3 \end{bmatrix}
=
\begin{bmatrix}
s_{11}^E & s_{12}^E & s_{13}^E & 0 & 0 & 0 & 0 & 0 & d_{31} \\
s_{12}^E & s_{11}^E & s_{13}^E & 0 & 0 & 0 & 0 & 0 & d_{31} \\
s_{13}^E & s_{13}^E & s_{33}^E & 0 & 0 & 0 & 0 & 0 & d_{33} \\
0 & 0 & 0 & s_{55}^E & 0 & 0 & 0 & d_{15} & 0 \\
0 & 0 & 0 & 0 & s_{55}^E & 0 & d_{15} & 0 & 0 \\
0 & 0 & 0 & 0 & 0 & s_{66}^E & 0 & 0 & 0 \\
0 & 0 & 0 & 0 & d_{15} & 0 & \varepsilon_{11}^T & 0 & 0 \\
0 & 0 & 0 & d_{15} & 0 & 0 & 0 & \varepsilon_{11}^T & 0 \\
d_{31} & d_{31} & d_{33} & 0 & 0 & 0 & 0 & 0 & \varepsilon_{33}^T
\end{bmatrix}
\begin{bmatrix} T_1 \\ T_2 \\ T_3 \\ T_4 \\ T_5 \\ T_6 \\ E_1 \\ E_2 \\ E_3 \end{bmatrix}
\tag{A.4}
$$

where the contracted notation (i.e., Voigt's notation: $11 \rightarrow 1$, $22 \rightarrow 2$, $33 \rightarrow 3$, $23 \rightarrow 4$, $13 \rightarrow 5$, $12 \rightarrow 6$) is used so that the vectors of strain and stress components are

$$
\begin{bmatrix} S_1 \\ S_2 \\ S_3 \\ S_4 \\ S_5 \\ S_6 \end{bmatrix}
=
\begin{bmatrix} S_{11} \\ S_{22} \\ S_{33} \\ 2S_{23} \\ 2S_{13} \\ 2S_{12} \end{bmatrix},
\qquad
\begin{bmatrix} T_1 \\ T_2 \\ T_3 \\ T_4 \\ T_5 \\ T_6 \end{bmatrix}
=
\begin{bmatrix} T_{11} \\ T_{22} \\ T_{33} \\ T_{23} \\ T_{13} \\ T_{12} \end{bmatrix}
\tag{A.5}
$$

Therefore the shear strain components in the contracted notation are the engineering shear strains. It should be noted from the elastic, piezoelectric, and dielectric constants in Equation (A.4) that the symmetries of transversely isotropic material behavior ($s_{11}^E = s_{22}^E$, $d_{31} = d_{32}$, etc.) are directly applied.

A.2   Reduced Equations for a Thin Beam

If the piezoelastic behavior of the thin structure is to be modeled as a thin beam based on the Euler–Bernoulli beam theory or Rayleigh beam theory, the stress components other than the one-dimensional bending stress T_1 are negligible so that

$$
T_2 = T_3 = T_4 = T_5 = T_6 = 0
\tag{A.6}
$$

Along with this simplification, if an electrode pair covers the faces perpendicular to the 3-direction, Equation (A.4) becomes

$$
\begin{bmatrix} S_1 \\ D_3 \end{bmatrix}
=
\begin{bmatrix} s_{11}^E & d_{31} \\ d_{31} & \varepsilon_{33}^T \end{bmatrix}
\begin{bmatrix} T_1 \\ E_3 \end{bmatrix}
\tag{A.7}
$$

which can be written as

$$\begin{bmatrix} s_{11}^E & 0 \\ -d_{31} & 1 \end{bmatrix} \begin{bmatrix} T_1 \\ D_3 \end{bmatrix} = \begin{bmatrix} 1 & -d_{31} \\ 0 & \varepsilon_{33}^T \end{bmatrix} \begin{bmatrix} S_1 \\ E_3 \end{bmatrix} \tag{A.8}$$

Therefore the stress–electric displacement form of the reduced constitutive equations for a thin beam is

$$\begin{bmatrix} T_1 \\ D_3 \end{bmatrix} = \begin{bmatrix} \bar{c}_{11}^E & -\bar{e}_{31} \\ \bar{e}_{31} & \bar{\varepsilon}_{33}^S \end{bmatrix} \begin{bmatrix} S_1 \\ E_3 \end{bmatrix} \tag{A.9}$$

where the reduced matrix of the elastic, piezoelectric, and dielectric constants is

$$\bar{C} = \begin{bmatrix} \bar{c}_{11}^E & -\bar{e}_{31} \\ \bar{e}_{31} & \bar{\varepsilon}_{33}^S \end{bmatrix} = \begin{bmatrix} s_{11}^E & 0 \\ -d_{31} & 1 \end{bmatrix}^{-1} \begin{bmatrix} 1 & -d_{31} \\ 0 & \varepsilon_{33}^T \end{bmatrix} \tag{A.10}$$

Here and hereafter, an overbar denotes that the respective constant is reduced from the three-dimensional form to the plane-stress condition. In Equation (A.10),

$$\bar{c}_{11}^E = \frac{1}{s_{11}^E}, \quad \bar{e}_{31} = \frac{d_{31}}{s_{11}^E}, \quad \bar{\varepsilon}_{33}^S = \varepsilon_{33}^T - \frac{d_{31}^2}{s_{11}^E} \tag{A.11}$$

where the superscript S denotes that the respective constant is evaluated at constant strain.

A.3   Reduced Equations for a Moderately Thick Beam

If the piezoelasticity of the structure is to be modeled as a moderately thick beam based on the Timoshenko beam theory, the stress components other than T_1 (the stress component in the axial direction) and T_5 (the transverse shear stress) are negligible so that

$$T_2 = T_3 = T_4 = T_6 = 0 \tag{A.12}$$

is applied in Equation (A.4). Then,

$$\begin{bmatrix} S_1 \\ S_5 \\ D_3 \end{bmatrix} = \begin{bmatrix} s_{11}^E & 0 & d_{31} \\ 0 & s_{55}^E & 0 \\ d_{31} & 0 & \varepsilon_{33}^T \end{bmatrix} \begin{bmatrix} T_1 \\ T_5 \\ E_3 \end{bmatrix} \tag{A.13}$$

which can be written as

$$\begin{bmatrix} s_{11}^E & 0 & 0 \\ 0 & s_{55}^E & 0 \\ -d_{31} & 0 & 1 \end{bmatrix} \begin{bmatrix} T_1 \\ T_5 \\ D_3 \end{bmatrix} = \begin{bmatrix} 1 & 0 & -d_{31} \\ 0 & 1 & 0 \\ 0 & 0 & \varepsilon_{33}^T \end{bmatrix} \begin{bmatrix} S_1 \\ S_5 \\ E_3 \end{bmatrix} \tag{A.14}$$

Therefore the stress–electric displacement form of the reduced constitutive equations is

$$
\begin{bmatrix} T_1 \\ T_5 \\ D_3 \end{bmatrix} = \begin{bmatrix} \bar{c}_{11}^E & 0 & -\bar{e}_{31} \\ 0 & \bar{c}_{55}^E & 0 \\ \bar{e}_{31} & 0 & \bar{\varepsilon}_{33}^S \end{bmatrix} \begin{bmatrix} S_1 \\ S_5 \\ E_3 \end{bmatrix} \tag{A.15}
$$

Here, the reduced matrix of the elastic, piezoelectric, and permittivity constants is

$$
\bar{C} = \begin{bmatrix} \bar{c}_{11}^E & 0 & -\bar{e}_{31} \\ 0 & \bar{c}_{55}^E & 0 \\ \bar{e}_{31} & 0 & \bar{\varepsilon}_{33}^S \end{bmatrix} = \begin{bmatrix} s_{11}^E & 0 & 0 \\ 0 & s_{55}^E & 0 \\ -d_{31} & 0 & 1 \end{bmatrix}^{-1} \begin{bmatrix} 1 & 0 & -d_{31} \\ 0 & 1 & 0 \\ 0 & 0 & \varepsilon_{33}^T \end{bmatrix} \tag{A.16}
$$

where

$$
\bar{c}_{11}^E = \frac{1}{s_{11}^E}, \quad \bar{c}_{55}^E = \frac{1}{s_{55}^E}, \quad \bar{e}_{31} = \frac{d_{31}}{s_{11}^E}, \quad \bar{\varepsilon}_{33}^S = \varepsilon_{33}^T - \frac{d_{31}^2}{s_{11}^E} \tag{A.17}
$$

Note that the transverse shear stress in Equation (A.15) is corrected due to Timoshenko [2,3]

$$
T_5 = \kappa \bar{c}_{55}^E S_5 \tag{A.18}
$$

where κ is the shear correction factor [2–12].

A.4 Reduced Equations for a Thin Plate

If the thin structure is to be modeled as a thin plate (i.e., Kirchhoff plate) due to two-dimensional strain fluctuations, the normal stress in the thickness direction of the piezoceramic and the respective transverse shear stress components are negligible:

$$
T_3 = T_4 = T_5 = 0 \tag{A.19}
$$

Equation (A.4) becomes

$$
\begin{bmatrix} S_1 \\ S_2 \\ S_6 \\ D_3 \end{bmatrix} = \begin{bmatrix} s_{11}^E & s_{12}^E & 0 & d_{31} \\ s_{12}^E & s_{11}^E & 0 & d_{31} \\ 0 & 0 & s_{66}^E & 0 \\ d_{31} & d_{31} & 0 & \varepsilon_{33}^T \end{bmatrix} \begin{bmatrix} T_1 \\ T_2 \\ T_6 \\ E_3 \end{bmatrix} \tag{A.20}
$$

which can be rearranged to give

$$
\begin{bmatrix} s_{11}^E & s_{12}^E & 0 & 0 \\ s_{12}^E & s_{11}^E & 0 & 0 \\ 0 & 0 & s_{66}^E & 0 \\ -d_{31} & -d_{31} & 0 & 1 \end{bmatrix} \begin{bmatrix} T_1 \\ T_2 \\ T_6 \\ D_3 \end{bmatrix} = \begin{bmatrix} 1 & 0 & 0 & -d_{31} \\ 0 & 1 & 0 & -d_{31} \\ 0 & 0 & 1 & 0 \\ 0 & 0 & 0 & \varepsilon_{33}^T \end{bmatrix} \begin{bmatrix} S_1 \\ S_2 \\ S_6 \\ E_3 \end{bmatrix} \tag{A.21}
$$

The stress–electric displacement form of the reduced constitutive equations becomes

$$
\begin{bmatrix} T_1 \\ T_2 \\ T_6 \\ D_3 \end{bmatrix} = \begin{bmatrix} \bar{c}_{11}^E & \bar{c}_{12}^E & 0 & -\bar{e}_{31} \\ \bar{c}_{12}^E & \bar{c}_{11}^E & 0 & -\bar{e}_{31} \\ 0 & 0 & \bar{c}_{66}^E & 0 \\ \bar{e}_{31} & \bar{e}_{31} & 0 & \bar{\varepsilon}_{33}^S \end{bmatrix} \begin{bmatrix} S_1 \\ S_2 \\ S_6 \\ E_3 \end{bmatrix}
\tag{A.22}
$$

where

$$
\bar{\mathbf{C}} = \begin{bmatrix} \bar{c}_{11}^E & \bar{c}_{12}^E & 0 & -\bar{e}_{31} \\ \bar{c}_{12}^E & \bar{c}_{11}^E & 0 & -\bar{e}_{31} \\ 0 & 0 & \bar{c}_{66}^E & 0 \\ \bar{e}_{31} & \bar{e}_{31} & 0 & \bar{\varepsilon}_{33}^S \end{bmatrix} = \begin{bmatrix} s_{11}^E & s_{12}^E & 0 & 0 \\ s_{12}^E & s_{11}^E & 0 & 0 \\ 0 & 0 & s_{66}^E & 0 \\ -d_{31} & -d_{31} & 0 & 1 \end{bmatrix}^{-1} \begin{bmatrix} 1 & 0 & 0 & -d_{31} \\ 0 & 1 & 0 & -d_{31} \\ 0 & 0 & 1 & 0 \\ 0 & 0 & 0 & \varepsilon_{33}^T \end{bmatrix}
\tag{A.23}
$$

Here, the reduced elastic, piezoelectric, and permittivity constants are

$$
\bar{c}_{11}^E = \frac{s_{11}^E}{\left(s_{11}^E + s_{12}^E\right)\left(s_{11}^E - s_{12}^E\right)}
\tag{A.24}
$$

$$
\bar{c}_{12}^E = \frac{-s_{12}^E}{\left(s_{11}^E + s_{12}^E\right)\left(s_{11}^E - s_{12}^E\right)}
\tag{A.25}
$$

$$
\bar{c}_{66}^E = \frac{1}{s_{66}^E}
\tag{A.26}
$$

$$
\bar{e}_{31} = \frac{d_{31}}{s_{11}^E + s_{12}^E}
\tag{A.27}
$$

$$
\bar{\varepsilon}_{33}^S = \bar{\varepsilon}_{33}^T - \frac{2d_{31}^2}{s_{11}^E + s_{12}^E}
\tag{A.28}
$$

where the first term (Equation (A.24)) is the elastic constant that is related to the bending stiffness of a piezoelectric plate (accounting for the Poisson effect) in the absence of torsion.

References

1. Standards Committee of the IEEE Ultrasonics, Ferroelectrics, and Frequency Control Society (1987) *IEEE Standard on Piezoelectricity*, IEEE, New York.
2. Timoshenko, S.P. (1921) On the correction for shear of the differential equation for transverse vibrations of prismatic bars. *Philosophical Magazine*, **41**, 744–746.
3. Timoshenko, S.P. (1922) On the transverse vibrations of bars of uniform cross-section. *Philosophical Magazine*, **43**, 125–131.
4. Mindlin, R.D. (1951) Thickness-shear and flexural vibrations of crystal plates. *Journal of Applied Physics*, **22**, 316–323.

5. Mindlin, R.D. (1952) Forced thickness-shear and flexural vibrations of piezoelectric crystal plates. *Journal of Applied Physics*, **23**, 83–88.

6. Cowper, G.R. (1966) The shear coefficient in Timoshenko beam theory. *ASME Journal of Applied Mechanics*, **33**, 335–340.

7. Kaneko, T. (1975) On Timoshenko's correction for shear in vibrating beams. *Journal of Physics D: Applied Physics*, **8**, 1927–1936.

8. Stephen, N.G. (1978) On the variation of Timoshenko's shear coefficient with frequency. *ASME Journal of Applied Mechanics*, **45**, 695–697.

9. Stephen, N.G. (1980) Timoshenko's shear coefficient from a beam subjected to gravity loading. *ASME Journal of Applied Mechanics*, **47**, 121–127.

10. Stephen, N.G. and Hutchinson, J.R. (2001) Discussion: shear coefficients for Timoshenko beam theory. *ASME Journal of Applied Mechanics*, **68**, 959–961.

11. Hutchinson, J.R. (2001) Shear coefficients for Timoshenko beam theory. *ASME Journal of Applied Mechanics*, **68**, 87–92.

12. Puchegger, S., Bauer, S., Loidl, D., Kromp, K., and Peterlik, H. (2003) Experimental validation of the shear correction factor. *Journal of Sound and Vibration*, **261**, 177–184.

Appendix B

Modeling of the Excitation Force in Support Motion Problems of Beams and Bars

B.1 Transverse Vibrations

Consider a uniform Euler–Bernoulli beam with two supports at $x = 0$ (the left support) and at $x = L$ (the right support) in its general form (Figure B.1). If the left and the right supports translate in the transverse direction with $g_L(t)$ and $g_R(t)$, respectively, and rotate with small angles $h_L(t)$ and $h_R(t)$, respectively, the equation of motion for the transverse forced vibrations of the beam due to support excitation can be written as

$$YI \frac{\partial^4 w_{rel}(x, t)}{\partial x^4} + c_s I \frac{\partial^5 w_{rel}(x, t)}{\partial x^4 \partial t} + c_a \frac{\partial w_{rel}(x, t)}{\partial t} + m \frac{\partial^2 w_{rel}(x, t)}{\partial t^2}$$
$$= -m \frac{\partial^2 w_b(x, t)}{\partial t^2} - c_a \frac{\partial w_b(x, t)}{\partial t} \tag{B.1}$$

where

$$w_b(x, t) = \delta_1(x) g_L(t) + \delta_2(x) h_L(t) + \delta_3(x) g_R(t) + \delta_4(x) h_R(t) \tag{B.2}$$

is the effective base displacement, $w_{rel}(x, t)$ is the vibratory (elastic) displacement of the beam relative to the moving base, and $\delta_1(x)$, $\delta_2(x)$, $\delta_3(x)$, and $\delta_4(x)$ are the boundary condition-dependent *displacement influence functions* given by Timoshenko *et al.* [1] (the remaining parameters in Equation (B.1) can be found in Section 2.1.1). The displacement influence functions for different boundary condition pairs are listed in Table B.1. The empty cells in Table B.1 are either because the beam does not have a support at a respective boundary or because it cannot be excited by the respective support motion at that boundary. For instance, there is no support at the right end of a clamped–free beam to define a support motion, while it is not possible to excite the beam by support rotation at the pinned boundaries.

Piezoelectric Energy Harvesting, First Edition. Alper Erturk and Daniel J. Inman.
© 2011 John Wiley & Sons, Ltd. Published 2011 by John Wiley & Sons, Ltd.

Figure B.1   Uniform beam excited transversely by the motion of its boundaries

Table B.1   Displacement influence functions for the transverse support excitation of beams with different boundary conditions

$x = 0$	$x = L$	$\delta_1(x)$	$\delta_2(x)$	$\delta_3(x)$	$\delta_4(x)$
Clamped	Free	1	x	—	—
Pinned	Pinned	$1 - \dfrac{x}{L}$	—	$\dfrac{x}{L}$	—
Clamped	Clamped	$1 - \dfrac{3x^2}{L^2} + \dfrac{2x^3}{L^3}$	$x - \dfrac{2x^2}{L} + \dfrac{x^3}{L^2}$	$\dfrac{3x^2}{L^2} - \dfrac{2x^3}{L^3}$	$-\dfrac{x^2}{L} + \dfrac{x^3}{L^2}$
Clamped	Pinned	$1 - \dfrac{3x^2}{2L^2} + \dfrac{x^3}{2L^3}$	$x - \dfrac{3x^2}{2L} + \dfrac{x^3}{2L^2}$	$\dfrac{3x^2}{2L^2} - \dfrac{x^3}{2L^3}$	—

Given the boundary condition pair and the support displacement input, the right hand side of Equation (B.1) becomes a known function of x and t (and the standard modal analysis procedure in the presence of a general external force can be followed to solve for $w_{rel}(x, t)$). The absolute displacement response of the beam relative to the fixed reference frame, $w(x, t)$, is simply the linear superposition of the base displacement and the vibratory displacement relative to the base: $w(x, t) = w_b(x, t) + w_{rel}(x, t)$.

B.2   Longitudinal Vibrations

Figure B.2 shows a uniform bar with two supports at $x = 0$ (the left support) and at $x = L$ (the right support). If the left and the right supports translate in the longitudinal direction with

Figure B.2   Uniform bar excited longitudinally by the motion of its boundaries

Table B.2   Displacement influence functions for the longitudinal support excitation of bars with different boundary conditions

$x = 0$	$x = L$	$\delta_1(x)$	$\delta_2(x)$
Clamped	Free	1	—
Clamped	Clamped	$1 - \dfrac{x}{L}$	$\dfrac{x}{L}$

$g_L(t)$ and $g_R(t)$, respectively, the equation of motion for the longitudinal forced vibrations of the bar due to support motion can be written as

$$-YA\frac{\partial^2 u_{rel}(x, t)}{\partial x^2} - c_s A\frac{\partial^3 u_{rel}(x, t)}{\partial x^2 \partial t} + c_a\frac{\partial u_{rel}(x, t)}{\partial t} + m\frac{\partial^2 u_{rel}(x, t)}{\partial t^2}$$
$$= -m\frac{\partial^2 u_b(x, t)}{\partial t^2} - c_a\frac{\partial u_b(x, t)}{\partial t} \tag{B.3}$$

where

$$u_b(x, t) = \delta_1(x)g_L(t) + \delta_2(x)g_R(t) \tag{B.4}$$

is the effective base displacement, $u_{rel}(x, t)$ is the vibratory displacement of the bar relative to the moving base, $\delta_1(x)$ and $\delta_2(x)$ are the boundary condition-dependent displacement influence functions, and the remaining parameters in Equation (B.3) can be found in Section 2.4.1. The displacement influence functions for the two possible boundary condition pairs are listed in Table B.2. Given the boundary condition pair and the support displacement input, one can express the right-hand-side forcing function in Equation (B.3) and solve for $u_{rel}(x, t)$. The displacement response of the bar relative to the fixed reference frame is simply $u(x, t) = u_b(x, t) + u_{rel}(x, t)$.

Reference

1. Timoshenko, S., Young, D.H., and Weaver, W. (1974) *Vibration Problems in Engineering*, John Wiley & Sons, Inc., New York.

Appendix C

Modal Analysis of a Uniform Cantilever with a Tip Mass

C.1 Transverse Vibrations

The following analytical modal analysis is given for the linear transverse vibrations of an undamped Euler–Bernoulli beam with clamped–free boundary conditions and a tip mass rigidly attached at the free end. The expressions for the undamped natural frequencies and mode shapes are obtained and the normalization conditions of the eigenfunctions are given. The procedure for reducing the partial differential equation of motion to an infinite set of ordinary differential equations is summarized, which is applicable to proportionally damped systems as well.

C.1.1 Boundary-Value Problem

Using the Newtonian or the Hamiltonian approach, the governing equation of motion for undamped free vibrations of a uniform Euler–Bernoulli beam can be obtained as [1]

$$YI\frac{\partial^4 w(x,t)}{\partial x^4} + m\frac{\partial^2 w(x,t)}{\partial t^2} = 0 \qquad (C.1)$$

where $w(x,t)$ is the transverse displacement of the neutral axis (at point x and time t) due to bending, YI is the bending stiffness, and m is the mass per unit length of the beam.[1]

The clamped–free boundary conditions with a tip mass attachment (Figure C.1) can be expressed as

$$w(0,t) = 0, \qquad \left.\frac{\partial w(x,t)}{\partial x}\right|_{x=0} = 0 \qquad (C.2)$$

---

[1] Note that the discussion here is given for free vibrations. Hence the clamped end does not move and $w(x,t)$ is identical to $w_{rel}(x,t)$ as far as the notation of Chapters 2 and 3 is concerned.

Piezoelectric Energy Harvesting, First Edition. Alper Erturk and Daniel J. Inman.
© 2011 John Wiley & Sons, Ltd. Published 2011 by John Wiley & Sons, Ltd.

Figure C.1   Cantilevered uniform beam with a tip mass attachment

$$\left[YI\frac{\partial^2 w(x,t)}{\partial x^2}+I_t\frac{\partial^3 w(x,t)}{\partial t^2 \partial x}\right]_{x=L}=0,\quad \left[YI\frac{\partial^3 w(x,t)}{\partial x^3}-M_t\frac{\partial^2 w(x,t)}{\partial t^2}\right]_{x=L}=0 \quad (C.3)$$

where M_t is the tip mass and I_t is the mass moment of inertia of the tip mass about $x=L$.[2] Therefore the geometric boundary conditions at $x=0$ are given by Equations (C.2) while the natural boundary conditions at $x=L$ are given by Equations (C.3). Equations (C.1)–(C.3) define the *boundary-value problem* for the transverse vibrations of a cantilevered uniform Euler–Bernoulli beam with a tip mass attachment.

C.1.2   Solution Using the Method of Separation of Variables

The method of separation of variables [3] can be used to solve Equation (C.1) by separating the spatial and temporal functions as

$$w(x,t)=\phi(x)\eta(t) \tag{C.4}$$

which can be substituted into Equation (C.1) to give

$$\frac{YI}{m}\frac{1}{\phi(x)}\frac{d^4\phi(x)}{dx^4}=-\frac{1}{\eta(t)}\frac{d^2\eta(t)}{dt^2} \tag{C.5}$$

The left hand side of Equation (C.5) depends on x alone while the right hand side depends on t alone. Since x and t are independent variables, the standard argument in the method of separation of variables states that both sides of Equation (C.5) must be equal to the same constant γ:

$$\frac{YI}{m}\frac{1}{\phi(x)}\frac{d^4\phi(x)}{dx^4}=-\frac{1}{\eta(t)}\frac{d^2\eta(t)}{dt^2}=\gamma \tag{C.6}$$

---

[2] Knowing the mass moment of inertia of the tip mass attachment about its centroid, one can use the *parallel axis theorem* [2] to obtain its mass moment of inertia about the end point of the elastic beam where the boundary condition is expressed.

yielding

$$\frac{d^4\phi(x)}{dx^4} - \gamma \frac{m}{YI}\phi(x) = 0 \tag{C.7}$$

$$\frac{d^2\eta(t)}{dt^2} + \gamma\eta(t) = 0 \tag{C.8}$$

It follows from Equation (C.8) that γ is a positive constant so that the response is oscillatory (not growing or decaying) since one is after the vibratory response with small oscillations. Therefore this positive constant can be expressed as the square of another constant: $\gamma = \omega^2$.
The solution forms of Equations (C.7) and (C.8) are then

$$\phi(x) = A\cos\left(\frac{\lambda}{L}x\right) + B\cosh\left(\frac{\lambda}{L}x\right) + C\sin\left(\frac{\lambda}{L}x\right) + D\sinh\left(\frac{\lambda}{L}x\right) \tag{C.9}$$

$$\eta(t) = E\cos\omega t + F\sin\omega t \tag{C.10}$$

where A, B, C, D, E, and F are unknown constants and

$$\lambda^4 = \omega^2\frac{mL^4}{YI} \tag{C.11}$$

Equation (C.4) can be employed in Equations (C.2) and (C.3) to obtain

$$\phi(0) = 0, \quad \left.\frac{d\phi_r(x)}{dx}\right|_{x=0} = 0 \tag{C.12}$$

$$\left[YI\frac{d^2\phi(x)}{dx^2} - \omega^2 I_t\frac{d\phi(x)}{dx}\right]_{x=L} = 0, \quad \left[YI\frac{d^3\phi(x)}{dx^3} + \omega^2 M_t\phi(x)\right]_{x=L} = 0 \tag{C.13}$$

The spatial form of the boundary conditions, Equations (C.12) and (C.13), should then be used to find the values of λ which give non-trivial $\phi(x)$. This process is called the *differential eigenvalue problem* and it is covered next.

C.1.3 Differential Eigenvalue Problem

Using Equation (C.9) in Equations (C.12) gives

$$A + B = 0 \tag{C.14}$$

$$C + D = 0 \tag{C.15}$$

Equation (C.9) then becomes

$$\phi(x) = A\left[\cos\left(\frac{\lambda}{L}x\right) - \cosh\left(\frac{\lambda}{L}x\right)\right] + C\left[\sin\left(\frac{\lambda}{L}x\right) - \sinh\left(\frac{\lambda}{L}x\right)\right] \qquad (C.16)$$

Hence the unknown constants are A and C only. Using Equation (C.16) in the remaining two boundary conditions given by Equations (C.13) yields

$$\begin{bmatrix} \cos\lambda + \cosh\lambda - \dfrac{\lambda^3 I_t}{mL^3}(\sin\lambda + \sinh\lambda) & \sin\lambda + \sinh\lambda + \dfrac{\lambda^3 I_t}{mL^3}(\cos\lambda - \cosh\lambda) \\[2mm] \sin\lambda - \sinh\lambda + \dfrac{\lambda M_t}{mL}(\cos\lambda - \cosh\lambda) & -\cos\lambda - \cosh\lambda + \dfrac{\lambda M_t}{mL}(\sin\lambda - \sinh\lambda) \end{bmatrix}\begin{Bmatrix} A \\ C \end{Bmatrix} = \begin{Bmatrix} 0 \\ 0 \end{Bmatrix}$$

$$(C.17)$$

The coefficient matrix in Equation (C.17) has to be singular in order to obtain non-trivial values of A and C (hence non-trivial $\phi(x)$). Setting the determinant of the above coefficient matrix gives the *characteristic equation* of the differential eigenvalue problem as

$$1 + \cos\lambda\cosh\lambda + \lambda\frac{M_t}{mL}(\cos\lambda\sinh\lambda - \sin\lambda\cosh\lambda) - \frac{\lambda^3 I_t}{mL^3}(\cosh\lambda\sin\lambda + \sinh\lambda\cos\lambda)$$

$$+ \frac{\lambda^4 M_t I_t}{m^2 L^4}(1 - \cos\lambda\cosh\lambda) = 0 \qquad (C.18)$$

For the given system parameters m, L, M_t, and I_t, one can solve for the roots of Equation (C.18), which are the *eigenvalues* of the system (note that M_t/mL and I_t/mL^3 are dimensionless in the above equation).[3] The boundary-value problem defined in Section C.1.1 is positive definite [1], hence the system has infinitely many positive eigenvalues for the infinitely many *natural modes* of vibration. The eigenvalue (or the dimensionless *frequency parameter*) of the rth vibration mode is denoted here as λ_r (where r is a positive integer), and it is associated with the rth *eigenfunction* denoted by $\phi_r(x)$:

$$\phi_r(x) = A_r\left[\cos\frac{\lambda_r}{L}x - \cosh\frac{\lambda_r}{L}x + \varsigma_r\left(\sin\frac{\lambda_r}{L}x - \sinh\frac{\lambda_r}{L}x\right)\right] \qquad (C.19)$$

where the second row of Equation (C.17) is used in Equation (C.16) to keep only a single modal constant (A_r) while ς_r is obtained from

$$\varsigma_r = \frac{\sin\lambda_r - \sinh\lambda_r + \lambda_r\dfrac{M_t}{mL}(\cos\lambda_r - \cosh\lambda_r)}{\cos\lambda_r + \cosh\lambda_r - \lambda_r\dfrac{M_t}{mL}(\sin\lambda_r - \sinh\lambda_r)} \qquad (C.20)$$

---

[3] If the tip mass can be modeled as a point mass, that is, if the rotary inertia of the tip mass is negligible, $I_t = 0$ can be used in these expressions. In the absence of a tip mass, simply $M_t = I_t = 0$.

The *undamped natural frequency* (or the eigenfrequency) of free oscillations for the *r*th vibration mode is obtained from Equation (C.11) as

$$\omega_r = \lambda_r^2 \sqrt{\frac{YI}{mL^4}} \qquad\qquad (C.21)$$

and it is associated with the natural *mode shape* of the *r*th vibration mode given by Equation (C.19).

C.1.4 Response to Initial Conditions

From Equation (C.4), the natural motion of the *r*th vibration mode becomes

$$w_r(x, t) = \phi_r(x)\eta_r(t) \qquad\qquad (C.22)$$

where $\eta_r(t)$ is called the *modal coordinate* (or the *modal response*). Since the distributed-parameter system has infinitely many vibration modes, the general response is a linear combination of the contributions from all vibration modes:

$$w(x, t) = \sum_{r=1}^{\infty} \phi_r(x)\eta_r(t) \qquad\qquad (C.23)$$

Using Equations (C.10) and (C.19) in (C.23), the response to initial conditions becomes

$$w(x, t) = \sum_{r=1}^{\infty} \left[\cos \frac{\lambda_r}{L}x - \cosh \frac{\lambda_r}{L}x + \varsigma_r \left(\sin \frac{\lambda_r}{L}x - \sinh \frac{\lambda_r}{L}x \right) \right] (A_r \cos \omega_r t + B_r \sin \omega_r t)$$

$$(C.24)$$

where A_r and B_r are the unknown constants which can be solved for using the initial conditions $w(x, 0)$ and $\partial w(x, t)/\partial t|_{t=0}$.

C.1.5 Orthogonality of the Eigenfunctions

Substituting two distinct solutions (modes *r* and *s*) of the eigenvalue problem into the spatial Equation (C.7) gives [1]

$$YI\frac{d^4\phi_r(x)}{dx^4} = \omega_r^2 m\phi_r(x) \qquad\qquad (C.25)$$

$$YI\frac{d^4\phi_s(x)}{dx^4} = \omega_s^2 m\phi_s(x) \qquad\qquad (C.26)$$

Multiplying Equation (C.25) by $\phi_s(x)$ and integrating over the domain gives

$$\int_0^L \phi_s(x) YI \frac{d^4\phi_r(x)}{dx^4} dx = \omega_r^2 \int_0^L \phi_s(x) m\phi_r(x) \, dx \tag{C.27}$$

Applying integration by parts twice to the left hand side of Equation (C.27) and using Equations (C.12) and (C.13), one can obtain

$$\int_0^L \frac{d^2\phi_s(x)}{dx^2} YI \frac{d^2\phi_r(x)}{dx^2} dx$$

$$= \omega_r^2 \left\{ \int_0^L \phi_s(x) m\phi_r(x) \, dx + \phi_s(L) M_t \phi_r(L) + \left[\frac{d\phi_s(x)}{dx} I_t \frac{d\phi_r(x)}{dx} \right]_{x=L} \right\} \tag{C.28}$$

Multiplying Equation (C.26) by $\phi_r(x)$ and following a similar procedure gives

$$\int_0^L \frac{d^2\phi_s(x)}{dx^2} YI \frac{d^2\phi_r(x)}{dx^2} dx$$

$$= \omega_s^2 \left\{ \int_0^L \phi_s(x) m\phi_r(x) \, dx + \phi_s(L) M_t \phi_r(L) + \left[\frac{d\phi_s(x)}{dx} I_t \frac{d\phi_r(x)}{dx} \right]_{x=L} \right\} \tag{C.29}$$

Subtracting Equation (C.29) from (C.28) and recalling that r and s are distinct modes ($\omega_r^2 \neq \omega_s^2$), the orthogonality condition of the eigenfunctions is obtained as

$$\int_0^L \phi_s(x) m\phi_r(x) \, dx + \phi_s(L) M_t \phi_r(L) + \left[\frac{d\phi_s(x)}{dx} I_t \frac{d\phi_r(x)}{dx} \right]_{x=L} = 0, \quad \omega_r^2 \neq \omega_s^2 \tag{C.30}$$

C.1.6   Normalization of the Eigenfunctions

It follows from the orthogonality condition given by Equation (C.30) that

$$\int_0^L \phi_s(x) m\phi_r(x) \, dx + \phi_s(L) M_t \phi_r(L) + \left[\frac{d\phi_s(x)}{dx} I_t \frac{d\phi_r(x)}{dx} \right]_{x=L} = \delta_{rs} \tag{C.31}$$

where δ_{rs} is the Kronecker delta. Equation (C.31) is the orthogonality condition of the eigenfunctions and $s = r$ can be used to solve for the modal amplitude A_r of the eigenfunctions (i.e., to normalize the eigenfunctions). The eigenfunctions normalized according to Equation (C.31) are called the *mass-normalized* eigenfunctions.

Equations (C.27)–(C.29) can be used along with Equations (C.12) and (C.13) to give the companion orthogonality condition as

$$\int_0^L \phi_s(x) YI \frac{d^4\phi_r(x)}{dx^4} dx - \left[\phi_s(x) YI \frac{d^3\phi_r(x)}{dx^3}\right]_{x=L} + \left[\frac{d\phi_s(x)}{dx} YI \frac{d^2\phi_r(x)}{dx^2}\right]_{x=L} = \omega_r^2 \delta_{rs}$$

(C.32)

Finally, from Equations (C.29) and (C.31), one extracts

$$\int_0^L \frac{d^2\phi_s(x)}{dx^2} YI \frac{d^2\phi_r(x)}{dx^2} dx = \omega_r^2 \delta_{rs}$$

(C.33)

which is the alternative companion form. Once any one of Equations (C.31)–(C.33) is satisfied, the other two are automatically satisfied. Therefore, either one of these three expressions can be used for mass normalizing the eigenfunctions (i.e., to solve for A_r).

C.1.7 Response to External Forcing

Consider the undamped forced vibrations of the beam shown in Figure C.1 under the excitation of the distributed transverse force per length $f(x, t)$ acting in the positive z-direction. The governing equation of motion is then

$$YI \frac{\partial^4 w(x, t)}{\partial x^4} + m \frac{\partial^2 w(x, t)}{\partial t^2} = f(x, t)$$

(C.34)

The solution of Equation (C.34) can be expressed in the form of Equation (C.23). Substituting Equation (C.23) into Equation (C.34) gives

$$YI \frac{\partial^4}{\partial x^4} \left[\sum_{r=1}^{\infty} \phi_r(x) \eta_r(t)\right] + m \frac{\partial^2}{\partial t^2} \left[\sum_{r=1}^{\infty} \phi_r(x) \eta_r(t)\right] = f(x, t)$$

(C.35)

Multiplying both sides of Equation (C.35) by the mass-normalized eigenfunction $\phi_s(x)$ and integrating over the length of the beam, one obtains

$$\sum_{r=1}^{\infty} \left[\eta_r(t) \int_0^L \phi_s(x) YI \frac{d^4\phi_r(x)}{dx^4} dx\right] + \sum_{r=1}^{\infty} \left[\frac{d^2\eta_r(t)}{dt^2} \int_0^L \phi_s(x) m \phi_r(x) dx\right] = \int_0^L \phi_s(x) f(x, t) dx$$

(C.36)

Using the orthogonality conditions given by Equations (C.31) and (C.32), one can then express the following:

$$\frac{d^2\eta_s(t)}{dt^2} + \omega_s^2\eta_s(t) - \sum_{r=1}^{\infty}\left\{\phi_s(x)\left[M_t\phi_r(x)\frac{d^2\eta_r(t)}{dt^2} - YI\frac{d^3\phi_r(x)}{dx^3}\eta_r(t)\right]\right\}_{x=L}$$

$$-\sum_{r=1}^{\infty}\left\{\frac{d\phi_s(x)}{dx}\left[I_t\frac{d\phi_r(x)}{dx}\frac{d^2\eta_r(t)}{dt^2} + YI\frac{d^2\phi_r(x)}{dx^2}\eta_r(t)\right]\right\}_{x=L}$$

$$= \int_0^L \phi_s(x)f(x,t)\,dx \tag{C.37}$$

In view of the boundary conditions given by Equations (C.2) and (C.3), Equation (C.37) reduces to

$$\frac{d^2\eta_r(t)}{dt^2} + \omega_r^2\eta_r(t) = F_r(t) \tag{C.38}$$

where

$$F_r(t) = \int_0^L \phi_r(x)f(x,t)\,dx \tag{C.39}$$

If the applied force is arbitrary, the Duhamel integral (undamped version) can be used to solve for the modal coordinate, which can then be used in Equation (C.23) to give the physical response for zero initial conditions as

$$w(x,t) = \sum_{r=1}^{\infty}\frac{\phi_r(x)}{\omega_r}\int_0^t F_r(\tau)\sin\omega_r(t-\tau)d\tau \tag{C.40}$$

If the applied force is harmonic of the form

$$f(x,t) = F(x)\cos\omega t \tag{C.41}$$

The steady-state response can be obtained as

$$w(x,t) = \sum_{r=1}^{\infty}\frac{\phi_r(x)}{\omega_r^2 - \omega^2}F_r\cos\omega t \tag{C.42}$$

where

$$F_r = \int_0^L \phi_r(x)F(x)\,dx \tag{C.43}$$

C.2 Longitudinal Vibrations

The analytical modal analysis for the linear longitudinal vibrations of a uniform clamped–free bar[4] with a tip mass attachment is summarized in this section. The steps are identical to those followed in the transverse vibrations case; however, the governing equation of motion, hence the resulting expressions, are completely different.

C.2.1 Boundary-Value Problem

Consider the free vibrations of the uniform bar shown in Figure C.1 in the longitudinal direction (i.e., x-direction). The governing equation of motion can be obtained as [1]

$$YA\frac{\partial^2 u(x,t)}{\partial x^2} - m\frac{\partial^2 u(x,t)}{\partial t^2} = 0 \qquad (C.44)$$

where $u(x,t)$ is the longitudinal displacement at any point x and time t, YA is the axial stiffness (Y is the elastic modulus and A is the cross-sectional area), and m is the mass per unit length of the bar. Equation (C.44) is in the one-dimensional form of the celebrated *wave equation*.

The displacement at the clamped end ($x = 0$) is zero, while the force resultant of the internal dynamic stresses is in equilibrium with the inertia of the tip mass attachment at the free end ($x = L$). Therefore, the boundary conditions are

$$u(0,t) = 0, \qquad \left[YA\frac{\partial u(x,t)}{\partial x} + M_t \frac{\partial^2 u(x,t)}{\partial t^2} \right]_{x=L} = 0 \qquad (C.45)$$

Equations (C.44) and (C.45) define the boundary-value problem for the longitudinal vibrations of a uniform clamped–free bar with a tip mass attachment.

C.2.2 Solution Using the Method of Separation of Variables

Separating the space and the time variables yields

$$u(x,t) = \varphi(x)\chi(t) \qquad (C.46)$$

which can be employed in Equation (C.44) to give

$$\frac{YA}{m}\frac{1}{\varphi(x)}\frac{d^2\varphi(x)}{dx^2} = \frac{1}{\chi(t)}\frac{d^2\chi(t)}{dt^2} = \upsilon \qquad (C.47)$$

---

[4] Following the conventional technical jargon, the structure is called a bar (or a rod) when longitudinal vibrations are discussed.

where the space-dependent and time-dependent sides are equal to each other and, following the standard argument of the separation of variables solution procedure, they are equal to a constant, which is denoted in Equation (C.47) by υ, yielding

$$\frac{d^2\varphi(x)}{dx^2} - \upsilon\frac{m}{YA}\varphi(x) = 0 \tag{C.48}$$

$$\frac{d^2\chi(t)}{dt^2} - \upsilon\chi(t) = 0 \tag{C.49}$$

It is clear from Equation (C.49) that non-growing or non-decaying solutions in time require that $\upsilon < 0$; therefore, one can set $\upsilon = -\omega^2$, where ω is another constant.

The solution forms of Equations (C.48) and (C.49) are then

$$\varphi(x) = A\cos\left(\frac{\alpha}{L}x\right) + B\sin\left(\frac{\alpha}{L}x\right) \tag{C.50}$$

$$\chi(t) = C\cos\omega t + D\sin\omega t \tag{C.51}$$

where A, B, C, and D are unknown constants and

$$\alpha^2 = \omega^2\frac{mL^2}{YA} \tag{C.52}$$

Equation (C.46) can be employed in Equation (C.45) to obtain the spatial form of the boundary conditions as

$$\varphi(0) = 0, \quad \left[YA\frac{d\varphi(x)}{dx} - \omega^2 M_t\varphi(x)\right]_{x=L} = 0 \tag{C.53}$$

C.2.3   Differential Eigenvalue Problem

Using Equation (C.50) in the first one of Equations (C.53) gives

$$A = 0 \tag{C.54}$$

which reduces Equation (C.50) to

$$\varphi(x) = B\sin\left(\frac{\alpha}{L}x\right) \tag{C.55}$$

Equation (C.55) can be used in the second boundary condition to give the characteristic equation as

$$\frac{M_t}{mL}\alpha\sin\alpha - \cos\alpha = 0 \tag{C.56}$$

Given the bar mass (mL) and the tip mass (M_t), Equation (C.56) can be used in order to solve for the eigenvalue (or the dimensionless frequency parameter) α_r of the rth vibration mode (the system is again positive definite and one seeks positive-valued eigenvalues). From Equation (C.55), the eigenfunction of the rth vibration mode is

$$\varphi_r(x) = B_r \sin\left(\frac{\alpha_r}{L}x\right) \tag{C.57}$$

where B_r is the modal amplitude constant (obtained from the normalization conditions in the solution process). Rearranging Equation (C.52), the undamped natural frequency of the rth vibration mode is obtained from[5]

$$\omega_r = \alpha_r\sqrt{\frac{YA}{mL^2}} \tag{C.58}$$

C.2.4   Response to Initial Conditions

The natural motion of the rth vibration mode can be expressed using Equation (C.46) as

$$u_r(x, t) = \varphi_r(x)\chi_r(t) \tag{C.59}$$

where $\chi_r(t)$ is the modal coordinate of the rth vibration mode. The general response is then expressed as a linear combination of the contributions from all vibration modes:

$$u(x, t) = \sum_{r=1}^{\infty} \varphi_r(x)\chi_r(t) \tag{C.60}$$

Using Equations (C.51) and (C.57) in Equation (C.60), the response to initial conditions becomes

$$u(x, t) = \sum_{r=1}^{\infty} (C_r \cos \omega_r t + D_r \sin \omega_r t) \sin\left(\frac{\alpha_r}{L}x\right) \tag{C.61}$$

where C_r and D_r are the unknown constants which can be solved for using the initial conditions $u(x, 0)$ and $\partial u(x, t)/\partial t|_{t=0}$.

C.2.5   Orthogonality of the Eigenfunctions

Two distinct solutions of the differential eigenvalue problem (for modes r and s) can be substituted into Equation (C.48) to give

$$-YA\frac{d^2\varphi_r(x)}{dx^2} = \omega_r^2 m\varphi_r(x) \tag{C.62}$$

$$-YA\frac{d^2\varphi_s(x)}{dx^2} = \omega_s^2 m\varphi_s(x) \tag{C.63}$$

---

[5] The undamped natural frequency is denoted by ω_r in both the transverse and the longitudinal vibration problems for ease of notation, but they are not identical.

Multiplying Equation (C.62) by $\varphi_s(x)$ and integrating the resulting equation over the length of the bar, one can obtain

$$-\int_0^L \varphi_s(x)YA\frac{d^2\varphi_r(x)}{dx^2}dx = \omega_r^2 \int_0^L \varphi_s(x)m\varphi_r(x)\,dx \qquad (C.64)$$

Integrating by parts the left hand side of Equation (C.64) and using Equations (C.53) leads to

$$\int_0^L \frac{d\varphi_s(x)}{dx}YA\frac{d\varphi_r(x)}{dx}dx = \omega_s^2 \left[\int_0^L \varphi_s(x)m\varphi_r(x)\,dx + \varphi_s(L)M_t\varphi_r(L)\right] \qquad (C.65)$$

Likewise, multiplying Equation (C.63) by $\varphi_r(x)$ and following the same steps, one can obtain

$$\int_0^L \frac{d\varphi_r(x)}{dx}YA\frac{d\varphi_s(x)}{dx}dx = \omega_r^2 \left[\int_0^L \varphi_r(x)m\varphi_s(x)\,dx + \varphi_r(L)M_t\varphi_s(L)\right] \qquad (C.66)$$

Subtracting Equation (C.66) from Equation (C.65) and recalling that r and s are distinct modes (hence $\omega_r^2 \neq \omega_s^2$), one concludes

$$\int_0^L \varphi_s(x)m\varphi_r(x)\,dx + \varphi_s(L)M_t\varphi_r(L) = 0, \quad \omega_r^2 \neq \omega_s^2 \qquad (C.67)$$

C.2.6 Normalization of the Eigenfunctions

The foregoing orthogonality condition can then be used to express the following mass normalization condition of the eigenfunctions:

$$\int_0^L \varphi_s(x)m\varphi_r(x)\,dx + \varphi_s(L)M_t\varphi_r(L) = \delta_{rs} \qquad (C.68)$$

Equations (C.64)–(C.66) can be used with Equations (C.53) to give the companion orthogonality relation as

$$-\int_0^L \varphi_s(x)YA\frac{d^2\varphi_r(x)}{dx^2}dx + \left[\varphi_s(x)YA\frac{d\varphi_r(x)}{dx}\right]_{x=L} = \omega_r^2\delta_{rs} \qquad (C.69)$$

and the alternative form is

$$\int_0^L \frac{d\varphi_r(x)}{dx} YA \frac{d\varphi_s(x)}{dx} dx = \omega_r^2 \delta_{rs} \tag{C.70}$$

which is obtained simply from Equations (C.66) and (C.68).

Equation (C.57) can be substituted into Equation (C.68) to solve for B_r (by setting $s = r$). Then the mass-normalized form of the eigenfunctions becomes

$$\varphi_r(x) = \frac{1}{\sqrt{\frac{mL}{2}\left(1 - \frac{\sin 2\alpha_r}{2\alpha_r}\right) + M_t \sin^2 \alpha_r}} \sin \frac{\alpha_r}{L} x \tag{C.71}$$

which satisfies Equations (C.69) and (C.70) as well.

C.2.7   Response to External Forcing

If the undamped bar shown in Figure C.1 is excited by the distributed axial force per length $p(x, t)$ acting in the positive x-direction, the governing equation of motion becomes

$$-YA \frac{\partial^2 u(x, t)}{\partial x^2} + m \frac{\partial^2 u(x, t)}{\partial t^2} = p(x, t) \tag{C.72}$$

The solution of Equation (C.72) can be expressed in the form of Equation (C.60) and substituting the latter into the former one obtains

$$-YA \frac{\partial^2}{\partial x^2}\left[\sum_{r=1}^\infty \varphi_r(x)\chi_r(t)\right] + m \frac{\partial^2}{\partial t^2}\left[\sum_{r=1}^\infty \varphi_r(x)\chi_r(t)\right] = p(x, t) \tag{C.73}$$

Multiplying both sides of Equation (C.73) by the mass-normalized eigenfunctions $\varphi_s(x)$ and integrating over the length of the bar gives

$$-\sum_{r=1}^\infty \left[\chi_r(t)\int_0^L \varphi_s(x) YA \frac{d^2\varphi_r(x)}{dx^2} dx\right] + \sum_{r=1}^\infty \left[\frac{d^2\chi_r(t)}{dt^2}\int_0^L \varphi_s(x) m\varphi_r(x) dx\right]$$

$$= \int_0^L \varphi_s(x) p(x, t) dx \tag{C.74}$$

Then, using Equations (C.68) and (C.69), Equation (C.74) can be rearranged to give

$$
\frac{d^2\chi_s(t)}{dt^2} + \omega_s^2\chi_s(t) - \sum_{r=1}^{\infty} \left\{ \varphi_s(x) \left[M_t\varphi_r(x)\frac{d^2\chi_r(t)}{dt^2} + YA\frac{d\varphi_r(x)}{dx}\chi_r(t) \right] \right\}_{x=L}
$$

$$
= \int_0^L \varphi_s(x)p(x,t)\,dx \tag{C.75}
$$

Considering the boundary conditions given by Equations (C.45), Equation (C.75) reduces to

$$
\frac{d^2\eta_r(t)}{dt^2} + \omega_r^2\eta_r(t) = P_r(t) \tag{C.76}
$$

where

$$
P_r(t) = \int_0^L \varphi_r(x)p(x,t)\,dx \tag{C.77}
$$

For an arbitrary forcing function, the physical response $u(x,t)$ can be obtained using the undamped version of the Duhamel integral (with zero initial conditions) as

$$
u(x,t) = \sum_{r=1}^{\infty} \frac{\varphi_r(x)}{\omega_r} \int_0^t P_r(\tau)\sin\omega_r(t-\tau)\,d\tau \tag{C.78}
$$

If the applied force is harmonic of the form

$$
p(x,t) = P(x)\cos\omega t \tag{C.79}
$$

The steady-state response can be obtained as

$$
u(x,t) = \sum_{r=1}^{\infty} \frac{\varphi_r(x)}{\omega_r^2 - \omega^2} P_r\cos\omega t \tag{C.80}
$$

where

$$
P_r = \int_0^L \varphi_r(x)P(x)\,dx \tag{C.81}
$$

References

1. Meirovitch, L. (2001) *Fundamentals of Vibrations*, McGraw-Hill, New York.
2. Meriam, J.L. and Kraige, L.G. (2001) *Engineering Mechanics: Dynamics*, John Wiley & Sons, Inc., New York.
3. Greenberg, M.D. (1998) *Advanced Engineering Mathematics*, Prentice Hall, Englewood Cliffs, NJ.

Appendix D

Strain Nodes of a Uniform Thin Beam for Cantilevered and Other Boundary Conditions

D.1 Strain Nodes of a Uniform Thin Cantilever without a Tip Mass

For modal vibrations, the *curvature eigenfunction* is simply the second derivative of the *displacement eigenfunction* given by Equation (C.19). In the absence of a tip mass ($M_t = I_t = 0$),

$$\frac{d^2\phi_r(x)}{dx^2} = -\left(\frac{\lambda_r}{L}\right)^2 A_r \left[\cos\frac{\lambda_r}{L}x + \cosh\frac{\lambda_r}{L}x + \frac{\sin\lambda_r - \sinh\lambda_r}{\cos\lambda_r + \cosh\lambda_r}\left(\sin\frac{\lambda_r}{L}x + \sinh\frac{\lambda_r}{L}x\right)\right] \tag{D.1}$$

Since the system is positive definite ($\lambda_r > 0$, hence $\lambda_r \neq 0$), the dimensionless positions of the strain nodes (i.e., the inflection points of the eigenfunctions) are roots of the following equation in the interval $0 < \bar{x} < 1$:

$$\cos\lambda_r\bar{x} + \cos\lambda_r\bar{x} + \frac{\sin\lambda_r - \sinh\lambda_r}{\cos\lambda_r + \cosh\lambda_r}(\sin\lambda_r\bar{x} + \sin\lambda_r\bar{x}) = 0 \tag{D.2}$$

where $\bar{x} = x/L$ is the dimensionless position along the beam axis and the eigenvalues are obtained from

$$1 + \cos\lambda\cosh\lambda = 0 \tag{D.3}$$

which is reduced from Equation (C.18). The dimensionless positions of the strain nodes over the length of the beam for the first five modes are listed in Table D.1. For convenience, the frequency parameters λ_r of the first five vibration modes are also provided in the table and they can be used in Equation (C.21) to predict the undamped natural frequencies of the cantilever (in short-circuit conditions if the constant electric field elastic modulus of the piezoceramic

Piezoelectric Energy Harvesting, First Edition. Alper Erturk and Daniel J. Inman.
© 2011 John Wiley & Sons, Ltd. Published 2011 by John Wiley & Sons, Ltd.

Table D.1 Frequency parameters and dimensionless positions of the strain nodes for a uniform Euler–Bernoulli beam with clamped–free boundary conditions (without a tip mass)

Mode	Frequency parameter (λ_r)	Strain node positions on x-axis ($\bar{x} = x/L$)			
1	1.87510407	—	—	—	—
2	4.69409113	0.2165	—	—	—
3	7.85475744	0.1323	0.4965	—	—
4	10.9955407	0.0944	0.3559	0.6417	—
5	14.1371684	0.0735	0.2768	0.5001	0.7212

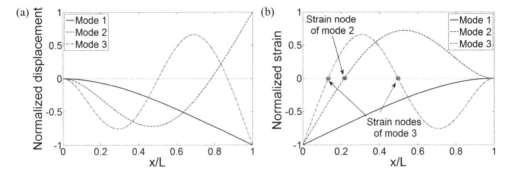

Figure D.1 Normalized (a) displacement and (b) strain mode shapes of a cantilevered beam without a tip mass for the first three vibration modes

is used in calculating YI). As can be seen from Table D.1, the rth vibration mode has $r - 1$ strain nodes and the only vibration mode of a cantilevered beam without strain nodes is the fundamental mode.

The normalized displacement mode shapes and the strain mode shapes of the first three vibration modes of a cantilevered beam without a tip mass are displayed in Figure D.1. In agreement with Table D.1, the strain node of the second vibration mode is located at $\bar{x} = 0.2165$ while the third vibration mode has two strain nodes at $\bar{x} = 0.1323$ and $\bar{x} = 0.4965$. These positions give an idea about how to locate the segmented electrodes (or how to etch the full electrodes) for harvesting energy from these modes without cancellation of the electrical outputs. For instance, to avoid cancellation in harvesting energy from the second vibration mode, two electrode pairs can be used to cover the regions $0 \leq \bar{x} < 0.2165$ and $0.2165 < \bar{x} \leq 1$ separately.[1] The voltage outputs of these electrode pairs will be out of phase with each other by $180°$ and they can be combined accordingly in the electrical circuit [1]. Otherwise, if continuous electrodes cover the entire length of the beam ($0 \leq \bar{x} \leq 1$), the negative area ($0 \leq \bar{x} < 0.2165$) under the strain curve cancels a considerable portion of the positive area ($0.2165 < \bar{x} \leq 1$), resulting in a significant amount of reduction in the voltage output.

---

[1] The idea of using "etched" electrodes to avoid cancellation of the piezoelectric reaction in vibrations of crystals dates back to Cady [2]. An alternative is to apply *patterned polarization* as in Kim *et al.* [3,4], which requires partial repolarization of the piezoceramic according to the strain node lines (so that full electrodes can be used).

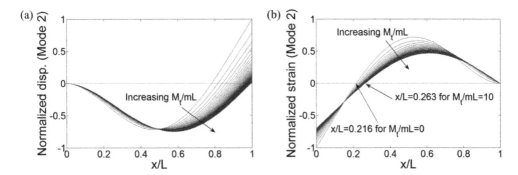

Figure D.2 Variation of the normalized (a) displacement and (b) strain mode shapes of the second vibration mode for a cantilevered beam with the ratio of tip mass to beam mass

D.2 Effect of a Tip Mass on the Strain Nodes

If a tip mass M_t of negligible mass moment of inertia ($I_t = 0$), that is, a point mass, is attached rigidly at $x = L$ to the cantilevered beam as shown in Figure C.1, the displacement and the strain mode shapes are affected. Figure D.2a shows the variation of the second mode shape whereas Figure D.2b displays the variation of the strain distribution of the second vibration mode with increasing M_t/mL. The shift in the strain node position due to increasing M_t/mL can be seen in Figure D.2b. As the M_t/mL ratio goes from 0 to 10, the strain node of the second mode moves from $\bar{x} = 0.2165$ to $\bar{x} = 0.2632$.

Figure D.3a shows the strain node positions of the second and the third modes versus M_t/mL ratio. As the M_t/mL ratio is increased from 0 to 10, the only strain node of the second mode moves from $\bar{x} = 0.2165$ to $\bar{x} = 0.2632$ whereas the two strain nodes of the third vibration mode move from $\bar{x} = 0.1323$ and $\bar{x} = 0.4965$ to $\bar{x} = 0.1468$ and $\bar{x} = 0.5530$, respectively. It is also useful to investigate the variation of the frequency parameters with the M_t/mL ratio, which is given in Figure D.3b for the first five vibration modes (recall that these frequency parameters give the undamped natural frequencies when they are used in Equation (C.21)).

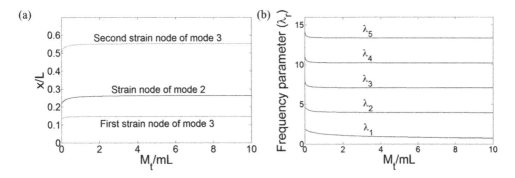

Figure D.3 (a) Variations of the strain node positions of the second and third vibration modes and (b) variations of the frequency parameters of the first five vibration modes with the ratio of tip mass to beam mass

The strain nodes move slightly toward the free end of the beam while the frequency parameters decrease with increasing M_t/mL ratio. The positions of the strain nodes are more sensitive to the variations in M_t/mL ratio in the relatively low M_t/mL region (i.e., for $M_t/mL \leq 1$). As far as the frequency parameters are concerned, other than that of the first vibration mode, all frequency parameters converge to a non-zero value as $M_t/mL \to \infty$. Careful investigation shows that, as $M_t/mL \to \infty$, the frequency parameter of the rth mode of a clamped–free beam with a tip mass converges to that of the $(r - 1)$th mode of a clamped–pinned beam without a tip mass[2] (except for λ_1, which goes to zero with a very slow rate). The fact that the boundary conditions of a cantilevered harvester beam with a tip mass shift from clamped–free to clamped–pinned as $M_t/mL \to \infty$ makes sense as the mass moment of inertia of the tip mass is neglected by setting $I_t = 0$ in Equation (C.18). Therefore, direct consideration of the strain nodes of a clamped–pinned beam should give a good estimate of the strain nodes for very large values of M_t/mL in modes other than the first mode. However, if the rotary inertia of the tip mass is not negligible and as it increases as $M_t/mL \to \infty$, the boundary conditions shift from clamped–free to clamped–clamped and it becomes more reasonable to estimate the strain node positions (of modes $r \geq 2$) from the eigenfunctions of a clamped–clamped beam for large values of M_t/mL.

If the mass moment of inertia at the tip (I_t) is not negligible, it can be normalized with respect to the rotary inertia of the free rigid beam about a certain point and a similar dimensionless analysis can be performed using Equation (C.18). The form of the eigenfunction expression given by Equation (C.19) is still the same; however, the eigenvalues to be used in this expression should be extracted from Equation (C.18) without neglecting the I_t term. As mentioned previously, for a large tip mass and mass moment of inertia, the free end of the beam also acts as a clamped end for modes $r \geq 2$. This fact is also evident from the dominating term in Equation (C.18) for $(M_t/mL)(I_t/mL^3) \to \infty$, which is the characteristic equation of a uniform Euler–Bernoulli beam with clamped–clamped boundary conditions $(1 - \cos \lambda \cosh \lambda = 0)$. Clamped–clamped boundary conditions may cause strong cancellations in the electrical outputs if full electrodes are used for covering the piezoceramic layer(s). Therefore, employing a large tip mass for reducing the natural frequencies of a cantilevered harvester has the side effect of reducing the electrical response of the vibration modes other than the fundamental mode. If higher vibrations of a cantilevered energy harvester are to be excited in a practical application, the addition of a tip mass results in a trade-off, as it has an undesired effect on the vibration modes other than the first mode (which can be avoided by using segmented electrodes [1]).

D.3 Strain Nodes for Other Boundary Conditions

Since the literature of energy harvesting [5] and the literature dealing with piezoelectric beams [6] have considered boundary conditions other than clamped–free as well, the numerical data of the strain node positions for some other practical boundary conditions are tabulated in this section. Tables D.2–D.4 display the positions of the strain nodes for the first five vibration modes of uniform Euler–Bernoulli beams with pinned–pinned, clamped–clamped,

---

[2] The dominating term in Equation (C.18) for $M_t/mL \to \infty$ is the characteristic equation of a clamped–pinned beam: $\tanh \lambda - \tan \lambda = 0$ (because the system is positive definite, $\lambda \neq 0$ in the dominating term).

Table D.2 Frequency parameters and dimensionless positions of the strain nodes for a uniform
Euler–Bernoulli beam with pinned–pinned boundary conditions

Mode	Frequency parameter (λ_r)	Strain node positions on x-axis ($\bar{x} = x/L$)			
1	π	—	—	—	—
2	2π	1/2	—	—	—
3	3π	1/3	2/3	—	—
4	4π	1/4	1/2	3/4	—
5	5π	1/5	2/5	3/5	4/5

and clamped–pinned boundary conditions. The frequency parameters are also provided and
they can be used in Equation (C.21) to predict the undamped natural frequencies.

Since the pinned–pinned (Table D.2) and clamped–clamped (Table D.3) boundary conditions are symmetric boundary conditions (yielding symmetric and anti-symmetric mode
shapes for odd and even modes, respectively), the positions of the strain nodes are symmetric
with respect to the center $\bar{x} = 0.5$ of the beam. However, for the clamped–pinned boundary
conditions (Table D.4), no such symmetry exists. It should be noted that it is safe to cover
the entire surface with continuous electrodes for harvesting energy from the first vibration
mode of a pinned–pinned beam. The rule for the pinned–pinned case is the same as the
clamped–free case (Table D.1): that is, the rth vibration mode has $r - 1$ strain nodes. On
the other hand, a beam with clamped–clamped boundary conditions has two strain nodes in
the first vibration mode. According to Table D.3, the rth vibration mode of a clamped–clamped
beam has $r + 1$ strain nodes. Hence, three electrode pairs (with discontinuities at $\bar{x} = 0.2242$
and at $\bar{x} = 0.7758$) can be used to extract the electrical outputs of a clamped–clamped energy
harvester without cancellation for vibrations with the first mode shape. Table D.4 shows that
the rth vibration mode has r strain nodes for a clamped–pinned beam. Therefore two electrode
pairs (with a discontinuity at $\bar{x} = 0.2642$) can handle the cancellation issue for the fundamental mode excitation of a harvester beam with clamped–pinned boundary conditions (note that
the clamped boundary is $\bar{x} = 0$ in Table D.4).

Among the boundary conditions investigated here, the clamped–clamped boundary condition pair constitutes a unique case. Theoretically, for all vibration modes of a clamped–clamped
beam, the modal coupling term is *zero* if full (continuous) electrodes cover the entire beam
surface (see Equation (3.27)) and the circuit equation cannot be excited. A similar issue (total
cancellation) is expected for the even vibration modes ($r = 2, 4, 6, \ldots, 2n$ where n is an

Table D.3 Frequency parameters and dimensionless positions of the strain nodes for a uniform
Euler–Bernoulli beam with clamped–clamped boundary conditions

Mode	Frequency parameter (λ_r)	Strain node positions on x-axis ($\bar{x} = x/L$)				
1	4.73004074	0.2242	0.7758	—	—	—
2	7.85320462	0.1321	0.5000	0.8679	—	—
3	10.9956079	0.0944	0.3558	0.6442	0.9056	—
4	14.1371655	0.0735	0.2768	0.5000	0.7232	0.9265
5	17.2787597	0.0601	0.2265	0.4091	0.5909	0.7735	0.9399

Table D.4 Frequency parameters and dimensionless positions of the strain nodes for a uniform Euler–Bernoulli beam with clamped–pinned boundary conditions

Mode	Frequency parameter (λ_r)	Strain node positions on x-axis ($\bar{x} = x/L$)				
1	3.92660231	0.2642	—	—	—	—
2	7.06858275	0.1469	0.5536	—	—	—
3	10.2101761	0.1017	0.3832	0.6924	—	—
4	13.3517688	0.0778	0.2931	0.5295	0.7647	—
5	16.4933614	0.0630	0.2372	0.4286	0.6190	0.8095

integer) of the pinned–pinned case if full electrodes cover the entire beam length. For the even modes of a beam with pinned–pinned boundary conditions, the slopes at the pinned boundaries are not zero, but they are equal to each other, yielding a total cancellation at these modes due to Equation (3.27).

The data provided in Tables D.1–D.4 are also useful for modal actuation of beams (see Equations (3.31) and (3.44)). From Table D.1, the fundamental mode of a clamped–free beam can be excited by locating the piezoelectric actuator(s) anywhere along the beam. One should prefer a location close to the clamped end (see the strain distribution of mode 1 in Figure D.1b) in order to minimize the required actuation input as formerly discussed by Crawley and de Luis [7]. However, excitation of the fundamental mode of a clamped–clamped beam is more critical as the piezoelectric actuator(s) should not cover the positions $\bar{x} = 0.2242$ and $\bar{x} = 0.7758$ (Table D.3). Covering one of these strain nodes with a piezoceramic actuator may require dramatically high voltage inputs compared to a smarter choice of actuation location.

References

1. Erturk, A., Tarazaga, P., Farmer, J.R., and Inman, D.J. (2009) Effect of strain nodes and electrode configuration on piezoelectric energy harvesting from cantilevered beams. *ASME Journal of Vibration and Acoustics*, **131**, 011010.
2. Cady, W.G. (1946) *Piezoelectricity: An Introduction to the Theory and Applications of Electromechanical Phenomena in Crystals*, McGraw-Hill, New York.
3. Kim, S., Clark, W.W., and Wang, Q.M. (2005) Piezoelectric energy harvesting with a clamped circular plate: analysis. *Journal of Intelligent Material Systems and Structures*, **16**, 847–854.
4. Kim, S., Clark, W.W., and Wang, Q.M. (2005) Piezoelectric energy harvesting with a clamped circular plate: experimental study. *Journal of Intelligent Material Systems and Structures*, **16**, 855–863.
5. Guan, M.J. and Liao, W.H. (2007) On the efficiencies of piezoelectric energy harvesting circuits towards storage device voltages. *Smart Materials and Structures*, **16**, 498–505.
6. Lesieutre, G.A. and Davis, C.L. (1997) Can a coupling coefficient of a piezoelectric device be higher than those of its active material? *Journal of Intelligent Material Systems and Structures*, **8**, 859–867.
7. Crawley, E.F. and de Luis, J. (1987) Use of piezoelectric actuators as elements of intelligent structures. *AIAA Journal*, **25**, 1373–1385.

Appendix E

Numerical Data for PZT-5A and PZT-5H Piezoceramics

PZT-5A and PZT-5H piezoceramics are the most commonly used engineering piezoceramics. Their typical three-dimensional piezoelastic properties can be found on the Web [1] or in the literature [2] as given in Table E.1.

In addition to these data, the mass densities of PZT-5A and PZT-5H are reported [1,2] as 7750 and 7500 kg/m^3, respectively. The permittivity components at constant stress are given in Table E.1 in the form of constant-stress dielectric constants ($\varepsilon_{11}^T/\varepsilon_0$, $\varepsilon_{33}^T/\varepsilon_0$) where the permittivity of free space is $\varepsilon_0 = 8.854\,\text{pF/m}$ [3]. The reduced constants for the Euler–Bernoulli and Rayleigh beam theories are obtained using Equations (A.11) as listed in Table E.2 whereas those for the Timoshenko beam theory are obtained from Equations (A.17) as in Table E.3. The reduced constants for the Kirchhoff plate theory are listed in Table E.4 (based on Equations (A.24)–(A.28)).

Table E.1 Three-dimensional properties of PZT-5A and PZT-5H

	PZT-5A	PZT-5H
s_{11}^E (pm^2/N)	16.4	16.5
s_{12}^E (pm^2/N)	−5.74	−4.78
s_{13}^E (pm^2/N)	−7.22	−8.45
s_{33}^E (pm^2/N)	18.8	20.7
s_{55}^E (pm^2/N)	47.5	43.5
s_{66}^E (pm^2/N)	44.3	42.6
d_{31} (pm/V)	−171	−274
d_{33} (pm/V)	374	593
d_{15} (pm/V)	584	741
$\varepsilon_{11}^T/\varepsilon_0$	1730	3130
$\varepsilon_{33}^T/\varepsilon_0$	1700	3400

Piezoelectric Energy Harvesting, First Edition. Alper Erturk and Daniel J. Inman.
© 2011 John Wiley & Sons, Ltd. Published 2011 by John Wiley & Sons, Ltd.

Table E.2 Reduced properties of PZT-5A and PZT-5H for the Euler–Bernoulli and Rayleigh beam theories

	PZT-5A	PZT-5H
$\bar{c}_{11}^{E}$ (MPa)	61.0	60.6
$\bar{e}_{31}$ (C/m^2)	−10.4	−16.6
$\bar{\varepsilon}_{33}^{S}$ (nF/m)	13.3	25.55

Table E.3 Reduced properties of PZT-5A and PZT-5H for the Timoshenko beam theory

	PZT-5A	PZT-5H
$\bar{c}_{11}^{E}$ (MPa)	61.0	60.6
$\bar{c}_{55}^{E}$ (MPa)	21.1	23.0
$\bar{e}_{31}$ (C/m^2)	−10.4	−16.6
$\bar{\varepsilon}_{33}^{S}$ (nF/m)	13.3	25.6

Table E.4 Reduced properties of PZT-5A and PZT-5H for the Kirchhoff plate theory

	PZT-5A	PZT-5H
$\bar{c}_{11}^{E}$ (MPa)	69.5	66.2
$\bar{c}_{12}^{E}$ (MPa)	24.3	19.2
$\bar{c}_{66}^{E}$ (MPa)	22.6	23.5
$\bar{e}_{31}$ (C/m^2)	−16.0	−23.4
$\bar{\varepsilon}_{33}^{S}$ (nF/m)	9.57	17.3

References

1. Engineering Fundamentals, Inc. www.efunda.com.
2. Heinonen, E., Juuti, J., and Leppavuori, S. (2005) Characterization and modelling of 3D piezoelectric ceramic structures with ATILA software. *Journal of European Ceramic Society*, **25**, 2467–2470.
3. Standards Committee of the IEEE Ultrasonics, Ferroelectrics, and Frequency Control Society (1987) *IEEE Standard on Piezoelectricity*, IEEE, New York.

Appendix F

Constitutive Equations for an Isotropic Substructure

F.1 Three-Dimensional Form of the Constitutive Equations for an Isotropic Material

The tensorial representation of the three-dimensional constitutive law for an isotropic substructure material is [1]

$$S_{ij} = \frac{1}{Y_s} \left[(1 + v_s) T_{ij} - v_s T_{kk} \delta_{ij} \right] \tag{F.1}$$

where Y_s is the elastic modulus and v_s is Poisson's ratio (subscript s denotes the substructure). The expanded form of Equation (F.1) can be given as

$$
\begin{bmatrix} S_1 \\ S_2 \\ S_3 \\ S_4 \\ S_5 \\ S_6 \end{bmatrix}
= \frac{1}{Y_s}
\begin{bmatrix}
1 & -v_s & -v_s & 0 & 0 & 0 \\
-v_s & 1 & -v_s & 0 & 0 & 0 \\
-v_s & -v_s & 1 & 0 & 0 & 0 \\
0 & 0 & 0 & 2(1+v_s) & 0 & 0 \\
0 & 0 & 0 & 0 & 2(1+v_s) & 0 \\
0 & 0 & 0 & 0 & 0 & 2(1+v_s)
\end{bmatrix}
\begin{bmatrix} T_1 \\ T_2 \\ T_3 \\ T_4 \\ T_5 \\ T_6 \end{bmatrix}
\tag{F.2}
$$

where the contracted notation is used and the shear strains are the engineering shear strains defined in Equation (A.5).

F.2 Reduced Equations for a Thin Beam

If the elastic behavior of the thin substructure is to be modeled as a thin beam based on the Euler–Bernoulli or Rayleigh beam theory, the stress components other than T_1 are negligible so that

$$T_2 = T_3 = T_4 = T_5 = T_6 = 0 \tag{F.3}$$

Piezoelectric Energy Harvesting, First Edition. Alper Erturk and Daniel J. Inman.
© 2011 John Wiley & Sons, Ltd. Published 2011 by John Wiley & Sons, Ltd.

Therefore, Equation (F.2) simply reduces to

$$T_1 = Y_s S_1$$

(F.4)

which is known as Hooke's law [1] in elementary mechanics.

F.3 Reduced Equations for a Moderately Thick Beam

If the elasticity of the thin substructure is to be modeled as a moderately thick beam (based on the Timoshenko beam theory), the negligible stress components are

$$T_2 = T_3 = T_4 = T_6 = 0$$

(F.5)

Equation (F.2) can be rearranged to give

$$\begin{bmatrix} T_1 \\ T_5 \end{bmatrix} = \begin{bmatrix} Y_s & 0 \\ 0 & G_s \end{bmatrix} \begin{bmatrix} S_1 \\ S_5 \end{bmatrix}$$

(F.6)

Here, the shear modulus is defined in terms of the elastic modulus and Poisson's ratio as

$$G_s = \frac{Y}{2(1 + v_s)}$$

(F.7)

and the transverse shear stress is corrected to

$$T_5 = \kappa G_s S_5$$

(F.8)

where κ is the shear correction factor.

F.4 Reduced Equations for a Thin Plate

If the elastic behavior of the substructure is modeled as a thin plate (i.e. Kirchhoff plate), the following stress components are negligible:

$$T_3 = T_4 = T_5 = 0$$

(F.9)

Equation (F.2) reduces to

$$\begin{bmatrix} S_1 \\ S_2 \\ S_6 \end{bmatrix} = Y_s \begin{bmatrix} 1 & -v_s & 0 \\ -v_s & 1 & 0 \\ 0 & 0 & 2(1 + v_s) \end{bmatrix} \begin{bmatrix} T_1 \\ T_2 \\ T_6 \end{bmatrix}$$

(F.10)

which can be rearranged to give the stress components in terms of the strain components as

$$
\begin{bmatrix} T_1 \\ T_2 \\ T_6 \end{bmatrix} = \frac{Y_s}{1 - v_s^2} \begin{bmatrix} 1 & v_s & 0 \\ v_s & 1 & 0 \\ 0 & 0 & (1 - v_s)/2 \end{bmatrix} \begin{bmatrix} S_1 \\ S_2 \\ S_6 \end{bmatrix}
\tag{F.11}
$$

where the first element of the right-hand-side coefficient matrix of elastic components is the elastic constant that is related to the bending stiffness of an isotropic plate (accounting for the Poisson effect) in the absence of torsion.

Reference

1. Sokolnikoff, I.S. (1946) *Mathematical Theory of Elasticity*, McGraw-Hill, New York.

Appendix G

Essential Boundary Conditions for Cantilevered Beams

G.1   Euler–Bernoulli and Rayleigh Beam Theories

If the beam is clamped at $x = 0$ and free at $x = L$, the essential boundary conditions are defined at the clamped boundary as

$$w^0(0, t) = 0 \tag{G.1}$$

$$\frac{\partial w^0(x, t)}{\partial x}\bigg|_{x=0} = 0 \tag{G.2}$$

$$u^0(0, t) = 0 \tag{G.3}$$

where $w^0(x, t)$ is the transverse displacement and $u^0(x, t)$ is the axial displacement of the neutral axis.

G.2   Timoshenko Beam Theory

The essential boundary conditions are given at the clamped end ($x = 0$) of a cantilevered Timoshenko beam as

$$w^0(0, t) = 0 \tag{G.4}$$
$$\psi^0(0, t) = 0 \tag{G.5}$$
$$u^0(0, t) = 0 \tag{G.6}$$

where $\psi^0(x, t)$ is the cross-section rotation.

Piezoelectric Energy Harvesting, First Edition. Alper Erturk and Daniel J. Inman.
© 2011 John Wiley & Sons, Ltd. Published 2011 by John Wiley & Sons, Ltd.

Appendix H

Electromechanical Lagrange Equations Based on the Extended Hamilton's Principle

In the absence of mechanical dissipative effects, the extended Hamilton's principle can be given for an electromechanical system as

$$\int_{t_1}^{t_2} (\delta T - \delta U + \delta W_{ie} + \delta W_{nc}) \, dt = 0 \tag{H.1}$$

where δT, δU, and δW_{ie} are the first variations [1] of the total kinetic energy, the total potential energy, and the internal electrical energy, and δW_{nc} is the virtual work of the non-conservative mechanical force and electric charge components.

The total kinetic energy can be given as a function of the generalized coordinates and their time derivatives:

$$T = T(q_1, q_2, \ldots, q_n, \dot{q}_1, \dot{q}_2, \ldots, \dot{q}_n) \tag{H.2}$$

The total potential energy and the internal electrical energy are functions of generalized coordinates only[1]

$$U = U(q_1, q_2, \ldots, q_n) \tag{H.3}$$

$$W_{ie} = W_{ie}(q_1, q_2, \ldots, q_n) \tag{H.4}$$

---

[1] One of the generalized coordinates is the electrical across variable, i.e., the voltage output across the load.

Piezoelectric Energy Harvesting, First Edition. Alper Erturk and Daniel J. Inman.
© 2011 John Wiley & Sons, Ltd. Published 2011 by John Wiley & Sons, Ltd.

The first variations of Equations (H.2)–(H.4) are

$$\delta T = \sum_{k=1}^{n} \left(\frac{\partial T}{\partial q_k} \delta q_k + \frac{\partial T}{\partial \dot{q}_k} \delta \dot{q}_k \right) \tag{H.5}$$

$$\delta U = \sum_{k=1}^{n} \frac{\partial U}{\partial q_k} \delta q_k \tag{H.6}$$

$$\delta W_{ie} = \sum_{k=1}^{n} \frac{\partial W_{ie}}{\partial q_k} \delta q_k \tag{H.7}$$

The virtual work done by the generalized non-conservative forces (Q_k) is

$$\delta W_{nc} = \sum_{k=1}^{n} Q_k \delta q_k \tag{H.8}$$

The extended Hamilton's principle becomes

$$\int_{t_1}^{t_2} (\delta T - \delta U + \delta W_{ie} + \delta W_{nc}) \, dt$$

$$= \int_{t_1}^{t_2} \sum_{k=1}^{n} \left[\left(\frac{\partial T}{\partial q_k} - \frac{\partial U}{\partial q_k} + \frac{\partial W_{ie}}{\partial q_k} + Q_k \right) \delta q_k + \frac{\partial T}{\partial \dot{q}_k} \delta \dot{q}_k \right] dt = 0 \tag{H.9}$$

with the auxiliary conditions $\delta q_k = 0$ ($k = 1, 2, \ldots, n$) at $t = t_1$ and $t = t_2$ [2]. Integration by parts is applied to the last term to give

$$\int_{t_1}^{t_2} \frac{\partial T}{\partial \dot{q}_k} \delta \dot{q}_k \, dt = \int_{t_1}^{t_2} \frac{\partial T}{\partial \dot{q}_k} \frac{d(\delta q_k)}{dt} dt$$

$$= \frac{\partial T}{\partial \dot{q}_k} \delta q_k \Big|_{t_1}^{t_2} - \int_{t_1}^{t_2} \frac{d}{dt} \left(\frac{\partial T}{\partial \dot{q}_k} \right) \delta q_k \, dt = - \int_{t_1}^{t_2} \frac{d}{dt} \left(\frac{\partial T}{\partial \dot{q}_k} \right) \delta q_k \, dt \tag{H.10}$$

Using Equation (H.10) in Equation (H.9) gives

$$\int_{t_1}^{t_2} \sum_{k=1}^{n} \left[\frac{\partial T}{\partial q_k} - \frac{\partial U}{\partial q_k} + \frac{\partial W_{ie}}{\partial q_k} + Q_k - \frac{d}{dt} \left(\frac{\partial T}{\partial \dot{q}_k} \right) \right] \delta q_k \, dt = 0 \tag{H.11}$$

For the extended Hamilton's principle to hold for arbitrary virtual displacements, Equation (H.11) reduces to the electromechanical Lagrange equations:

$$\frac{d}{dt}\left(\frac{\partial T}{\partial \dot{q}_k}\right) - \frac{\partial T}{\partial q_k} + \frac{\partial U}{\partial q_k} - \frac{\partial W_{ie}}{\partial q_k} = Q_k \tag{H.12}$$

where the mechanical dissipative effects can be introduced as generalized non-conservative forces.

References

1. Dym, C.L. and Shames, I.H. (1973) *Solid Mechanics: A Variational Approach*, McGraw-Hill, New York.
2. Meirovitch, L. (1970) *Methods of Analytical Dynamics*, McGraw-Hill, New York.

Index

Piezoelectric Energy Harvesting, First Edition. Alper Erturk and Daniel J. Inman.
© 2011 John Wiley & Sons, Ltd. Published 2011 by John Wiley & Sons, Ltd.

Printed and bound by CPI Group (UK) Ltd, Croydon, CR0 4YY

07/03/2024

14467102-0001